Fuzzy Mathematical Techniques with Applications

"Indocti discant, et ament meminisse periti."

CHARLES JEAN FRANÇOIS HÉNAULT
1749

Fuzzy Mathematical Techniques with Applications

Abraham Kandel
Florida State University

ADDISON-WESLEY PUBLISHING COMPANY

Reading, Massachusetts • Menlo Park, California
Don Mills, Ontario • Wokingham, England
Amsterdam • Sydney • Singapore • Tokyo
Mexico City • Bogotá • Santiago • San Juan

Contains substantial material reprinted from
Fuzzy Techniques in Pattern Recognition,
copyright 1982 by John Wiley & Sons, Inc.
Reprinted by permission of John Wiley & Sons, Inc.

Reprinted with corrections July, 1986

Library of Congress Cataloging in Publication Data

Kandel, Abraham.
Fuzzy mathematical techniques with applications.
Bibliography: p.
Includes index.
1. Fuzzy sets. 2. Fuzzy systems. I. Title.
QA248.K35 1986 511.3′2 85-3892
ISBN 0-201-11752-5

CDEFGHIJ-MA-898

In honor of my father
JACOB KANDEL
may he live 120 years full
with happiness and health,
and my late mother
DINA KANDEL
who passed away while this
work was in progress.

Prologue

Although there is extensive literature on the theory of fuzzy sets and its applications, it is difficult for one who wishes to acquire a familiarity with the theory to find a text that both provides a readable introduction and presents an up-to-date exposition of some of the main applications of the theory. Professor Kandel's text serves this purpose with authority, and his treatment of the subject matter reflects his many important contributions to both the theory and its applications.

As is pointed out by Professor Kandel, the theory of fuzzy sets has been, and will continue to be, controversial for some time to come. In essence, the controversy stems from the deep-seated scientific tradition of respect for what is quantitative and precise, and disdain for what is qualitative and fuzzy. And yet, it is a fact that much of human reasoning is imprecise in nature, and most of it is not amenable to formalization within the framework of classical logic and probability theory. This is particularly true of commonsense reasoning—a mode of reasoning that is so natural for humans and so difficult for machines. In coming years, the ability of the theory of fuzzy sets to provide a basis for a formalization of commonsense reasoning may well be an acid test of its usefulness in artificial intelligence. It is my conviction that the test will be passed and the theory of fuzzy sets will eventually become a standard tool for the management of uncertainty in expert systems. Many of the techniques needed for this purpose are the techniques described in Professor Kandel's book.

A frequently raised question that is addressed by Professor Kandel is: Can any significant problems in the realm of uncertainty be dealt with more effectively through the use of the theory of fuzzy sets than through classical probability theory? The wide spectrum of problems treated by Professor Kandel includes those not amenable to analysis within the framework of probability theory. The important issue here is not the axiomatics of probability theory but its expressiveness as a language of uncertainty. Thus the main limitation of probability theory stems from its failure to provide three basic concepts that are a part of the theory of fuzzy sets, namely, the concept of a fuzzy event, the concept of a fuzzy probability, and the concept of a fuzzy syllogism. Viewed in this perspective, Professor Kandel's authoritative exposition of these and related concepts, drawn from the theory of fuzzy logic and possibility theory, provides the reader with an extensive background for applying the theory of fuzzy sets to systems and phenomena that do not lend themselves to analysis by conventional methods.

University of California
Berkeley, CA

L. A. Zadeh

Foreword

As researchers and practitioners continue their efforts to build intelligent systems, they must come to grips with the issue of uncertainty in human knowledge. As we learn more and more about uncertainty, we realize that randomness is not the only form it takes. Philosophers have discussed these other forms of uncertainty, but it is only now that we are starting to get quantitative tools to describe and manipulate these additional forms of uncertainty. Beginning in 1965, L. A. Zadeh of the University of California at Berkeley introduced the theory of fuzzy subsets as a means for representing uncertainty. The type of uncertainty that this theory was meant to handle has as its roots the type of imprecision and ambiguity which is so prevalent in human discourse and thought. Zadeh's original paper sparked the interest of many researchers worldwide, and this has resulted in the rapid development of the field. The theory of fuzzy subsets, while based upon multivalued logic, extends this logic in the same way that ordinary set theory extends binary logic. In particular, fuzzy set theory frees us from the so-called law of contradiction and allows us to entertain conflicting propositions.

Kandel has provided in this text a comprehensive survey of this field. The material herein presented will provide an invaluable reference for those interested in acquainting themselves with this field for the purpose of doing research. Perhaps more importantly however, this book will be of use to researchers in artificial intelligence and other related areas. As they begin to look for methods to include uncertainty in their models, they will find that fuzzy sets provide a natural mechanism for this process. This book provides a comprehensive source to enable these system builders to quickly learn the necessary fundamentals of this theory in one place, rather than having to work their way through the voluminous literature in the field.

Iona College
New Rochelle, NY

Ronald R. Yager

To the Reader

Since the publication of the now classic paper on Fuzzy Sets by Professor Lotfi A. Zadeh in 1965, the theory of fuzzy mathematics has gained more and more recognition from many researchers in a wide range of scientific fields. Like other theories that have broken away from tradition, the theories of fuzzy mathematics have been and will continue to be controversial for some time to come. The theories of fuzzy mathematics are attractive not only because they are based on the very intuitive idea of fuzzy sets, but because they are capable of generating many intellectually appealing structures that provide today's scientists and engineers with new insights into interesting, significant, and often-debated problems in both science and engineering.

The present volume will be of great value to those who are interested in acquainting themselves with the basic theories of fuzzy mathematics and in exploring their potential as a methodology for dealing with phenomena that are vague, imprecise, or too complex or too ill-defined to be susceptible of analysis by conventional mathematical means.

This volume is intended to be a rather exhaustive research monograph on the theories of fuzzy mathematics, and is based on a large compilation of the literature on the subject. The first six chapters discuss in detail the different aspects of fuzzy mathematics; in Chapter Seven we present several research contributions to the various applications of the mathematical theories discussed in the first six chapters. The core bibliography on the theories of fuzzy mathematics and their applications is quite useful as a research compendium.

The monograph is aimed at readers at the graduate level with a strong mathematical background, and can be used as a textbook in graduate-level seminars in mathematics, computer science, engineering, human-centered systems, and related subjects. There is a need for a comprehensive monograph describing the mathematical theories that challenge the traditional reliance on two-valued logic and classical set theory as a basis for scientific inquiry. That this need existed became clear to me when I observed the tremendous interest shown by researchers and students in a wide range of scientific areas, in the mathematically oriented, concise article by myself and Byatt (1978) and in the early versions of the bibliography on fuzzy sets by myself and Davis (1976) and by myself and Yager (1979). This extensive presentation of the mathematical notions that have been developed in the framework of fuzzy set theory, as well as the bibliography, should satisfy the present demand by members of the scientific community.

Some of the material contained in this book has been used in courses at the Florida State University and at Ben Gurion University of the Negev in Israel, where I have spent part of the academic year 1984/1985. I am indebted to my students and peers at both institutions for their help and encouragement throughout the development of this book. Especially I want to express my lasting gratitude to Douglas H. Schlak for assistance in putting together the extensive bibliography, and to Lawrence O. Hall, Christie Wilcox, Karen L. Johnson, John R. Van Wingen, and all the students in my graduate seminar, for proofreading the manuscript. Portions of the material in this text are derived from many early investigations listed in the bibliography.

I would like to acknowledge my indebtedness to all those researchers and the many scientists who have contributed to the fast-growing field of research known as "fuzzy set theory and its applications." It is my pleasure to particularly acknowledge the constant encouragement and consistent support of Lotfi A. Zadeh, Madan M. Gupta, Arnold Kaufmann, Ronald R. Yager, Hans J. Zimmermann, Wyllis Bandler, and the late King-Sun Fu. I owe a debt of gratitude to all of them. This work was partly supported by NSF grant IST 8405953.

Tallahassee, FL Abe Kandel

Contents

CHAPTER ONE

Fuzzy Sets

1.1 INTRODUCTION

The "hard" sciences, such as engineering, chemistry, or physics, construct exact mathematical models of empirical phenomena, and then use these models to make predictions. Some aspects of the "real world" always escape such precise mathematical models, and usually there is an elusive inexactness as part of the original model.

A central idea in the Platonic philosophy is that, in the real world, elements are perturbed by imperfection and thus, for example, there exists no element that is perfectly round. "Perfect notions" or "exact concepts" correspond to the sort of things envisaged in pure mathematics, while "inexact structures" prevail in real life. It is our belief that inexact structures are rich enough in operations and properties to be of genuine use in constructing models for a wide variety of situations; further, these mathematical properties will provide a practical guide for both philosophical and technical reasoning.

The purpose of this chapter is to introduce a unifying point of view to the notion of inexactness, based on the theory of fuzzy sets introduced by Zadeh. The term Fuzzy in the sense used here seems to have been first introduced in Zadeh [1962]. In that paper Zadeh called for a "mathematics of fuzzy or cloudy quantities which are not describable in terms of probability distributions." This paper was followed in 1965 by a technical exposition of just such a mathematics now termed the theory of FUZZY SETS (Zadeh [1965]). The reasons supporting the representation of inexact concepts by fuzzy sets has been given by Goguen [1967]. Perhaps his most convincing argument is a *representation theorem* which says that any system satisfying certain axioms is equivalent to a system of fuzzy sets. Since the axioms are intuitively plausible for the system of all inexact concepts, the theorem allows us to conclude that inexact concepts can be represented by fuzzy sets. The representation theorem is a precise mathematical result in the theory of categories, so that a very precise meaning is given to the concepts "system," "equivalent," and "represented."

Essentially, fuzziness is a type of imprecision that stems from a grouping of elements into classes that do not have sharply defined boundaries. Such classes — called *fuzzy sets* — arise, for example, whenever we describe ambiguity, vagueness, and ambivalence in mathematical models of empirical phenomena. Since certain aspects of reality always escape such models, the strictly binary (and even the ternary) approach to the treatment of physical phenomena is not always adequate to describe systems in the real world; and the attributes of the system variables often emerge from an elusive fuzziness, a readjustment to context, or an effect of human imprecision. In many cases, however, even if the model is precise, fuzziness may be a concomitant of complexity. Systems of high cardinality are rampant in real life and their computer simulations require some kind of mathematical formulation to deal with the imprecise descriptions.

The theory of fuzzy sets has as one of its aims the development of a methodology for the formulation and solution of problems that are too complex or too ill-defined to be susceptible to analysis by conventional techniques. Because of its unorthodoxy, it has been and will continue to be controversial for some time to come. Eventually, though, the theory of fuzzy sets is likely to be recognized as a natural development in the evolution of scientific thinking; and the skepticism about its usefulness will be viewed, in retrospect, as a manifestation of the human attachment to tradition and resistance to innovation.

In what follows, our attention is focused primarily on defining some of the basic notions within the conceptual framework of fuzzy set theory and exploring some of their elementary implications.

1.2 FUZZY SETS AND OPERATIONS ON FUZZY SETS

The theory of fuzzy sets deals with a subset A of the universe of discourse X, where the transition between full membership and no membership is gradual rather than abrupt. The "fuzzy subset" has no well-defined boundaries whereas the universe of discourse (the universe X) covers a definite range of objects. Fuzzy classes of objects are often encountered in the real world. For instance, A may be the set of beautiful women in a town X, or A may be the set of long streets in town X. Traditionally, the grade of membership 1 is assigned to those objects that fully and completely belong to A, while 0 is assigned to objects that do not belong to A at all. The more an object x belongs to A, the closer to 1 is its grade of membership $\chi_A(x)$.

In abstract (or conventional, or nonfuzzy) set theory, the sets considered are defined as collections of objects having some very general property P;

nothing special is assumed or considered about the nature of the individual objects. For example, we define a set X as the set of streets. Symbolically,

$$X = \{x \mid x \quad \text{is a street}\}.$$

Now what about the "class of long streets"? First of all, is it a set in the ordinary sense? Before we answer that, we may first ask: "How long is a long street? Is a one-mile street a long street? If so, then is there any difference between a half-mile street and a one-mile long street, etc.?" Frankly, we do not know how to answer these questions adequately from the information "long street," because the "class of long streets" does not constitute a set in the usual sense. In fact, most of the classes of objects encountered in the real physical world are of this fuzzy, not sharply defined type. They do not have precisely defined criteria of membership. In such classes, it is not necessary for an object to belong or not belong to a class; there may be intermediate grades of membership. This is the concept of a fuzzy set, which is a "class" with a continuum of grades of membership.

Fuzzy-set theory, introduced by Zadeh in 1965, is a generalization of abstract set theory. In other words, the former always includes the latter as a special case; definitions, theorems, and proofs of fuzzy-set theory always hold for nonfuzzy sets. Because of this generalization, fuzzy-set theory has a wider scope of applicability than abstract set theory in solving problems that involve, to some degree, subjective evaluation.

Intuitively, a fuzzy set is a class that admits the possibility of *partial membership* in it. Let $X = \{x\}$ denote a space of objects. Then a fuzzy set A in X is a set of ordered pairs

$$A = \{x, \chi_A(x)\}, \qquad x \in X,$$

where $\chi_A(x)$ is termed "the grade of membership of x in A." We shall assume for simplicity that $\chi_A(x)$ is a number in the interval $[0, 1]$, with the grades 1 and 0 representing, respectively, *full membership* and *nonmembership* in a fuzzy set, as discussed before. We have assumed that an exact comparison is possible for the truths of any two inexact statements "$x \in A$" and "$y \in A$," and that the exact relation so obtained satisfies the minimal consistency requirements of *transitivity* and *reflexivity*; the ordering $x \geq y$ means "x is at least as true as y" with $x \leq y$ denoting "x is not truer than y."

The grades of membership reflect an "ordering" of the objects in the universe; it is interesting to note that the grade-of-membership value $\chi_A(x)$ of an object x in A can be interpreted as the degree of compatibility of the predicate associated with A and the object x. As will be seen later, it is also possible to interpret $\chi_A(x)$ as the degree of possibility that x is the value of a parameter fuzzily restricted by A.

In general, we distinguish three kinds of inexactness: *generality*, that a concept applies to a variety of situations; *ambiguity*, that it describes more than one distinguishable subconcept; and *vagueness*, that precise boundaries are not defined. All three types of inexactness are represented by a fuzzy set:

Generality occurs when the universe is not just one point; ambiguity occurs when there is more than one local maximum of a membership function; and vagueness occurs when the function takes values other than just 0 and 1.

We now consider several examples of fuzzy sets.

■ *Example 1.2.1*

In this example we consider the class of all real numbers that are much greater than 1. We can define this set as

$$A = \{x | x \text{ is a real number} \quad \text{and} \quad x \gg 1\}.$$

But it is not a well-defined set for the reasons mentioned before. This set may be defined subjectively by a membership function such as

$$\chi_A(x) = 0 \qquad \text{for } x \leq 1,$$

$$\chi_A(x) = \frac{x-1}{x} \qquad \text{for } x > 1. \qquad ■$$

The assignment of the membership function of a fuzzy set is subjective in nature and, in general, reflects the context in which the problem is viewed. Although the assignment of the membership function of a fuzzy set A is "subjective," it cannot be assigned arbitrarily. For example, it would be totally wrong to assign the membership function of Example 1.2.1 as

$$\chi_A(x) = \begin{cases} \dfrac{x-1}{x} & \text{for } x \leq 1, \\ 0 & \text{for } x > 1. \end{cases}$$

A function χ_A such as

$$\chi_A(x) = \begin{cases} 0 & \text{for } x \leq 1, \\ e^{-(x-1)} & \text{for } x > 1, \end{cases}$$

which monotonically decreases as x increases, for $x > 1$, or

$$\chi_A(x) = \begin{cases} 0 & \text{for } x \leq 1, \\ 1 - e^{-1000(x-1)} & \text{for } x > 1, \end{cases}$$

which increases monotonically, but is approximately equal to 1 for $x = 1.1$, should not be considered, as they describe other classes of objects rather than the one required by Example 1.2.1. Functions such as these will be called nonadmissible functions of the fuzzy set A. The function $\chi_A(x)$ as defined in Example 1.2.1 and other functions such as

$$\chi_A(x) = \begin{cases} 0 & \text{for } x \leq 1, \\ 1 - e^{-0.1(x-1)} & \text{for } x > 1; \end{cases} \tag{1.2.1}$$

$$\chi_A(x) = \begin{cases} 0 & \text{for } x \leq 1, \\ 1 - [\cosh(x-1)]^{-1} & \text{for } x > 1, \end{cases} \tag{1.2.2}$$

which satisfy the condition $0 \leq \chi_A(x) \leq 1$ for all $x \in X$ and are consistent with the specification of the set, will be called the *admissible functions of A*.

The *support* of A is the set of points in X at which $\chi_A(x) > 0$. The *height* of A is the supremum of $\chi_A(x)$ over X. A *crossover point* of A is the point in X whose grade of membership in A is 0.5. We say that A is *normal* if its height is 1; otherwise it is subnormal.

■ ***Example 1.2.2***

Let the universe be the interval [0, 120], with x interpreted as age. A fuzzy subset A of X labeled *old* may be defined by a grade of membership function such as

$$\chi_A(x) = \begin{cases} 0 & \text{for } 0 \le x \le 40, \\ \left(1 + \left(\dfrac{x-40}{5}\right)^{-2}\right)^{-1} & \text{for } 40 < x \le 120. \end{cases}$$

In this example, the support of *old* is the interval (40, 120], the height of *old* is effectively 1, and the crossover point of *old* is 45.

To simplify the representation of fuzzy sets, it is convenient to use the following notation:

A nonfuzzy finite set such as $X = \{x_1, x_2, \ldots, x_n\}$ *is expressed as*

$$X = \sum_{j=1}^{n} x_j = x_1 + x_2 + \cdots + x_n,$$

with the understanding that this is a representation of X as the union of its constituent singletons, with the plus sign (+) *playing the role of "union" rather than the arithmetic sum. Thus,*

$$x_j + x_k = x_k + x_j$$

and

$$x_j + x_j = x_j$$

for $j, k = 1, 2, \ldots, n$.

As a simple extension of this notation, a finite fuzzy set A on X is expressed as

$$A = \chi_A(x_1)/x_1 + \cdots + \chi_A(x_n)/x_n = \sum_{j=1}^{n} \chi_A(x_j)/x_j.$$

When X is not finite, we can use the notation

$$A = \int_X \chi_A(x)/x.$$

■ ***Example 1.2.3***

In the universe of discourse comprising $\mathbb{N}$ = {positive integers}, the fuzzy set A, labeled "integers approximately equal to 5," may be defined as

$$A = 0.1/2 + 0.4/3 + 0.9/4 + 1.0/5 + 0.9/6 + 0.4/7 + 0.1/8.$$ ■

■ ***Example 1.2.4***

In the universe of discourse comprised of $\mathbb{R}$ = {real numbers}, the fuzzy set A, labeled "real numbers clustered around 5," may be defined by the grade-of-membership function

$$\chi_A(x) = \left\{1 + \left[\tfrac{1}{4}(x-5)\right]^2\right\}^{-1}$$

or as

$$A = \int_{\mathbb{R}} \left\{1 + \left[\tfrac{1}{4}(x-5)\right]^2\right\}^{-1} / x.$$ ■

In many cases it is convenient to express the membership function of a fuzzy subset of the real line in terms of a standard function, whose parameters may be adjusted to fit a specified membership function in an approximate fashion. Two such functions, the S-function and the π-function, are defined by

$$S(u;\alpha,\beta,\gamma) = \begin{cases} 0 & \text{for } u \le \alpha, \\ 2\left(\dfrac{u-\alpha}{\gamma-\alpha}\right)^2 & \text{for } \alpha \le u \le \beta, \\ 1 - 2\left(\dfrac{u-\gamma}{\gamma-\alpha}\right)^2 & \text{for } \beta \le u \le \gamma, \\ 1 & \text{for } u \ge \gamma; \end{cases} \tag{1.2.3}$$

$$\pi(u;\beta,\gamma) = \begin{cases} S\left(u;\gamma-\beta,\gamma-\dfrac{\beta}{2},\gamma\right) & \text{for } u \le \gamma, \\ 1 - S\left(u;\gamma,\gamma+\dfrac{\beta}{2},\gamma+\beta\right) & \text{for } u \ge \gamma, \end{cases} \tag{1.2.4}$$

In $S(u;\alpha,\beta,\gamma)$, the parameter β, $\beta = (\alpha+\gamma)/2$, is the crossover point. In $\pi(u;\beta,\gamma)$, β is the bandwidth, that is, the separation between the crossover points of π, while γ is the point at which π is unity.

In some cases, the assumption that χ_A is a mapping from X to $[0,1]$ may be too restrictive, and it may be desirable to allow χ_A to take values in a lattice or, more particularly, in a Boolean algebra. For most purposes, however, it is sufficient to deal with the first two of the following hierarchy of fuzzy sets.

Definition 1.2.1

A fuzzy subset, A, of X is of Type 1 if its membership function, χ_A, is a mapping from X to $[0,1]$; and A is of Type K, $K = 2, 3, \ldots,$ if χ_A is mapping from X to the set of fuzzy subsets of Type $(K-1)$. For simplicity, it will always be understood that A is of Type 1 if it is not specified to be of a higher type.

■ ***Example 1.2.5***

Suppose that X is the set of all nonnegative integers and A is a fuzzy subset of X labeled "small integers." Then A is of Type 1 if the grade of membership of a generic element x in A is a number in the interval [0, 1], for example,

$$\chi_{\text{small integers}}(x) = \left[1 + \left(\frac{x}{5}\right)^2\right]^{-1}, \qquad x = 0, 1, 2, \ldots$$

On the other hand, A is of Type 2 if for each x in X, $\chi_A(x)$ is a fuzzy subset of [0, 1] of Type 1, for example, for $x = 10$,

$$\chi_{\text{small integers}}(10) = \text{low}$$

where low is a fuzzy subset of [0, 1] whose membership function is defined by, say,

$$\chi_{\text{low}}(v) = 1 - S(v; 0, 0.25, 0.5), \qquad v \in [0, 1]$$

which implies that

$$\chi_{\text{low}} = \int_0^1 (1 - S(v; 0, 0.25, 0.5))/v. \qquad \blacksquare$$

α-Cuts

When we want to exhibit an element $x \in X$ that typically belongs to a fuzzy set A, we may demand that its membership value be greater than some threshold $\alpha \in (0, 1]$. The ordinary set of such elements is the α-cut A_α of A,

$$A_\alpha = \{x \in X, \qquad \chi_A(x) \geq \alpha\}.$$

One also defines the *strong α-cut*

$$A_{\bar{\alpha}} = \{x \in X, \qquad \chi_A(x) > \alpha\}.$$

The membership function of a fuzzy set A can be expressed in terms of the characteristic functions of its α-cuts according to the formula

$$\chi_A(x) = \sup_{\alpha \in (0,1]} \min[\alpha, \chi_{A_\alpha}(x)],$$

where

$$\chi_{A_\alpha}(x) = \begin{cases} 1 & \text{iff } x \in A_\alpha, \\ 0 & \text{otherwise.} \end{cases}$$

It is easily checked that the following properties hold:

$$(A \cup B)_\alpha = A_\alpha \cup B_\alpha, \qquad (A \cap B)_\alpha = A_\alpha \cap B_\alpha.$$

However, $(\bar{A})_{\bar{\alpha}} = \overline{(A_{1-\alpha})} \neq \overline{(A_\alpha)}$ if $\alpha \neq \frac{1}{2}$. This result stems from the fact that generally there are elements that belong neither to A_α nor to $(\bar{A})_\alpha$, that is,

$$(A_\alpha \cup (\bar{A})_\alpha \neq X).$$

The *level fuzzy sets* of a fuzzy set A are defined as the fuzzy sets $\tilde{A}_\alpha$, $\alpha \in (0, 1)$, such that

$$\tilde{A}_\alpha = \{(x, \chi_A(x)), x \in A_\alpha\}.$$

The rationale behind this definition is the fact that in practical applications it is sufficient to consider fuzzy sets defined in only one part of their support — the most significant part — in order to save computing time and computer memory storage.

A fuzzy set A may be decomposed into its level sets through the *resolution identity*

$$A = \int_0^1 \alpha A_\alpha$$

or

$$A = \sum_\alpha \alpha A_\alpha,$$

where αA_α is the product of a scalar α with the set A_α and $\int_0^1$ (or Σ_α) is the union of the A_α, with α ranging from 0 to 1.

The resolution identity may be viewed as the result of combining those elements in A that fall into the same level set. More specifically, suppose that A is represented in the form

$$A = 0.1/2 + 0.3/4 + 0.5/7 + 0.9/8 + 1/9.$$

Then A can be rewritten as

$$\begin{aligned} A = 0.1/2 + 0.1/4 + 0.1/7 + 0.1/8 + 0.1/9 \\ + 0.3/4 + 0.3/7 + 0.3/8 + 0.3/9 \\ + 0.5/7 + 0.5/8 + 0.5/9 \\ + 0.9/8 + 0.9/9 \\ + \quad 1/9 \end{aligned}$$

or

$$\begin{aligned} A = 0.1(1/2 + 1/4 + 1/7 + 1/8 + 1/9) \\ +0.3(1/4 + 1/7 + 1/8 + 1/9) \\ +0.5(1/7 + 1/8 + 1/9) \\ +0.9(1/8 + 1/9) \\ +1/9, \end{aligned}$$

with the level sets given by

$$\begin{aligned} A_{0.1} &= 2 + 4 + 7 + 8 + 9, \\ A_{0.3} &= 4 + 7 + 8 + 9, \\ A_{0.5} &= 7 + 8 + 9, \\ A_{0.9} &= 8 + 9, \\ A_1 &= 9. \end{aligned}$$

As can be seen easily, the resolution principle may provide a convenient way of generalizing various concepts associated with nonfuzzy sets to fuzzy sets. For example, let X be a linear vector space. Then A is convex if and only if for all $\lambda \in [0, 1]$ and all x_1, x_2 in X,

$$\chi_A(\lambda x_1 + (1 - \lambda) x_2) \geq \min(\chi_A(x_1), \chi_A(x_2)).$$

In terms of the level sets of A, A is convex if and only if the A_α are convex for all $\alpha \in [0, 1]$. Dually, A is concave if and only if

$$\chi_A(\lambda x_1 + (1-\lambda)x_2) \leq \max(\chi_A(x_1), \chi_A(x_2)).$$

In many cases the fuzzy α-cuts can replace the classical grade-of-membership function as described above. For example, let A be a fuzzy set on X and A_α its α-cut. A_α can be written $\chi_A^{-1}([\alpha, 1])$, that is, the inverse image of the interval $[\alpha, 1]$. Let $\chi_{[\alpha,1]}$ be the characteristic function of the interval $[\alpha, 1]$ in the universe $[0, 1]$:

$$\chi_{A_\alpha}(x) = \chi_{[\alpha,1]}(\chi_A(x)), \qquad \forall x \in X.$$

A fuzzy α-cut can be understood as the set of elements whose membership values are greater than "approximately α," that is, belong to a fuzzy interval $(\tilde{\alpha}, 1]$, where $\chi_{(\tilde{\alpha},1]}$ is a continuous nondecreasing function from $[0, 1]$ to $[0, 1]$ and $\chi_{(\tilde{\alpha},1]}(1) = 1$. It is natural to extend $\chi_{A_\alpha}(x)$ into

$$\chi_{A_{\tilde{\alpha}}}(x) = \chi_{(\tilde{\alpha},1]}(\chi_A(x)), \qquad \forall x \in X,$$

where $A_{\tilde{\alpha}}$ is the fuzzy α-cut of A.

Set-Theoretic Operations

Let A and B be fuzzy subsets of X. We can now discuss some basic operations performed on fuzzy sets.

1. Two fuzzy sets, A and B, are said to be *equal* $(A = B)$ iff

$$\int_X \chi_A(x)/x = \int_X \chi_B(x)/x \qquad \text{or} \qquad \forall x \in X, \qquad \chi_A(x) = \chi_B(x).$$

2. A is *contained* in $B (A \subseteq B)$ iff $\int_X \chi_A(x)/x \leq \int_X \chi_B(x)/x$.
3. The *union* of fuzzy sets A and B is denoted by $A \cup B$ and is defined by

$$A \cup B \triangleq \int_X (\chi_A(x) \vee \chi_B(x))/x,$$

where $\vee$ is the symbol for max.

4. The *intersection* of A and B is denoted by $A \cap B$ and is defined by

$$A \cap B \triangleq \int_X (\chi_A(x) \wedge \chi_B(x))/x,$$

where $\wedge$ is the symbol for min.

A justification of the choice of max and min was given by Bellman and Giertz [1973]; max and min are the only operators f and g that meet the following requirements:

(i) The membership value of x in a compound fuzzy set depends on the membership value of x in the elementary fuzzy sets that form it, but not on anything else:

$$\forall x \in X, \qquad \chi_{A \cup B}(x) = f(\chi_A(x), \chi_B(x)),$$

$$\chi_{A \cap B}(x) = g(\chi_A(x), \chi_B(x)).$$

(ii) f and g are commutative, associative, and mutually distributive operators.

(iii) f and g are continuous and nondecreasing with respect to each of their arguments. Intuitively, the membership of x in $A \cup B$ or $A \cap B$ cannot decrease when the membership of x in A or B increases. A small increase of $\chi_A(x)$ or $\chi_B(x)$ cannot induce a strong increase of $\chi_{A \cup B}(x)$ or $\chi_{A \cap B}(x)$.

(iv) $f(u, u)$ and $g(u, u)$ are strictly increasing. If $\chi_A(x_1) = \chi_B(x_1) > \chi_A(x_2) = \chi_B(x_2)$, then the membership of x_1 in $A \cup B$ or $A \cap B$ is certainly strictly greater than that of x_2.

(v) Membership in $A \cap B$ requires more, and membership in $A \cup B$ less, than the membership in one of A or B:

$$\forall x \in X, \qquad \chi_{A \cap B}(x) \leq \min(\chi_A(x), \chi_B(x)),$$

$$\chi_{A \cup B}(x) \geq \max(\chi_A(x), \chi_B(x)).$$

(vi) Complete membership in A and in B implies complete membership in $A \cap B$. Complete lack of membership in A and in B implies complete lack of membership in $A \cup B$:

$$g(1, 1) = 1, \qquad f(0, 0) = 0.$$

The above assumptions are consistent and sufficient to ensure the uniqueness of the choice of union and intersection operators.

Fung and Fu [1974] also found max and min to be the only possible operators. They use a slightly different set of assumptions. They kept (i) and added the following:

(ii′) f and g are commutative, associative, and idempotent.
(iii′) f and g are nondecreasing.
(vii) f and g can be recursively extended to $m \geq 3$ arguments.
(viii) $\forall x \in X$, $f(1, \chi_A(x)) = 1$, $g(0, \chi_A(x)) = 0$.

The interpretation of these axioms was given in the framework of group decision-making with a slightly more general valuation set.

5. The *complement* of A is denoted by $\bar{A}$ and is defined by

$$\bar{A} \triangleq \int_X (1 - \chi_A(x))/x.$$

Namely,

$$\forall x \in X, \qquad \chi_{\bar{A}}(x) = 1 - \chi_A(x).$$

The justification of this negation is more difficult than that of min and max. Natural conditions to impose on a complementation function h were proposed by Bellman and Giertz [1973]:

(i) $\chi_{\bar{A}}(x)$ depends only on $\chi_A(x)$: $\quad \chi_{\bar{A}}(x) = h(\chi_A(x))$.
(ii) $h(0) = 1$ and $h(1) = 0$, to recover the usual complementation when A is an ordinary subset.
(iii) h is continuous and strictly monotonically decreasing, since membership in $\bar{A}$ should become smaller when membership in A increases.
(iv) h is involutive: $\quad h(h(\chi_{\bar{A}}(x))) = \chi_A(x)$.

The above assumptions do not determine h uniquely, not even if we require in addition $h(\frac{1}{2}) = \frac{1}{2}$. However, $h(u) = 1 - u$ if the following requirement is introduced (Gaines [1978]):

(v) $\forall x_1 \in X, \forall x_2 \in X$, if $\chi_A(x_1) + \chi_A(x_2) = 1$, then $\chi_{\bar{A}}(x_1) + \chi_{\bar{A}}(x_2) = 1$.

Instead of (v), Bellman and Giertz have proposed the following very strong condition:

(vi) $\forall x_1 \in X, \forall x_2 \in X, \chi_A(x_1) - \chi_A(x_2) = \chi_{\bar{A}}(x_2) - \chi_{\bar{A}}(x_1)$, which means that a certain change in the membership value in A should have the same effect on the membership in $\bar{A}$.

Assumptions (i), (ii), and (vi) entail $h(u) = 1 - u$.

However, there may be situations where (v) and (vi) may appear to be not really necessary assumptions. Sugeno [1974] defines the λ-complement $\bar{A}^\lambda$ of A via the fraction given by

$$\chi_{\bar{A}^\lambda}(x) = (1 - \chi_A(x))/(1 + \lambda_{\chi_A}(x)), \qquad \lambda \in (-1, +\infty).$$

λ-complementation satisfies (i), (ii), (iii), and (iv).

6. The *product* of A and B is denoted by AB and is defined by

$$AB \triangleq \int_X \chi_A(x)\chi_B(x)/x.$$

Thus, A^α, where α is any positive number, should be interpreted as

$$A^\alpha \triangleq \int_X (\chi_A(x))^\alpha / x.$$

Similarly, if α is any nonnegative real number such that $\alpha \operatorname{Sup}_x \chi_A(x) \leq 1$, then

$$\alpha A \triangleq \int_X \alpha \chi_A(x) / x.$$

As a special case, the operation of *concentration* can be defined as

$$\mathrm{CON}(A) \triangleq A^2,$$

while that of *dilation* can be expressed by

$$\mathrm{DIL}(A) \triangleq A^{0.5}.$$

7. The *bounded sum* of A and B is denoted by $A \oplus B$ and is defined by

$$A \oplus B \triangleq \int_X 1 \wedge (\chi_A(x) + \chi_B(x)) / x$$

where $+$ is the arithmetic sum.

8. The *bounded difference* of A and B is denoted by $A \ominus B$ and is defined by

$$A \ominus B \triangleq \int_X 0 \vee (\chi_A(x) - \chi_B(x)) / x$$

where $-$ is the arithmetic difference.

9. The *left-square* of A is denoted by ${}^2A$ and is defined by

$${}^2A \triangleq \int_V \chi_A(x) / x^2$$

where $V \triangleq \{x^2 \mid x \in X\}$. More generally,

$${}^\alpha A \triangleq \int_V \chi_A(x) / x^\alpha$$

where $V \triangleq \{x^\alpha \mid x \in X\}$.

10. If $A_1, \ldots, A_k$ are fuzzy subsets of X, and $w_1, \ldots, w_k$ are nonnegative weights adding up to unity, then a *convex combination* of $A_1, \ldots, A_k$ is a fuzzy set A whose membership function is expressed by

$$\chi_A = w_1 \chi_{A_1} + \cdots + w_k \chi_{A_k} = \sum_{j=1}^{k} w_j \chi_{A_j},$$

where in this case, $+$ (Σ) denotes the arithmetic sum. The concept of a convex combination is useful in the representation of linguistic modifiers such as "essentially" and "typically," which modify the weights associated with the components of a fuzzy set.

11. If $A_1, \ldots, A_k$ are fuzzy subsets of $X_1, \ldots, X_k$, respectively, the Cartesian product of $A_1, \ldots, A_k$ is denoted by $A_1 \times \cdots \times A_k$ and is defined as a fuzzy subset of $X_1 \times \cdots \times X_k$ whose membership function is expressed by

$$\chi_{A_1 \times \cdots \times A_k}(x_1, \ldots, x_k) = \chi_{A_1}(x_1) \wedge \cdots \wedge \chi_{A_k}(x_k).$$

Equivalently,

$$A_1 \times \cdots \times A_k = \int_{X_1 \times \cdots \times X_k} (\chi_{A_1}(x_1) \wedge \cdots \wedge \chi_{A_k}(x_k))/(x_1, \ldots, x_k).$$

■ ***Example 1.2.6***

Let $X = \{1, 2, 3, 4, 5, 6, 7\}$ and let

$$A = 0.8/3 + 1/5 + 0.6/6,$$
$$B = 0.7/3 + 1/4 + 0.5/6;$$

then

$$A \cup B = 0.8/3 + 1/4 + 1/5 + 0.6/6,$$
$$A \cap B = 0.7/3 + 0.5/6;$$
$$\bar{A} = 1/1 + 1/2 + 0.2/3 + 1/4 + 0.4/6 + 1/7,$$
$$AB = 0.56/3 + 0.3/6,$$
$$A^2 = 0.64/3 + 1/5 + 0.36/6;$$
$$0.5A = 0.4/3 + 0.5/5 + 0.3/6;$$
$$\text{CON}(B) = 0.49/3 + 1/4 + 0.25/6,$$
$$\text{DIL}(B) = 0.84/3 + 1/4 + 0.7/6;$$
$$A \oplus B = 1/3 + 1/4 + 1/5 + 1/6,$$
$$A \ominus B = 0.1/3 + 1/5 + 0.1/6;$$
$${}^2A = 0.8/9 + 1/25 + 0.6/36,$$
$${}^3A = 0.8/27 + 1/125 + 0.6/216.$$ ■

■ ***Example 1.2.7***

Let $X_1 = X_2 = \{2, 4, 6\}$ and let

$$A_1 = 0.5/2 + 1/4 + 0.6/6,$$
$$A_2 = 1/2 + 0.6/4;$$

then

$$A_1 \times A_2 = 0.5/(2,2) + 1/(4,2) + 0.6/(6,2) + 0.5/(2,4) + 0.6/(4,4) + 0.6/(6,4).$$ ■

Following the excellent presentation by Dubois and Prade [1979b], suppose that the operations of union (U), intersection (I), and complement (C) are preserved in fuzzy sets. Considering elements a, b, and c, each with values in [0, 1] we get:

$$\forall a, b, c \in [0,1]^3: \quad U(a, I(b,c)) = I(U(a,b), U(a,c)); \quad (1.2.5)$$

$$I(a, U(b,c)) = U(I(a,b), I(a,c)); \quad (1.2.6)$$

$$U(a,a) = a; \qquad I(a,a) = a; \quad (1.2.7)$$

$$U(a, C(a)) = 1; \qquad I(a, C(a)) = 0. \quad (1.2.8)$$

Now the following propositions hold:

Proposition 1.2.1

Assume that the excluded-middle laws [Eqs. (1.2.8)] hold for fuzzy sets; then union and intersection cannot be idempotent.

Proof

Due to the strict decreasing nature and continuity of C,

$$\exists p \in (0,1): \quad C(p) = p.$$

Note that p cannot be 0 or 1 since $C(0) = 1$ and $C(1) = 0$. Now since the excluded-middle laws hold, $U(p,p) = 1$ and $I(p,p) = 0$, which is inconsistent with idempotency [Eqs. (1.2.7)] due to $p \in \{0,1\}$. Q.E.D.

Proposition 1.2.2

If the excluded-middle laws [Eqs. (1.2.8)] hold for fuzzy sets and both union and intersection satisfy

$$\forall a \in (0,1), U(0,a) \neq 1,$$

$$\forall a \in (0,1), I(1,a) \neq 0,$$

then $\cup$ *and* $\cap$ *are no longer mutually distributive.*

When we want to choose union and intersection to combine fuzzy sets, we have to give up either excluded-middle laws or distributivity and idempotency.

Operators that satisfy excluded-middle laws are defined by:

$$U(a,b) = \min(1, a+b),$$

$$I(a,b) = \max(0, a+b-1),$$

$$C(a) = 1 - a.$$

Now $\tilde{P}(X)$, the set of subsets of X, is a nondistributive complemented structure. Yager [1979h] has shown that $\max(0, a+b-1)$ was the only mapping generating an intersection operator that was associative, and

$$\forall k: \quad a - k \in [0,1], b + k \in [0,1];$$

$$\forall a, b \in [0,1], I(a,b) = I(a-k, b+k) \quad (1.2.9)$$

when [Eq. (1.2.9)] is interpreted as a linear compensation effect between membership values.

An interesting operator is an *averaging operator*; an example of an averaging operator is the arithmetic mean

$$M(a,b) = \frac{a+b}{2}.$$

But others are worth considering, such as:

$$M(a,b) = \frac{\min(a,b)}{1-|a-b|}$$

and

$$M(a,b) = \frac{\max(a,b)}{1+|a-b|}, \quad \text{where } a,b \in [0,1].$$

Another interesting combination of sets is the *symmetrical difference*, in which we keep only elements that belong to *only one* of two sets.

Consider $A \Delta B = B \Delta A$, where Δ denotes the symmetrical differences between fuzzy sets. The symmetrical differences can be generated by using mappings D from $[0,1]^2$ to $[0,1]$ such that

$$D(0,1) = D(1,0) = 1; \; D(0,0) = D(1,1) = 0, \tag{1.2.10}$$

$$\forall a,b \in [0,1], \quad D(a,b) = D(1-a, 1-b). \tag{1.2.11}$$

A rather general form of symmetrical difference is $D(a,b) = f(g(a,b), g(1-a, 1-b))$, where f and g are any continuous mappings from $[0,1]^2$ and $[0, 1]$, f being commutative and such that $f(g(0,0), g(1,1)) = 0$; $f(g(0,1), g(1,0)) = 1$. Examples of symmetrical differences of fuzzy sets are discussed by Dubois and Prade [1979b] and include:

a) $g(a,b) = \min(1-a, b)$ and $f(a,b) = \max(a,b)$ hence

$$D(a,b) = \max(\min(1-a,b), \min(a, 1-b)).$$

This is consistent with the usual definition of $A \Delta B = (\bar{A} \cap B) \cup (A \cap \bar{B})$ translating $\cap$ and $\cup$ into "min" and "max." Note that $A \Delta B = (A \cup B) \cap (\bar{A} \cup \bar{B})$ also holds with these fuzzy-set operators. Here, Δ is associative.

b) $g(a,b) = 1 - a + b$ and $f(a,b) = |a-b|/2$; hence $D(a,b) = |a-b|$, which is consistent with $A \Delta B = (\bar{A} \cap B) \cup (A \cap \bar{B})$ translating $\cap$ and $\cup$ into $\max(0, . + . - 1)$ and $\min(1, . + .)$. Here, Δ is no longer associative.

The Extension Principle

One of the basic ideas of fuzzy set theory, which provides a general extension of nonfuzzy mathematical concepts to fuzzy environments, is the *extension principle*. This is a basic identity that allows the domain of the definition of a mapping or a relation to be extended from points in X to fuzzy subsets of X.

More specifically, suppose that f is a mapping from X to Y and A is a fuzzy subset of X expressed as

$$A = \chi_1/x_1 + \cdots + \chi_n/x_n.$$

Then the extension principle asserts that

$$f(A) = f(\chi_1/x_1 + \cdots + \chi_n/x_n) \equiv \chi_1/f(x_1) + \cdots + \chi_n/f(x_n).$$

Thus the image of A under f can be deduced from the knowledge of the images of $x_1, \ldots, x_n$ under f.

If we have an n-ary function f, which is a mapping from the Cartesian product $X_1 \times X_2 \times \cdots \times X_n$ to a universe Y such that $y = f(x_1, \ldots, x_n)$, and $A_1, \ldots, A_n$, which are n fuzzy sets in $X_1, \ldots, X_n$, respectively, which are characterized by a set of membership functions $\{\chi_{A_i}(x_i)\}_{i=1}^n$, then the extension principle allows us to induce from n fuzzy sets $\{A_i\}_{i=1}^n$ a fuzzy set F on Y such that

$$\chi_F(y) = \operatorname*{Sup}_{\substack{x_1, \ldots, x_n \\ y = f(x_1, \ldots, x_n)}} \min\left[\chi_{A_1}(x_1), \ldots, \chi_{A_n}(x_n)\right],$$

$$\chi_F(y) = 0 \quad \text{if } f^{-1}(y) = \emptyset.$$

■ ***Example 1.2.8***

Let $X = 1 + 2 + \cdots + 7$ and "small" be a fuzzy subset on X defined by "small" $= 1/1 + 1/2 + 0.8/3 + 0.5/4$. If we take f as the operation of squaring, then

$$f(\text{small}) = {}^2\text{small} = (\text{small}^2) = 1/1 + 1/4 + 0.8/9 + 0.5/16. \quad ■$$

■ ***Example 1.2.9***

Let $X_1 = X_2 = 1 + 2 + \cdots + 7$ and let

$$A_1 = \text{approximately } 2 = 0.6/1 + 1/2 + 0.8/3,$$

$$A_2 = \text{approximately } 4 = 0.8/3 + 1/4 + 0.7/5.$$

Let $*$ be a binary operation defined on $X_1 \times X_2$ with values in Y, and let A_1 and A_2 be fuzzy subsets on X_1 and X_2, respectively, such that

$$A_1 = \chi_{A_{11}}/x_{11} + \cdots + \chi_{A_{1n}}/x_{1n},$$

and

$$A_2 = \chi_{A_{21}}/x_{21} + \cdots + \chi_{A_{2n}}/x_{2n}.$$

The operation $*$ may be extended to fuzzy subsets of X_1 and X_2 by defining

$$A_1 * A_2 = \left(\sum_i \chi_{A_{1i}}/x_{1i}\right) * \left(\sum_j \chi_{A_{2j}}/x_{2j}\right)$$

$$= \sum_{i,j} (\chi_{A_{1i}} \wedge \chi_{A_{2j}})/(x_{1i} * x_{2j}).$$

In our case we take $*$ as the arithmetic product of A_1 and A_2 and we get

$$\begin{aligned} A_1 * A_2 &= (\text{approximately } 2 \times \text{approximately } 4) \\ &= (0.6/1 + 1/2 + 0.7/3) \times (0.8/3 + 1/4 + 0.7/5) \\ &= 0.6/3 + 0.6/4 + 0.6/5 + 0.8/6 + 1/8 + 0.7/9 \\ &\quad + 0.7/10 + 0.7/12 + 0.7/15. \quad \blacksquare \end{aligned}$$

If the support of A is a continuum, that is,

$$A = \int_X \chi_A(x)/x,$$

then the statement of the extension principle assumes the following form:

$$f(A) \triangleq f\left(\int_X \chi_A(x)/x\right) \triangleq \int_Y \chi_A(x)/f(x),$$

with the understanding that $f(x)$ is a point in Y and $\chi_A(x)$ is its grade of membership in $f(A)$, which is a fuzzy subset of Y.

In some applications, it is convenient to use a modified form of the extension principle by decomposing A into its constituent level-sets rather than its fuzzy singletons. Thus, on writing

$$A = \int_0^1 \alpha A_\alpha,$$

where A_α is an α-level-set of A, the statement of the extension principle takes the form

$$f(A) = f\left(\int_0^1 \alpha A_\alpha\right) \equiv \int_0^1 \alpha f(A_\alpha)$$

when the support of A is a continuum, and

$$f(A) = f\left(\sum_\alpha \alpha A_\alpha\right) = \sum_\alpha \alpha f(A_\alpha),$$

when either the support of A is a countable set or the distinct level-sets of A form a countable collection.

Clearly, with the principle of extension, we can fuzzify any domain of mathematical reasoning using set theory. It should be clear, however, that another way of fuzzifying a structure is just to replace nonfuzzy sets by fuzzy sets or by the family of their α-cuts in the framework of the structure.

1.3 PROPERTIES OF FUZZY SETS

1.3.1 Cardinality

In the case of a crisp (nonfuzzy) subset, A, of a universe of discourse, U, the proposition "u is an element of A" is either true or false, and hence there is just one way in which the cardinality of A, that is, the count of elements of A, may be defined. However, even though the count may be defined uniquely, there may be some uncertainty about its value if there is an uncertainty regarding the membership status of points of U in A.

By contrast, in the case of a fuzzy subset, F, of U, the proposition "u is an element of F" is generally true, based upon the meaning of the concept of cardinality. Among the different definitions of cardinality, some associate with a fuzzy set F a real number, in which case the cardinality of a fuzzy set is nonfuzzy. Others associate with F a fuzzy number, since it may be argued that the cardinality of a fuzzy set should be a fuzzy number. For simplicity, we shall restrict our attention to finite universes of discourse, in which case a fuzzy subset, F, of $U = \{u_1, \ldots, u_n\}$ may be expressed symbolically as

$$F = \chi_1/u_1 + \cdots + \chi_n/u_n$$

or, more simply, as

$$F = \chi_1 u_1 + \cdots + \chi_n u_n,$$

in which the term χ_i/u_i, $i = 1, \ldots, n$, signifies that χ_i is the grade of membership of u_i in F, and the plus sign represents union.

A simple way of extending the concept of cardinality to fuzzy sets is to form the *sigma count*, which is the arithmetic sum of the grades of membership in F. Thus

$$\sum \mathrm{Count}(F) \triangleq \sum_i \chi_i, \quad i = 1, \ldots, n,$$

with the understanding that the sum may be rounded, if need be, to the nearest integer. Furthermore, one may stipulate that terms whose grade of membership falls below a specified threshold be excluded from the summation. The purpose of such an exclusion is to avoid a situation in which a large number of terms with low grades of membership become count-equivalent to a small number of terms with high membership.

As a simple illustration of the concept of sigma-count, assume that the fuzzy set of close friends of Ted is expressed as

$$F = 1/\mathrm{Joe} + 0.8/\mathrm{Sam} + 0.7/\mathrm{Eli} + 0.9/\mathrm{Ralph} + 0.9/\mathrm{Ron}.$$

In this case,

$$\begin{aligned} \sum \mathrm{Count}(F) &= 1 + 0.8 + 0.7 + 0.9 + 0.9 \\ &= 4.3. \end{aligned}$$

A sigma-count may be *weighted*, in the sense that if $w = (w_1, \ldots, w_n)$ is an n-tuple of nonnegative real numbers, then the *weighted sigma-count of F with respect to w* is defined by

$$\sum \mathrm{Count}(F; w) \triangleq \sum_i w_i \chi_i, \quad i = 1, \ldots, n.$$

This definition implies that $\Sigma\,\mathrm{Count}(F; w)$ may be interpreted as the sigma-count of a fuzzy multiset $'F$ in which the grade of membership and the multiplicity of u_i, $i = 1, \ldots, n$, are, respectively, χ_i and w_i. The sigma-count of a fuzzy set is clearly a real number, in this case. A fuzzy multiset $'F$ may be represented as

$$'F = \sum_i \chi_i / m_i \times u_i,$$

in which m_i is the *multiplicity* of u_i and χ_i is the grade of membership of u_i in the fuzzy set $F = \Sigma_i \chi_i / u_i$. The multiplicity, m_i, is a nonnegative real number that is usually, but not necessarily, an integer. Thus, a fuzzy multiset may have identical elements, or elements that differ only in their grade of membership. It may be argued, however, that the cardinality of a fuzzy set should be a fuzzy number. If one adopts such a philosophy, then a natural way of defining fuzzy cardinality is by defining F in terms of its level sets. Namely,

$$F = \sum_\alpha \alpha F_\alpha,$$

in which the α-level-sets F_α are nonfuzzy sets defined by

$$F_\alpha \triangleq \{u | \chi_F(u) \geq \alpha\}, \qquad 0 < \alpha \leq 1,$$
$$\chi_{\alpha F_\alpha}(u) = \alpha \chi_F(u), \qquad u \in U.$$

In terms of this representation, there are three fuzzy counts, *F* Counts, that may be associated with F. First, the *FG* Count is defined as the conjunctive fuzzy integer

$$FG\,\text{Count}(F) = 1/0 + \Sigma_\alpha \alpha / \text{Count}(F_\alpha), \qquad \alpha > 0.$$

Second, the *FL* Count is defined as

$$FL\,\text{Count}(F) = (FG\,\text{Count}(F))' \ominus 1,$$

where $'$ denotes the complement (the count of the complementary set), and $\ominus$ 1 means that 1 is subtracted from the fuzzy number *FG* Count(F). And finally, the *FE* Count(F) is defined as the intersection of *FG* Count(F) and *FL* Count(F), that is,

$$FE\,\text{Count}(F) = FG\,\text{Count}(F) \cap FL\,\text{Count}(F).$$

Equivalently — and more precisely — we may define the counts in question via the membership function of F, that is,

$$\chi_{FG\,\text{Count}(F)}(i) \triangleq \sup_\alpha \{\alpha | \text{Count}(F_\alpha) \geq i\}, \qquad i = 0, 1, \ldots, n,$$
$$\chi_{FL\,\text{Count}(F)}(i) \triangleq \sup_\alpha \{\alpha | \text{Count}(F_\alpha \geq n - i)\},$$
$$\chi_{FE\,\text{Count}(F)}(i) \triangleq \chi_{FG\,\text{Count}(F)}(i) \wedge \chi_{FL\,\text{Count}(F)}(i),$$

where $\wedge$ stands for min in infix position.

As a simple illustration, consider the fuzzy set expressed as

$$F = 0.6/u_1 + 0.9/u_2 + 1/u_3 + 0.7/u_4 + 0.3/u_5.$$

In this case,

$$F_{1.0} = u_3,$$
$$F_{0.9} = u_2 + u_3,$$
$$F_{0.7} = u_2 + u_3 + u_4,$$
$$F_{0.6} = u_1 + u_2 + u_3 + u_4,$$
$$F_{0.3} = u_1 + u_2 + u_3 + u_4 + u_5,$$

which implies that F may be expressed by

$$F = 1(u_3) + 0.9(u_2 + u_3) + 0.7(u_2 + u_3 + u_4) + 0.6(u_1 + u_2 + u_3 + u_4) + 0.3(u_1 + u_2 + u_3 + u_4 + u_5),$$

and hence that

$$FG\,\text{Count}(F) = 1/0 + 1/1 + 0.9/2 + 0.7/3 + 0.6/4 + 0.3/5$$
$$FL\,\text{Count}(F) = 0.1/2 + 0.3/3 + 0.4/4 + 0.7/5 + 1/6 + \cdots \ominus 1$$
$$= 0.1/1 + 0.3/2 + 0.4/3 + 0.7/4 + 1/5 + \cdots$$
$$FE\,\text{Count}(F) = 0.1/1 + 0.3/2 + 0.4/3 + 0.6/4 + 0.3/5,$$

while, by comparison,

$$\Sigma\,\text{Count}(F) = 0.6 + 0.9 + 1.0 + 0.7 + 0.3 = 3.5.$$

Hence,

(a) $\chi_{FG\,\text{Count}}(i)$ is the truth value of the proposition "F contains *at least* i elements."

(b) $\chi_{FL\,\text{Count}}(i)$ is the truth value of the proposition "F contains *at most* i elements."

(c) $\chi_{FE\,\text{Count}}(i)$ is the truth value of the proposition "F contains i *and only* i elements."

From (a), it follows that $FG\,\text{Count}(F)$ may readily be obtained from F by first sorting F in the order of decreasing grades of membership and then replacing the elements by consecutive integers and adding the term 1/0. For example, for F defined above, we have

$$F\downarrow\, = 1/u_3 + 0.9/u_2 + 0.7/u_4 + 0.6/u_1 + 0.3/u_5$$
$$NF\downarrow\, = 1/1 + 0.9/2 + 0.7/3 + 0.6/4 + 0.3/5,$$

and

$$FG\,\text{Count}(F) = 1/0 + 1/1 + 0.9/2 + 0.7/3 + 0.6/4 + 0.3/5,$$

where $F\downarrow$ denotes F sorted in descending order, and $NF\downarrow$ is $F\downarrow$ with the ith u replaced by i. An immediate consequence of this relation between $\Sigma\,\text{Count}(F)$ and $FG\,\text{Count}(F)$ is the identity

$$\Sigma\,\text{Count}(F) = \sum_i \chi_{FG\,\text{Count}}(i) - 1,$$

which shows that, as a real number, $\Sigma\,\text{Count}(F)$ may be regarded as a "summary" of the fuzzy number $FG\,\text{Count}(F)$.

A type of count that plays an important role in the representation of meaning is that of *relative count* (or *relative cardinality*). Specifically, if F and G are fuzzy sets, then the relative sigma-count of F in G is defined as the ratio:

$$\Sigma\,\text{Count}(F/G) = \frac{\Sigma\,\text{Count}(F \cap G)}{\Sigma\,\text{Count}(G)},$$

which represents the proportion of elements of F that are in G, with the intersection $F \cap G$ defined by

$$\chi_{F \cap G}(u) = \chi_F(u) \wedge \chi_G(u).$$

The corresponding definition for the FG Count is

$$FG\,\text{Count}(F/G) = \Sigma_\alpha \alpha/\text{Count}(F_\alpha \cap G_\alpha)/\text{Count}(G_\alpha),$$

where the F_α and G_α represent the α-sets of F and G, respectively. The terms of the form α_1/u and α_2/u should not be combined into a single term $(\alpha_1 \vee \alpha_2)/u$, as they would be in the case of a fuzzy set.

The Σ Count and F Counts of fuzzy sets have a number of basic properties of which only a few will be stated here. Specifically, if F and G are fuzzy sets, then from the identity

$$(a \vee b) + (a \wedge b) = a + b,$$

which holds for any real numbers, it follows at once that

$$\Sigma\,\text{Count}(F \cap G) + \Sigma\,\text{Count}(F \cup G) = \Sigma\,\text{Count}(F) + \Sigma\,\text{Count}(G),$$

since

$$\chi_{F \cap G}(u) = \chi_F(u) \wedge \chi_G(u), \qquad u \in U,$$

and

$$\chi_{F \cup G}(u) = \chi_F(u) \vee \chi_G(u).$$

Thus, if F and G are disjoint, then

$$\Sigma\,\text{Count}(F \cup G) = \Sigma\,\text{Count}(F) + \Sigma\,\text{Count}(G),$$

and, more generally,

$$\Sigma\,\text{Count}(F) \vee \Sigma\,\text{Count}(G) \leq \Sigma\,\text{Count}(F \cup G) \leq \Sigma\,\text{Count}(F) + \Sigma\,\text{Count}(G)$$

and

$$\Sigma\,\text{Count}(F) + \Sigma\,\text{Count}(G) - \Sigma\,\text{Count}(U) \leq \Sigma\,\text{Count}(F \cap G)$$
$$\leq \Sigma\,\text{Count}(F) \wedge \Sigma\,\text{Count}(G).$$

Clearly,

$$\Sigma\,\text{Count}(A \cap B) \leq \Sigma\,\text{Count}(A),$$
$$\Sigma\,\text{Count}(A \cap B) \leq \Sigma\,\text{Count}(B),$$
$$\Sigma\,\text{Count}(A \cup B) \leq \Sigma\,\text{Count}(U).$$

In the case of F Counts and, more specifically, the FG Count, we also have

$$FG\,\text{Count}(F \cap G) \oplus FG\,\text{Count}(F \cup G) = FG\,\text{Count}(F) \oplus FG\,\text{Count}(G),$$

where $\oplus$ denotes the addition of fuzzy numbers, which is defined by

$$\chi_{A \oplus B}(u) = \sup_v(\chi_A(v) \wedge \chi_B(u - v)), \qquad u, v \in (-\infty, \infty),$$

where A and B are fuzzy numbers, and χ_A and χ_B are their respective membership functions.

A basic identity that holds for relative counts may be expressed as:

$$\Sigma\,\text{Count}(F \cap G) = \Sigma\,\text{Count}(G)\,\Sigma\,\text{Count}(F/G)$$

for sigma-counts, and as

$$FG\,\text{Count}(F \cap G) = F\,\text{Count}(G) \otimes FG\,\text{Count}(F/G)$$

for FG Counts, where $\otimes$ denotes the multiplication of fuzzy numbers, which is defined by

$$\chi_{A \otimes B}(u) = \sup_v\left(\chi_A(v) \wedge \chi_B\left(\frac{u}{v}\right)\right), \qquad u, v \in (-\infty, \infty), \quad v \neq 0.$$

An inequality involving relative sigma-counts that is of relevance to the analysis of evidence in expert systems is the following:

$$\begin{aligned}\Sigma\,\text{Count}(F/G) + \Sigma\,\text{Count}(\neg F/G) &\geq 1\\ &= 1 \text{ if } G \text{ is nonfuzzy,}\end{aligned}$$

where $\neg F$ denotes the complement of F, that is,

$$\chi_{\neg F}(u) = 1 - \chi_F(u), \qquad u \in U.$$

It should be noted that, if the relative sigma-count $\Sigma\,\text{Count}(F/G)$ is identified with the conditional probability $\text{Prob}(F/G)$, then

$$\text{Prob}(\neg F/G) \geq 1 - \text{Prob}(F/G)$$

rather than

$$\text{Prob}(\neg F/G) = 1 - \text{Prob}(F/G),$$

which holds if G is nonfuzzy.

The inequality in question follows at once from

$$\begin{aligned}\Sigma\,\text{Count}(\neg F/G) &= \frac{\Sigma_i[(1 - \chi_F(u_i)) \wedge \chi_G(u_i)]}{\Sigma_i \chi_G(u_i)}\\ &\geq \frac{\Sigma_i(1 - \chi_F(u_i)\chi_G(u_i))}{\Sigma_i \chi_G(u_i)}\\ &\geq 1 - \frac{\Sigma_i \chi_F(u_i)\chi_G(u_i)}{\Sigma_i \chi_G(u_i)}\\ &\geq 1 - \frac{\Sigma_i \chi_F(u_i) \wedge \chi_G(u_i)}{\Sigma_i \chi_G(u_i)},\end{aligned}$$

since

$$\Sigma\,\mathrm{Count}(F/G) = \frac{\Sigma_i \chi_F(u_i) \wedge \chi_G(u_i)}{\Sigma_i \chi_G(u_i)}.$$

The concept of cardinality plays an essential role in representing the meaning of fuzzy quantifiers. This connection will be made more concrete in Chapter 2, and a basis for inference from propositions containing fuzzy quantifiers will be established.

1.3.2 Convex Fuzzy Sets

The notion of convexity can be generalized to fuzzy sets of a universe X, which we shall assume to be a real Euclidean N-dimensional space.

A fuzzy set A is *convex* iff its α-cuts are convex. An equivalent definition of convexity is: A is convex iff

$$\forall x_1 \in X, \quad \forall x_2 \in X, \quad \forall \lambda \in [0,1],$$

$$\chi_A[\lambda x_1 + (1-\lambda)x_2] \geq \min[\chi_A(x_1), \chi_A(x_2)].$$

It should be noted that this definition does not imply that χ_A is a convex function of x.

Clearly if A and B are convex, so is $A \cap B$. An element x of X can also be written $(x^1, x^2, \ldots, x^N)$ since X has N dimensions. The *projection* (*shadow*) of A on the hyperplane $H = \{x, x^i = 0\}$ is defined to be a fuzzy set $P_H(A)$ such that

$$\chi_{P_H(A)}(x^1, \ldots, x^{i-1}, x^{i+1}, \ldots, x^N) = \sup_{x^i} \chi_A(x^1, \ldots, x^N).$$

When A is a convex fuzzy set, so is $P_H(A)$. Moreover, if A and B are convex and if $\forall H$, $P_H(A) = P_H(B)$, then $A = B$.

1.4 *L*-FUZZY SETS

Let L be a set. An *L-fuzzy set A* is associated with a function χ_A from the universe X to L. If L has a given structure, such as lattice or group structure, $P_L(X)$, the set of L-fuzzy sets on X, will have the same structure too. For example, let L be a lattice. The intersection and the union of L-fuzzy sets can be induced in the following way:

$$\forall x \in X, \quad \chi_{A \cap B}(x) = \inf(\chi_A(x), \chi_B(x)),$$

$$\forall x \in X, \quad \chi_{A \cup B}(x) = \sup(\chi_A(x), \chi_B(x)),$$

where inf and sup denote respectively the greatest lower bound and the least upper bound. Note that membership values of L-fuzzy sets cannot always be compared unless L is linearly ordered. Moreover, distributivity and complementation require that a richer structure be defined.

There may occur some situations for which valuation sets different from [0, 1] are worth considering. For instance, if m ordinary fuzzy sets A_i $(i = 1, m)$ in X correspond to m properties, it is possible to associate with each $x \in X$ the vector of membership values $[\chi_{A_i}(x)]$ that represent the degree to which x satisfies the properties. A function from X to the set $L = [0, 1]^m$ can be built and L is a complete lattice that is not a linear ordering.

Now assume that each element x of X is described by means of only one property among $A_1, \ldots, A_m$, supposedly the most significant one for x. The property that best describes an element $x' \neq x$ may be different from that which describes x. We obtain in this way a partition of X into m classes. Obviously, it is meaningless to compare membership values of elements in different classes. Thus, the valuation set is here a collection of m disjoint linear orderings.

One can obviously extend the concept of fuzzy sets into a hierarchy of k types of fuzzy sets in a recursive way:

a type 1 fuzzy set is an ordinary fuzzy set in X;
a type k fuzzy set $(k > 1)$ in X is an L-fuzzy set whose membership values are type $(k - 1)$ fuzzy sets on [0, 1].

Let $\tilde{P}_k(X)$ be the set of type k fuzzy sets in X. Then $\tilde{P}_1(X) = \tilde{P}(X)$.

Union, intersection, and complementation of type k fuzzy sets can also be recursively defined by induction from the structure of the valuation set. Let us denote these operators by $\cup_k$, $\cap_k$, ${}^{-k}$ for instance,

$$\cup_1 = \cup; \ \chi_{A \cup_k B}(x) = \chi_A(x) \cup_{k-1} \chi_B(x), \quad k > 1.$$

We follow the excellent presentation of Mizumoto and Tanaka [1976d] in their discussion of fuzzy sets of type 2.

Fuzzy sets of type 2

A fuzzy set of type 2, A, in a universe of discourse X, is characterized by a *fuzzy membership function* χ_A as

$$\chi_A\colon \quad X \to [0, 1]^{[0,1]},$$

where the value $\chi_A(x)$ is the *fuzzy grade*, and is a fuzzy set in the unit interval [0, 1] represented by

$$\chi_A(x) = \int f(u)/u, \quad u \in [0, 1],$$

where f is a membership function for the fuzzy grade $\chi_A(x)$ and is defined as

$$f\colon \quad [0, 1] \to [0, 1].$$

Let $\chi_A(x)$ and $\chi_B(x)$ be fuzzy grades for fuzzy sets of type 2, A and B, represented as

$$\left.\begin{aligned} \chi_A(x) &= \int f(u)/u, \qquad u \in [0,1], \\ \chi_B(x) &= \int g(w)/w, \qquad w \in [0,1]. \end{aligned}\right\} \quad \begin{aligned} &f \text{ and } g \text{ depend on } x \\ &\text{as well as on } u \text{ (or } w\text{)}. \end{aligned}$$

Then the operations of algebraic product and algebraic sum for fuzzy grades $\chi_A(x)$ and $\chi_B(x)$ are defined as follows by using the extension principle:

Algebraic product

$$AB \Leftrightarrow \chi_{AB}(x) = \chi_A(x) \cdot \chi_B(x)$$
$$= \left(\int f(u)/u\right) \cdot \left(\int g(w)/w\right) = \int (f(u) \wedge g(w))/uw.$$

Algebraic sum

$$A \dotplus B \Leftrightarrow \chi_{A \dotplus B}(x) = \chi_A(x) \dotplus \chi_B(x) = \int (f(u) \wedge g(w))/(u \dotplus w)$$
$$= \int (f(u) \wedge g(w))/(u + w - uw).$$

The *complement* of a fuzzy set of type 2, A, is defined as

$$\bar{A} \Leftrightarrow \chi_{\bar{A}}(x) = \neg\chi_A(x) = \int f(u)/(1-u).$$

■ ***Example 1.4.1***

As a simple example, we shall execute the operation of algebraic product for discrete fuzzy grades $\chi_A(x)$ and $\chi_B(x)$. Let

$$\chi_A(x) = 0.5/0.2 + 1/0.4 + 0.8/0.6,$$

and let

$$\chi_B(x) = 1/0.2 + 0.9/0.4 + 0.4/0.6.$$

Then $\chi_{AB}(x)$ is given by

$$\begin{aligned} \chi_A(x) \cdot \chi_B(x) &= \frac{1 \wedge 0.5}{0.2 \times 0.2} + \frac{1 \wedge 1}{0.2 \times 0.4} + \frac{1 \wedge 0.8}{0.2 \times 0.6} \\ &\quad + \frac{0.9 \wedge 0.5}{0.4 \times 0.2} + \frac{0.9 \wedge 1}{0.4 \times 0.4} + \frac{0.9 \wedge 0.8}{0.4 \times 0.6} \\ &\quad + \frac{0.4 \wedge 0.5}{0.6 \times 0.2} + \frac{0.4 \wedge 1}{0.6 \times 0.4} + \frac{0.4 \wedge 0.8}{0.6 \times 0.6} \\ &= 0.5/0.04 + 1/0.08 + 0.8/0.12 \\ &\quad + 0.9/0.16 + 0.8/0.24 + 0.4/0.36. \end{aligned}$$

■

■ ***Example 1.4.2***

Let $\chi_A(x)$ and $\chi_B(x)$ be *continuous* fuzzy grades such that

$$\chi_A(x) = \chi_B(x) = \int_0^1 u/u.$$

We can obtain the algebraic product, algebraic sum, and negation of fuzzy grades $\chi_A(x)$ and $\chi_B(x)$ as follows:

$$\chi_A(x) \cdot \chi_B(x) = \int_0^1 \sqrt{u}\,/u,$$

$$\chi_A(x) \dotplus \chi_B(x) = \int_0^1 1 - \sqrt{1-u}\,/u,$$

$$\neg\chi_A(x) = \int_0^1 1 - u/u.$$

■

Join and meet

Join ($\sqcup$) *and meet* ($\sqcap$) *of fuzzy grades* χ_A *and* χ_B *are defined as follows by using the extension principle*:

$$\textit{Join}: \quad \chi_A \sqcup \chi_B = \int (f(u) \vee g(w))/(u \vee w),$$

$$\textit{Meet}: \quad \chi_A \sqcap \chi_B = \int (f(u) \wedge g(w))/(u \wedge w),$$

where $\vee$ *and* $\wedge$ *stand for max and min, respectively.*

Clearly, arbitrary fuzzy grades satisfy idempotent laws, commutative laws, and associative laws under join ($\sqcup$) and meet ($\sqcap$). Thus, they constitute a partially ordered set. Also, convex fuzzy grades are closed and also satisfy distributive laws under $\sqcup$ and $\sqcap$. Therefore, they form a commutative semiring, but do not form a lattice since they do not satisfy absorption laws.

It can be shown that normal convex fuzzy grades are closed and also satisfy absorption laws under $\sqcup$ and $\sqcap$. Thus, they form a distributive lattice under $\sqcup$ and $\sqcap$.

Therefore, if we let χ_A be convex fuzzy grade, and let χ_B and χ_C be arbitrary fuzzy grades, then we have the following results:

$$\chi_A \cdot (\chi_B \sqcup \chi_C) = (\chi_A \cdot \chi_B) \sqcup (\chi_A \cdot \chi_C),$$

$$\chi_A \cdot (\chi_B \sqcap \chi_C) = (\chi_A \cdot \chi_B) \sqcap (\chi_A \cdot \chi_C),$$

$$\chi_A \dotplus (\chi_B \sqcup \chi_C) = (\chi_A \dotplus \chi_B) \sqcup (\chi_A \dotplus \chi_C),$$

$$\chi_A \dotplus (\chi_B \sqcap \chi_C) = (\chi_A \dotplus \chi_B) \sqcap (\chi_A \dotplus \chi_C).$$

If, however, χ_A is not convex, the above identities do not hold even if χ_B and χ_C are convex, as the following example by Mizumoto and Tanaka [1976d] shows:

■ ***Example 1.4.3***

Let

$$\chi_A = \int_0^{0.5} 1 - 2u/u + \int_{0.5}^1 2u - 1/u, \qquad \chi_B = \int_0^{0.5} 2u/u + \int_{0.5}^1 1/u,$$

$$\chi_C = \int_0^{0.5} 2u/u + \int_{0.5}^1 2(1-u)/u.$$

■

Then we have

$$\chi_A \cdot (\chi_B \sqcap \chi_C) = \int_0^{3/16} [(3 - \sqrt{1 + 16u})/2]/u + \int_{3/16}^{0.5} [(\sqrt{1 + 16u} - 1)/2]/u + \int_{0.5}^{1} 2(1 - u)/u,$$

$$(\chi_A \cdot \chi_B) \sqcap (\chi_A \cdot \chi_C) = \int_0^{u_0} 1 - 2u/u + \int_{u_0}^{0.5} [(\sqrt{1 + 16u} - 1)/2]/u + \int_{0.5}^{1} 2(1 - u)/u, \quad u_0 = \frac{5 - \sqrt{17}}{4},$$

and clearly

$$\chi_A \cdot (\chi_B \sqcap \chi_C) \neq (\chi_A \cdot \chi_B) \sqcap (\chi_A \cdot \chi_C)$$

when χ_A is nonconvex. Similarly, if χ_A and χ_B are convex fuzzy grades, then

$$(\chi_A \sqcap \chi_B) \cdot (\chi_A \sqcup \chi_B) = \chi_A \cdot \chi_B$$

and

$$(\chi_A \sqcup \chi_B) \dot{+} (\chi_A \sqcap \chi_B) = \chi_A \dot{+} \chi_B. \quad ■$$

If χ_A and/or χ_B are nonconvex, the above identities are not satisfied as shown by the example below [Mizumoto and Tanaka [1976d]:

■ ***Example 1.4.4***

Let χ_A and χ_B be nonconvex fuzzy grades given by

$$\chi_A = \int_0^{0.5} 1 - 2u/u + \int_{0.5}^{1} 2u - 1/u,$$

and

$$\chi_B = \int_0^{0.5} 2u/u + \int_{0.5}^{1} 2(1 - u)/u.$$

Then

$$(\chi_A \sqcup \chi_B) \cdot (\chi_A \sqcap \chi_B) = \int_0^{u_0} 1 - 2u/u + \int_{u_0}^{1/4} 2\sqrt{u}/u + \int_{1/4}^{9/25} 2(1 - \sqrt{u})/u + \int_{9/25}^{0.5} [(-1 + \sqrt{1 + 16u})/2]/u + \int_{0.5}^{1} 2(1 - u)/u,$$

$$u_0 = 1 - \frac{\sqrt{3}}{2},$$

and

$$\chi_A \cdot \chi_B = \int_0^{3/16} [(3 - \sqrt{1 + 16u})/2]/u + \int_{3/16}^{0.5} [(-1 + \sqrt{1 + 16u})/2]/u + \int_{0.5}^{1} 2(1 - u)/u. \quad ■$$

CHAPTER TWO

Possibility Theory and Fuzzy Quantification

2.1 ESTIMATION OF MEMBERSHIP FUNCTIONS

In many applied problems, we have to estimate the membership function of some fuzzy variables. We follow the brilliant discussion by Dishkant [1981] in observing such an estimation technique. It has been shown by Giles [1976b] that fuzzy-set techniques result from Łukasiewicz' infinitely many-valued logic t_w. In classical logic the truth values are 0 and 1, but in t_w the truth values cover the whole segment [0, 1]. For all that, for both logics, a proposition is true if its truth value is 1.

The main connectives of t_w are: rigid conjunction $\wedge$, strong conjunction $\circ$, disjunction $\vee$, implication $\rightarrow$, negation $\sim$, equivalence $\leftrightarrow$, and quantifiers $\forall$ and $\exists$. They are defined with the help of equalities:

$$\begin{aligned}
\chi \wedge \lambda &\triangleq \min\{\chi, \lambda\}, & \chi \circ \lambda &\triangleq \max\{\chi + \lambda - 1, 0\}, \\
\chi \vee \lambda &\triangleq \max\{\chi, \lambda\}, & \chi \rightarrow \lambda &\triangleq \min\{1 - \chi + \lambda, 1\}, \\
\sim \chi &\triangleq 1 - \chi, & \chi \leftrightarrow \lambda &\triangleq 1 - |\chi - \lambda|, \\
\forall x_\chi(x) &\triangleq \inf_x \chi(x), & \exists x_\chi(x) &\triangleq \sup_x \chi(x).
\end{aligned}$$

The main difficulty, deriving from the recording of facts in the language of t_w, is the presence of many different connectives in t_w, corresponding to one connective of the classical logic. However, one must decide when it is appropriate to use the rigid conjunction and when the strong one. We now present the limit theorem due to Dishkant [1981].

Limit theorem

Let $\chi_k(x), k = 1, 2, \ldots,$ be the sequence of the membership functions of the fuzzy variables S_k and assume that the following conditions are fulfilled:

1. *For any k the bearer X_k of S_k is a finite interval and there is such M that for any k and any $x \in X_k$, $|\chi_k'''(x)| \leq M$.*
2. *For any k there is one and only one point a_k such that $\chi_k(a_k) = 1$.*
3. *There are A and B such that for any k, $0 < A \leq -\chi_k''(a_k) \leq B$.*
4. *There is $m > 0$ such that for any k, if b is a bending point of χ_k', $x \in X_k$ and $(b - a_k)(x - b) > 0$, then $|\chi_k'(x)| \geq m$.*
5. *For any k, $\chi_k(x)$ is continuous.*

Then for $N \to \infty$

$$\chi(z) - \rho(z) \to 0$$

uniformly for all $z \in R$. Here $'$ denotes the derivative,

$$\chi \triangleq \chi_{\Sigma S_k}, \quad a \triangleq \Sigma a_k, \quad c \triangleq \left(\Sigma c_k^{-1}\right)^{-1}, \quad c_k \triangleq -\chi_k''(a_k), \quad \Sigma \triangleq \Sigma_k^N = 1,$$

and

$$\rho(z) \triangleq \max\left\{1 - \tfrac{1}{2}c(z - a)^2, 0\right\}.$$

In a practical situation the limit theorem determines, at the best, only a type of membership function, but the parameters a and c remain unknown. One must solve the problem of the parameter estimation. The main peculiarity of this consideration is the interpretation of unknown parameters as fuzzy variables.

Let $\lambda_A(t)$ be a membership function of a fuzzy parameter A. Consider a general case and take λ_A: $T \to [0, 1]$, where T is an arbitrary set of the parameter values, and call the value $t_0 \in T$ an *optimal estimate* of the parameter A if

$$\exists t \lambda_A(t) \to \lambda_A(t_0)$$

is true in t_w. Namely, t_0 is an optimal estimate if the existence of any estimate involving t_0 is an estimate. Thus if one agrees that "t is an estimate for A," then he must agree that "t_0 is an estimate for A" also. This is a distinctive property of the optimal estimate, expressed in t_w.

Of course, in some cases there are one or many optimal estimates; in others there is not one. In the last case we may consider ε-optimal estimate, the definition of which is easy to give. In essence, t_0 is one of the values of t, for which $\lambda_A(t)$ is maximal. Hence,

$$t_0 = \arg\max \lambda_A(t).$$

Therefore, let $\chi(x, t)$ be the truth value of the assertion "If A has the value t, then the fuzzy variable S has the value x." Assume χ: $X \times T \to [0, 1]$ is *a priori* known. The problem is in finding an optimal estimate t_0 of A, with the aid of empirical data. Such data may be the results of experiment or some subjective estimate. Suppose the empirical data to be the truth values χ_k of sentences: "S has a value x_k," $k = 1, 2, \ldots, n$.

On the whole, the empirical situation may be expressed in t_w by the condition

$$\bigwedge_{1 \leq k \leq n} (\chi(x_k, t) \leftrightarrow \chi_k),$$

where $\bigwedge_{1 \leq k \leq n}$ denotes the multinomial rigid conjunction. This means that the parameter A becomes a fuzzy variable with a membership function, defined above: Thus the optimal estimate of A is given by

$$t_0 = \arg\min_t \max_{1 \leq k \leq n} |\chi(x_k, t) - \chi_k|.$$

■ ***Example 2.1.1***

Let S be a fuzzy real number; its membership function $\chi_s(x; a, c)$ depends on a pair $\langle a, c \rangle$ of unknown parameters. Let $\chi_s(x; a, c)$ be *a priori* given. Denote by χ_k the truth value of the sentence "x_k is the value of S." Let χ_k, $k = 1, \ldots, n$ be given. We have to find an estimate $\langle a_0, c_0 \rangle$ of the parameters.

The optimal estimate $\langle a_0, c_0 \rangle$ is the one minimizing

$$\max_{1 \leq k \leq n} |\chi_k - \chi(x_k; a, c)|.$$

As an example, let us take S to be a normal fuzzy variable. Then $\langle a_0, c_0 \rangle$ is such that the maximal vertical distance between the curve

$$\rho(z) = \max\left\{0, 1 - \tfrac{1}{2}c(z - a)^2\right\}$$

and the set of points $\langle x_k, \chi_k \rangle$, $k = 1, 2, \ldots, n$, is minimal. ■

■ ***Example 2.1.2***

Let X be a one-element set. It is therefore necessary to estimate the truth value χ_0 of a constant proposition. If the experiment gives values $\chi_1, \chi_2, \ldots, \chi_n$ for χ, then the optimal estimate χ_0 is such that $\max_k |\chi_0 - \chi_k|$ is minimal. Obviously, $\chi_0 = \frac{1}{2}(\chi_1 + \chi_n)$ if $\chi_1 \leq \chi_2 \leq \cdots \leq \chi_n$. That is, if experts give values $\chi_1 = 0.4$, $\chi_2 = 0.7$, $\chi_3 = 0.8$ for the truth of proposition p, then one must take $\chi_0 = \frac{1}{2}(0.4 + 0.8) = 0.6$ as the truth value of p. ■

■ ***Example 2.1.3***

Let X be some set of measurable subsets of a measure space, dependent on the parameter t. Let $\chi(x, t)$ be the probability of the event x if some fixed parameter takes the value $t \in T$. Assume fulfilled experiments in which the events $x_1, \ldots, x_n$ occur. The problem is in estimating the unknown value of the parameter. This is the formulation of the classical problem of mathematical statistics.

The simultaneous occurrence of all x_k is equivalent to the occurrence of the random sample $x_0 = x_1 \cap x_2 \cap \cdots \cap x_n$. Of course, the sentence "x_0 happens" has the truth value $\chi_0 = 1$, and thus

$$t_0 = \arg\max \chi(x_0, t).$$

In other words, as the estimate of the unknown parameter one must take that value by which the probability of the sample is maximal. We have just obtained the *principle of maximum likelihood* of the classical statistical theory. Hence, in the field of statistics, the fuzzy method coincides with the classical method of maximum likelihood. ■

2.2 TOWARD THE THEORY OF POSSIBILITY

This section is devoted to the subject of possibility developed by Zadeh [1978]. We now follow Zadeh's powerful presentation.

Let Y be a variable taking values in X; then a *possibility distribution*, Π_Y, associated with Y may be viewed as a fuzzy constraint on the values that may be assigned to Y. Such a distribution is characterized by a *possibility distribution function* π_Y: $X \to [0,1]$ which associates with each $x \in X$ the "degree of ease" or the possibility that Y may take x as a value.

In some cases, the constraint on the values of Y is physical in origin; in many cases, however, the possibility distribution that is associated with a variable is epistemic rather than physical. A basic assumption in fuzzy logic is that such epistemic possibility distributions are induced by propositions expressed in a natural language. In more concrete terms, this assumption may be stated as the following postulate.

Possibility Postulate

If F is a fuzzy subset of X characterized by its membership function χ_F: $X \to [0,1]$, then the proposition "Y is F" induces a possibility distribution Π_Y that is equal to F. Equivalently, "Y is F" translates into the possibility assignment equation $\Pi_Y = F$; that is,

$$Y \text{ is } F \to \Pi_Y = F,$$

which signifies that the proposition "Y is F" has the effect of constraining the values that may be assumed by Y, with the possibility distribution Π_Y identified with F.

An important aspect of the concept of a possibility distribution is that it is nonstatistical in nature. As a consequence, if P_Y is a probability distribution associated with Y, then the only connection between Π_Y and P_Y is that impossibility (i.e., zero possibility) implies improbability but not vice versa. Thus, Π_Y cannot be inferred from P_Y nor can P_Y be inferred from Π_Y.

As in the case of probabilities, one can define joint and conditional possibilities. Thus, if Y_1 and Y_2 are variables taking values in X_1 and X_2, respectively, then we can define the joint and conditional possibility distributions through their respective distribution functions:

$$\pi_{(Y_1,Y_2)}(x_1,x_2) = \text{Poss}\{Y_1 = x_1, Y_2 = x_2\}, \qquad x_1 \in X_1, \quad x_2 \in X_2,$$

and

$$\pi_{(Y_1|Y_2)}(x_1|x_2) = \text{Poss}\{Y_1 = x_1|Y_2 = x_2\},$$

where the last equation represents the conditional distribution function of Y_1 given Y_2.

If we know the distribution function of Y_1 and the conditional distribution function of Y_2 given Y_1, then we can construct the joint distribution function of Y_1 and Y_2 by forming the conjunction

$$\pi_{(Y_1, Y_2)}(x_1, x_2) = \pi_{Y_1}(x_1) \wedge \pi_{(Y_2|Y_1)}(x_2|x_1).$$

Several forms of the marginal possibility distribution function of Y_1 are:

(i) $\pi_{Y_1}(x_1) = \sup_{x_2 \in X_2} \pi_{(Y_1, Y_2)}(x_1, x_2)$;
(ii) $\pi_{Y_1}(x_1) = \sup_{x_2 \in X_2} \pi_{(Y_1|Y_2)}(x_1|x_2)$;
(iii) $\pi_{Y_1}(x_1) = \pi_{(Y_1|Y_2)}(x_1, \tilde{x}_2(x_1))$,

where $\tilde{x}_2(x_1)$ is the value of x_2 (for a given x_1) at which

$$\pi_{(Y_1|Y_2)}(x_1, x_2) = 1$$

if $\tilde{x}_2(x_1)$ is defined for every $x_1 \in X_1$.

Intuitively, (i) represents the possibility of assigning a value to Y_1 as perceived by an observer (Y_1, Y_2) observing the joint possibility distribution $\Pi_{(Y_1, Y_2)}$. Case (ii) represents the perception of an observer $(Y_1|Y_2)$ observing only the conditional possibility distribution $\Pi_{(Y_1|Y_2)}$ and is unconcerned with or unaware of $\Pi_{(Y_2|Y_1)}$. Similarly, (iii) expresses the perception of an observer who assumes that x_2 is assigned a value (if it exists) which makes $\pi_{(Y_1|Y_2)}(x_1, x_2)$ equal to unity.

In relating π_{Y_1} to $\pi_{(Y_1|Y_2)}$ through the supremum operator, we are tacitly invoking the principle of maximum restriction, which asserts that, in the absence of complete information about Π_{Y_1}, we should equate Π_{Y_1} to the maximal possibility distribution that is consistent with the partial information about Π_{Y_1}.

Let π be a possibility distribution induced by a fuzzy set F in X. Let G be a nonfuzzy set of X. The possibility that x belongs to G is $\Pi(G)$, where

$$\Pi(G) = \operatorname*{Sup}_{x \in G} \chi_F(x) = \operatorname*{Sup}_{x \in G} \pi(G).$$

■ ***Example 2.2.1***

If p is a proposition of the form $p \triangleq X$ is F, which translates into the possibility assignment equation

$$\Pi_{A(X)} = F,$$

where F is a fuzzy subset of U and $A(X)$ is an implied attribute of X taking values in U, then the information conveyed by p, $I(p)$, may be identified with the possibility distribution $\Pi_{A(X)}$ of the fuzzy variable $A(X)$. Thus, the connection between $I(p)$, $\Pi_{A(X)}$, and F is expressed by

$$I(p) \triangleq \Pi_{A(X)},$$

where $F = \Pi_{A(X)} = R(A(X))$ and R represents the restriction on $A(X)$. ■

The proposition $p \triangleq$ (Ted is young) translates into the possibility assignment equation

$$\Pi_{\text{Age(Ted)}} = \text{young},$$

where χ_{young} is given. Then

$$I(\text{Ted is young}) = \Pi_{\text{Age(Ted)}},$$

in which the possibility distribution function of Age(Ted) is given by

$$\pi_{\text{Age(Ted)}}(u) = 1 - S(u; 20, 30, 40), \qquad u \in [0, 100].$$

From the definition of $I(p)$ it follows that if $p \triangleq X$ is F and $q \triangleq X$ is G, then p is at least as informative as q, expressed as $I(p) \geqq I(q)$, if $F \subset G$. Thus, we have a partial ordering of the $I(p)$ defined by

$$F \subset G \Rightarrow I(X \text{ is } F) \geqq I(X \text{ is } G),$$

which implies that the more restrictive a possibility distribution is, the more informative is the proposition with which it is associated.

The following is a list of translation rules for fuzzy propositions:

1) Modifier rule

If

$$Y \textit{ is } F \rightarrow \Pi_Y = F,$$

then

$$Y \textit{ is } mF \rightarrow \Pi_Y = F^+,$$

where m is a modifier such as not, very, more or less, etc., and F^+ is a modification of F induced by m. More specifically: If m = not, then $F^+ = \bar{F}$ = complement of F, that is,

$$\chi_{F^+}(x) = 1 - \chi_F(x), \qquad x \in X.$$

If m = very, then $F^+ = F^2$, that is,

$$\chi_{F^+}(x) = \chi_F^2(x), \qquad x \in X.$$

If m = more or less, then $F^+ = \sqrt{F}$, that is,

$$\chi_{F^+}(x) = \sqrt{\chi_F(x)}, \qquad x \in X.$$

2) Conjunctive, disjunctive and implicational rules

If

$$Y \textit{ is } F \rightarrow \Pi_Y = F \qquad \textit{and} \qquad Z \textit{ is } G \rightarrow \Pi_Z = G,$$

where F and G are fuzzy subsets of X_1 and X_2, respectively, then

(a) $$Y \textit{ is } F \quad \textit{and} \quad Z \textit{ is } G \rightarrow \Pi_{(Y,Z)} = F \times G,$$

where

$$\chi_{F \times G}(x_1, x_2) \triangleq \chi_F(x_1) \wedge \chi_G(x_2).$$

(b) $$Y \textit{ is } F \textit{ or } Z \textit{ is } G \rightarrow \Pi_{(Y,Z)} = F^* \cup G^*,$$

where

$$F^* \triangleq F \times X_2, \qquad G^* \triangleq X_1 \times G$$

and

$$\chi_{F^* \cup G^*}(x_1, x_2) = \chi_F(x_1) \vee \chi_G(x_2).$$

(c) *If Y is F, then* $$Z \textit{ is } G \rightarrow \Pi_{(Z|Y)} = F^{*\prime} \oplus G^*,$$

where $\Pi_{(Z|Y)}$ *denotes the conditional possibility distribution of* Z *given* Y, *and the bounded sum* $\oplus$ *is defined by*

$$\chi_{F^{*\prime}\oplus G^{*}}(x_1, x_2) = 1 \wedge (1 - \chi_F(x_1) + \chi_G(x_2)).$$

3) Quantification rule

If $X = \{x_1, \ldots, x_n\}$, Q *is a quantifier such as many, few, several, all, some, most, etc., and*

$$Y \textit{ is } F \rightarrow \Pi_Y = F,$$

then the proposition "QY *are* F*" translates into*

$$\Pi_{\text{Count}(F)} = Q,$$

where

$$\text{Count}(F) = \sum_{i=1}^{n} \chi_i.$$

4) Truth qualification rule

Let τ *be a fuzzy truth value, e.g., very true, quite true, more or less true, etc. Such a truth value may be regarded as a fuzzy subset of the unit interval that is characterized by a membership function* χ_τ: $[0,1] \rightarrow [0,1]$.

A truth-qualified proposition can be expressed as "Y *is* F *is* τ." *The translation rule for such propositions can be given by*

$$Y \textit{ is } F \textit{ is } \tau \rightarrow \Pi_Y = F^+,$$

where

$$\chi_{F^+}(x) = \chi_\tau(\chi_F(x)).$$

In his approach to approximate reasoning, Zadeh [1974d] uses various rules of inference. These rules, when applied, yield some very interesting conclusions drawn from imprecise propositions. Some of the rules of inference are:

1) Projection rule

Consider a fuzzy proposition whose translation is expressed as

$$p \rightarrow \Pi_{(Y_1, \ldots, Y_n)} = F,$$

and let $Y_{(s)}$ *denote a subvariable of the variable* $Y \triangleq (Y_1, \ldots, Y_n)$.

Further, let $\Pi_{Y_{(s)}}$ *denote the marginal possibility distribution of* $Y_{(s)}$; *that is,*

$$\Pi_{Y_{(s)}} = \text{Proj}_{X_{(s)}} F,$$

where $X_i, i = 1, \ldots, n$, *is the universe of discourse associated with* Y_i;

$$X_{(s)} = X_{i_1} \times \cdots \times X_{i_k},$$

and the projection of F *on* $X_{(s)}$ *is defined by the possibility distribution function*

$$\pi_{Y_s}(x_{i_1}, \ldots, x_{i_k}) = \operatorname*{Sup}_{x_{j_1}, \ldots, x_{j_m}} \chi_F(x_1, \ldots, x_n)$$

where $\bar{s} \triangleq (j_1, \ldots, j_m)$ *is the index subsequence that is complementary to* s, *and* χ_F *is the membership function of* F.

Now let q be a retranslation of the possibility assignment equation

$$\Pi_{Y_{(s)}} = \mathrm{Proj}_{X_{(s)}} F.$$

Then, the projection rule asserts that q may be inferred from p; for example,

$$p \to \Pi_{(Y_1,\ldots,Y_n)} = F,$$

$$q \leftarrow \Pi_{Y_{(s)}} = \mathrm{Proj}_{X_{(s)}} F.$$

2) Conjunction rule

Consider a proposition p which is an assertion concerning the possible values of, say, two variables X and Y, which take values in U and V, respectively. Similarly, let q be an assertion concerning the possible values of the variables Y and Z, taking values in V and W. With these assumptions, the translations of p and q may be expressed as

$$\left.\begin{aligned} p &\to \Pi^p_{(X,Y)} = F \\ q &\to \Pi^q_{(Y,Z)} = G \end{aligned}\right\}.$$

Let F^* and G^* be, respectively, the cylindrical extensions of F and G in $U \times V \times W$. Thus,

$$F^* = F \times W$$

and

$$G^* = U \times G.$$

Using the conjunction rule, one can infer from p and q a proposition that is defined by the following scheme:

$$\begin{array}{l} p \to \Pi^p_{(X,Y)} = F, \\ q \to \Pi^q_{(Y,Z)} = G, \\ \hline r \leftarrow \Pi^r_{(X,Y,Z)} = F^* \cap G^*. \end{array}$$

On combining the projection and conjunction rules, Zadeh obtains the *compositional rule of inference*, which includes the classical *modus ponens* as a special case.

More specifically, on applying the projection rule, one obtains the following inference scheme

$$\begin{array}{l} p \to \Pi^p_{(X,Y)} = F, \\ q \to \Pi^q_{(Y,Z)} = G, \\ \hline r \leftarrow \Pi^r_{(X,Z)} = F \circ G, \end{array}$$

where the composition of F and G is defined by

$$\chi_{F \circ G}(u,w) = \sup_v \left(\chi_F(u,v) \wedge \chi_G(v,w)\right).$$

In particular, if p is a proposition of the form "X is F" and q is a proposition of the form "If X is G, then Y is H," then we get

$$\begin{array}{l} p \to \Pi_X = F, \\ q \to \Pi_{(Y|X)} = G^{*\prime} \oplus H^*, \\ \hline r \leftarrow \Pi_{(Y)} = F \circ (G^{*\prime} \oplus H^*). \end{array}$$

■ ***Example 2.2.2***

Premises: $p \triangleq$ Item x is near pattern D,

$q \triangleq$ Item y is near pattern C.

Question: What is the distance between items x and y?

Let (x_1, x_2) and (y_1, y_2) be the coordinates of items x and y, respectively. Let $\Pi_{(x_1, x_2)}$ and $\Pi_{(y_1, y_2)}$ be the possibility distributions induced by p and q; namely, derived from the definition of the binary relation NEAR.

Now, let d be the distance between items x and y, where

$$d = \sqrt{(x_1 - y_1)^2 + (x_2 - y_2)^2}.$$

Using the extension principle, the possibility distribution function of d is given by

$$\Pi_d(z) = \operatorname*{Sup}_{u_1, v_1, u_2, v_2} \left[\Pi_{(x_1, x_2)}(u_1, v_1) \wedge \Pi_{(y_1, y_2)}(u_2, v_2) \right],$$

where $z = \sqrt{(u_1 - u_2)^2 + (v_1 - v_2)^2}$ and the supremum is taken over all possible values of x_1, x_2, y_1, y_2 under the definition of z. ■

This is basically a construction of a conceptual framework for inference from propositions whose meaning is not sharply defined. Through the use of fuzzy logic, the answer to a query is usually expressed in the form of a possibility distribution of one or more variables. In contrast to the conventional techniques of inference, the standards of precision in fuzzy logic are generally not high; but through the use of linguistic variables and linguistic approximation, these standards can be adjusted to fit the imprecision and unreliability of the information that is resident in the data.

Although in principle there is no connection between probabilities and possibilities, in practice the knowledge of possibilities conveys some information about the probabilities, but not vice versa. Certainly, if an event is impossible, then it is also improbable. However, it is not true that an event that is possible is also probable. This rather weak connection between the two may be stated more precisely in the form of the *possibility/probability consistency principle*, namely: If Y is a variable that takes the values $y_1, \ldots, y_n$ with probabilities $p_1, \ldots, p_n$ and possibilities $\chi_1, \ldots, \chi_n$, respectively, then the *degree of consistency* of the probabilities $p_1, \ldots, p_n$ with the possibilities $\chi_1, \ldots, \chi_n$ is given by

$$\rho = \chi_1 p_1 + \chi_2 p_2 + \cdots + \chi_n p_n.$$

Intuitively, this means that, in order to be consistent with χ's, high probabilities should not be assigned to those values of Y that are associated with low degrees of possibility.

2.3 EXTENDED FUZZY SETS

We will now describe the extension principle that will enable us to use, in a more powerful way, some of the principles outlined before.

Let g be a function from U to V. The *extension principle* — as its name implies — serves to extend the domain of definition of g from U to the set of fuzzy subsets of U. In particular, if F is a finite fuzzy subset of U expressed as

$$F = \chi_1/u_1 + \cdots + \chi_n/u_n,$$

then $g(F)$ is a finite fuzzy subset of V defined as

$$\begin{aligned} g(F) &= g(\chi_1/u_1 + \cdots + \chi_n/u_n) \\ &= \chi_1/g(u_1) + \cdots + \chi_n/g(u_n). \end{aligned}$$

Furthermore, if U is the cartesian product of $U_1, \ldots, U_N$, so that $u = (u^1, \ldots, u^N)$, $u^i \in U_i$, and we know only the projections of F on $U_1, \ldots, U_N$, whose membership functions are, respectively, $\chi_{F1}, \ldots, \chi_{FN}$, then

$$g(F) = \Sigma_u \chi_{F1}(u^1) \wedge \cdots \wedge \chi_{FN}(u^N)/g(u^1, \ldots, u^N),$$

with the understanding that, in replacing $\chi_F(u^1, \ldots, u^N)$ with $\chi_{F1}(u^1) \wedge \cdots \wedge \chi_{FN}(u^N)$, we are now invoking the *principle of maximal possibility*. In short, this principle asserts that, in the absence of complete information about a possibility distribution Π, we should equate Π to the maximal (i.e., least restrictive) possibility distribution that is consistent with the partial information about Π.

As a simple illustration of the extension principle, assume that $U = \{1, 2, \ldots, 10\}$; g is the operation of squaring; and SMALL is a fuzzy subset of U defined by

$$\text{SMALL} = 1/1 + 1/2 + 0.8/3 + 0.6/4 + 0.4/5.$$

The *right square* of SMALL is given by

$$\text{SMALL}^2 \triangleq 1/1 + 1/2 + 0.64/3 + 0.36/4 + 0.16/5,$$

and, more generally, for a subset F of U and any real m, we have

$$\chi_{m_F}(u) \triangleq (\chi_F(u))^m, \qquad u \in U.$$

As another example, we can show that it is quite natural to extend the α-level-set of A, $\chi_{A_\alpha}(x)$, into

$$\chi_{A_{\tilde{\alpha}}}(x) = \chi_{(\tilde{\alpha},1]}(\chi_A(x)), \qquad \forall x \in X,$$

where $A_{\tilde{\alpha}}$ is the fuzzy α-cut of A discussed before. This can easily be derived from the extension principle; since $A_\alpha = \chi_A^{-1}([\alpha, 1])$, we also have $A_{\tilde{\alpha}} = \chi_A^{-1}((\tilde{\alpha}, 1])$. Hence, we can extend χ_A^{-1}, viewed as a multivalued function, from $P([0,1])$ in X, and the extension principle can be generalized to deal with such functions. For example,

$$\chi_{A_{\tilde{\alpha}}}(x) = \sup_{\substack{\alpha \\ x \in \chi_A^{-1}([\alpha,1])}} \chi^*_{(\tilde{\alpha},1]}([\alpha, 1]),$$

where $\chi^*_{(\tilde{\alpha},1]}([\alpha, 1]) = \chi_{(\tilde{\alpha},1]}(\alpha)$. Note that

$$\{\alpha, x \in \chi_A^{-1}([\alpha, 1])\} = [0, \chi_A(x)];$$

and since $\chi^*_{(\tilde{\alpha},1]}$ is nondecreasing and continuous, $\chi_{A_{\tilde{\alpha}}}(x) = \chi^*_{(\tilde{\alpha},1]}([\chi_A(x), 1])$, and thus $\chi_{A_{\tilde{\alpha}}}(x)$ is the greatest among the membership values of the sets $[\alpha, 1]$ whose images under χ_A^{-1} contain x.

2.4 FUZZY NUMBERS

possible fuzzy membership functions

This short exposition of the properties of fuzzy numbers follows the presentation in Zadeh [1982] and Dubois and Prade [1979b] and [1978c].

By a fuzzy number, we mean a number that is characterized by a possibility distribution or is a fuzzy subset of real numbers. Simple examples of fuzzy numbers are fuzzy subsets of the real line labeled *small*, *approximately* 8, *very close to* 5, *more or less large*, *much larger than* 6, *several*, etc. In general, a fuzzy number is either a convex or a concave fuzzy subset of the real line. A special case of a fuzzy number is an interval. Viewed in this perspective, fuzzy arithmetic may be viewed as a generalization of interval arithmetic.

Fuzzy arithmetic is not intended to be used in situations in which a high degree of precision is required. To take advantage of this assumption, it is expedient to represent the possibility distribution associated with a fuzzy number in a standardized form that involves a small number of parameters — usually two — which can be adjusted to fit the given distribution.

1. π-numbers

The possibility distribution of such numbers is bell-shaped and piecewise-quadratic. The distribution is characterized by two parameters; (a) the peak-point, i.e., the point at which $\pi = 1$, and (b) the bandwidth, β, which is defined as the distance between the cross-over points, i.e., the points at which $\pi = 0.5$. Thus, a fuzzy π-number, x, is expressed as (p, β), where p is the peak-point and β is the bandwidth; or, alternatively, as (p, β'), where β' is the normalized bandwidth, that is, $\beta' = \beta/p$. As a function of $u, u \in (-\infty, \infty)$, the values of $\pi_x(u)$ are defined by the equations

$$\pi_x(u) = 0 \qquad \text{for } u \le p - \beta$$

and

$$= \frac{2}{\beta^2}(u - p + \beta)^2 \qquad \text{for } p - \beta \le u \le p - \frac{\beta}{2}$$

$$= 1 - \frac{2}{\beta^2}(u - p)^2 \qquad \text{for } p - \frac{\beta}{2} \le u \le p + \frac{\beta}{2}$$

$$= \frac{2}{\beta^2}(u - p - \beta)^2 \qquad \text{for } p + \frac{\beta}{2} \le u \le p + \beta.$$

2. *s*-numbers

As its name implies, the possibility distribution of an *s*-number has the shape of an *s*. Thus, the equations defining an *s*-number expressed as (p/β), are:

$$\pi_x(u) = 0 \qquad \text{for } u \le p - \beta$$

$$= \frac{2}{\beta^2}(u - p + \beta)^2 \qquad \text{for } p - \beta \le u \le p - \frac{\beta}{2}$$

$$= 1 - \frac{2}{\beta^2}(u - p)^2 \qquad \text{for } p - \frac{\beta}{2} \le u \le p$$

$$= 1 \qquad \text{for } u \ge p,$$

where β (the bandwidth) is the length of the transition interval from $\pi_x = 0$ to $\pi_x = 1$, and p is the left peak-point, i.e., the right endpoint of the transition interval.

3. *z*-numbers

A *z*-number is a mirror image of an *s*-number. Thus, the defining equations for a *z*-number, expressed as $(p \setminus \beta)$, are:

$$\pi_x(u) = 1 \qquad \text{for } u \le p$$

$$= 1 - \frac{2}{\beta^2}(u - p)^2 \qquad \text{for } p \le u \le p + \frac{\beta}{2}$$

$$= \frac{2}{\beta^2}(u - p - \beta)^2 \qquad \text{for } p + \frac{\beta}{2} \le u \le p + \beta$$

$$= 0 \qquad \text{for } u \ge p + \beta,$$

where p is the right peak-point and β is the bandwidth.

4. *s / z*-numbers

An s/z-number has a flat-top possibility distribution which may be regarded as the intersection of the possibility distribution of an s-number and a z-number, with the understanding that the left peak-point of the s-number lies to the left of the right peak-point of the z-number. In some cases, however, it is expedient to disregard the latter restrictions and allow an s/z-number to have a sharp peak rather than a flat top. An s/z-number is represented as an ordered pair $(p_1/\beta_1; p_2 \backslash \beta_2)$ in which the first element is an s-number and the second element is a z-number.

5. *z/s*-numbers

The possibility distribution of a z/s number is the complement of that of an s/z-number. Thus, whereas an s/z-number is a convex fuzzy subset of the real line, a z/s-number is a concave fuzzy subset. Equivalently, the possibility distribution of a z/s-number may be regarded as the union of the possibility distributions of a z-number and an s-number. A z/s-number is represented as $(p_1 \backslash \beta_1; p_2/\beta_2)$.

Next we discuss the arithmetic operations on fuzzy numbers. Let $*$ denote an arithmetic operation such as addition, subtraction, multiplication, or division, and let $x * y$ be the result of applying $*$ to the fuzzy numbers x and y.

By the use of the extension principle, it can readily be established that the possibility distribution function of $x * y$ may be expressed in terms of those of x and y by the relation

$$\pi_{x*y}(w) = \vee_{u,v}(\pi_x(u) \wedge \pi_y(v)),$$

subject to the constraint

$$w = u * v | u, v, w \in (-\infty, \infty)$$

where $\vee_{u,v}$ denotes the supremum over u, v, and $\wedge \triangleq \min$.

It can easily be shown that if x and y are numbers of the same type (for example, π-numbers), then so are $x + y$ and $x - y$. Furthermore, the characterizing parameters of $x + y$ and $x - y$ depend in a very simple and natural way on those of x and y. More specifically, if $x = (p, \beta)$ and $y = (q, \gamma)$, then

$$(p, \beta) + (q, \gamma) = (p + q, \beta + \gamma)$$
$$(p/\beta) + (q/\gamma) = (p + q/\beta + \gamma)$$
$$(p \backslash \beta) + (q \backslash \gamma) = (p + q \backslash \beta + \gamma)$$
$$(p_1/\beta_1; p_2 \backslash \beta_2) + (q_1/\gamma_1; q_2 \backslash \gamma_2) = (p_1 + q_1/\beta_1 + \beta_2; p_2 + q_2 \backslash \gamma_1 + \gamma_2)$$
$$(p, \beta) - (q, \gamma) = (p - q, \beta + \gamma),$$

and similarly for other types of numbers.

In the case of multiplication, it is true only as an approximation that, if x and y are π-numbers, then so is $x \times y$. However, the relation between the peak-points and normalized bandwidths that is stated below is exact:

$$(p, \beta') \times (q, \gamma') = (p \times q, \beta' + \gamma').$$

The operation of *division*, $x \div y$, may be regarded as the composition of (a) forming the reciprocal of y, and (b) multiplying the result by x. In general, the operation $1 \div y$ does not preserve the type of y and hence the same applies to $x \div y$. However, if y is a π-number whose peak-point is much larger than 1 and whose normalized bandwidth is small, then $1 \div y$ is approximately a π-number defined by

$$1 \div (p, \beta) \cong (1 \div p, (\beta' \div p')),$$

and consequently

$$(p, \beta) \div (q, \gamma) \cong (p \div q, (\beta' \div p + \gamma' \div q')).$$

As a simple example of operations on fuzzy numbers, suppose that x is a π-number (p, β) and y is a number that is much larger than x. The question is: What is the possibility distribution of y?

Assume that the relation $y \gg x$ is characterized by a conditional possibility distribution $\Pi_{(y|x)}$ (that is, the conditional possibility distribution of y given x) which for real values of x is expressed as an s-number:

$$\Pi_{(y|x)} = (q(x)/\gamma(x))$$

whose peak-point, $q(x)$, and bandwidth, $\gamma(x)$, depend on x.

On applying the extension principle to the composition of the binary relation $\gg$ defined above with the unary relation x, it is readily found that y is an s-number that is approximately characterized by

$$y = (q(p)/[q(p) - q(p - \beta)]).$$

In this way, then, the possibility distribution of y may be expressed in terms of the possibility distribution of x and the conditional possibility distribution of y given x.

Because of the reproducibility property of possibility distributions, the computational effort involved in the manipulation of fuzzy numbers is generally not much greater than that required in interval arithmetic. The bounds on the results, however, are usually appreciably tighter because in the case of fuzzy numbers the possibility distribution functions are allowed to take intermediate values in the interval [0, 1], and not just 0 or 1, as in the case of intervals.

In what follows we adopt Yager's excellent analysis [1979e] in showing that equations that have fuzzy parameters cannot be solved in the usual manner. This is a result of the lack of multiplicative and additive inverses. A procedure will be presented that enables us to get the degree of truth associated with a given solution in this case. In essence, with each potential solution to a fuzzy equality, we can associate a degree of truth. The best solution would be the one with the highest degree of truth. Using Yager's notation, any crisp number $a \in \mathbb{R}$ can be expressed as a fuzzy number as follows:

$$a = \left\{ \frac{1}{a} \right\}.$$

We shall use the term *crisp number* to distinguish the special case of the ordinary numbers from the rest of the fuzzy numbers. Clearly, *for any noncrisp fuzzy number A, there exist no inverse numbers $\bar{A}$ and A^* under $+$ and $\times$ respectively, such that*

$$A + \bar{A} = 0, \qquad A \times A^* = 1.$$

This implies that

$$A + (-A) \neq 0$$

and

$$A \times \frac{1}{A} \neq 1.$$

This also means that, in general, algebraic equations involving fuzzy numbers cannot be solved. For example, if A is a fuzzy number, then, in general,

$$X + A = b$$

cannot be solved for X. This is due to the fact that we cannot isolate X by adding $(-A)$ to both sides. That is, for any C,

$$X + (A + C) = b + C;$$

since $A + C = D \neq 0$ we have $X + D = b + C$.

Similarly we cannot solve $AX = b$, where A is fuzzy.

Furthermore, this implies that there exist no fuzzy numbers $W_1, W_2, \ldots, W_m$ such that $F = W_1 + W_2 + \cdots + W_m$, where F is either a fuzzy or crisp number. This implies that in using fuzzy weights in multiobjective decision-making, it is impossible to get a normalized set of fuzzy weights. In general this implies that if f is a mathematical function involving fuzzy parameters, then one can't find a number x, such that $f(x) = F$.

A second manifestation of this result is that, even if a solution does exist, it becomes difficult to find. For example, if A and B are two fuzzy numbers and if we compute their sum, the fuzzy number C, by

$$C(z) = \max[A(x) \wedge B(y)] \qquad \text{for } z = x + y,$$

then, given the computed C and A, there exists a solution $(X = B)$ to the equation $C = A + X^2$. However, even though a solution $X = B$ exists, there is considerable difficulty in finding this value B. Since we can't subtract A from both sides and isolate X, the only available procedure is trial and error.

In either case we may need or be willing to settle for a solution that makes the given equation *only approximately true*. We shall present a procedure that will enable us to determine the degree to which a proposed solution satisfies a given equation. That is, given an equation and a proposed solution, we can see the goodness of this solution; and if it is good enough, we can accept it as an approximate solution.

Assume that x is a crisp variable, b is a number (fuzzy or crisp), and $F(x)$ is a mathematical function that has fuzzy parameters. That is, some of the coefficients in F are fuzzy numbers. The problem then is to find x such that

$$b = F(x).$$

Assume that x_0 is proposed as a solution to the equation. We can rephrase the problem as follows: "Find the compatibility of the statement $x = x_0$ in the light of the referenced fact that $b = F(x)$." The statement $x = x_0$ implies that $F(x) = F(x_0) = V$.

Therefore, we have the following two propositions:

$$P_1: \quad V = b \qquad \text{(our reference statement)},$$

and

$$P_2: \quad V = F(x_0).$$

Using the methods of the previous section, we can then determine the degree to which x_0 satisfies the equation $b = F(x)$, which is in turn the truth of the statement $F(x_0) = b$, by using Yager's procedure [1979e].

We shall summarize the procedure for finding the degree to which $x = x_0$ satisfies $b = F(x)$.

1. Calculate $F(x_0) = C$, a fuzzy number.
2. Obtain the compatibility of $P_1 = V$ is b and $P_2 = V$ is C as

$$\tau = \frac{b(x)}{C(x)}.$$

Then τ measures the degree to which x_0 solves $F(x) = b$.

■ ***Example 2.4.1***

Find the degree to which $X = 4$ solves

$$\text{"approx. 7"} = X + \text{"approx. 3"}$$

Let

$$\text{"approx. 7"} = \left\{ e^{-(x-7)^2}/x \right\} = b.$$

Let

$$\text{"approx. 3"} = \left\{ e^{-(y-3)^2}/y \right\}.$$

Then

$$4 + \text{"approx. 3"} = \{1/4\} + \left\{ e^{-(y-3)^2}/y \right\} = C,$$

$$C = \left\{ e^{-(y-3)^2}/(4 + y) \right\},$$

and, letting $x = y + 4$,

$$C = \left\{ e^{-(x-7)^2}/x \right\}.$$

■

We note that this procedure is valid for crisp numbers.

■ ***Example 2.4.2***

Solve $7 = X + 3$. Letting $7 = \{1/7\}$ and $3 = \{1/3\}$, we get

$$x + 3 = \{1/3 + x\}.$$

Trying as our solution $x = 4$, we get

$$4 + 3 = C = \{1/7\}.$$

Then $b(x) = \{1/7\}$ and $c(x) = \{1/7\}$. Therefore our value for compatibility τ is

$$\{b(x)/C(x)\} = \{1/1\} = 1,$$

which implies that $x = 4$ is absolutely true. However, if we try as our solution some crisp $x \neq 4$, then we get

$$X + 3 = C = \{1/x + 3\},$$

where $X + 3 \neq 7$; then

$$C = \{1/a\}, \qquad a \neq 7$$

and

$$b = \{1/7\}$$

and therefore

$$\tau = \{b(x)/C(x)\} = \{1/0, 0/1\} = 0,$$

which is the value for absolutely false. ■

We now can see the essential imprecision in fuzzy arithmetic. That is, given an equation we can only get a solution that imprecisely satisfies the equation. The measure of the truth of a solution would be some truth value τ. We can try to improve the solution by trying different values of x, and thus increase the value of τ. Work on this and related subjects was performed by Sanchez [1977, 1981].

2.5 FUZZY QUANTIFIERS

An important aspect of fuzzy quantifiers is that their occurrence in human discourse is, for the most part, implicit rather than explicit. For example, when we assert that "Basketball players are very tall," what we usually mean is that "Almost all basketball players are very tall," and similarly, the statement "Heavy smoking causes lung cancer," might be interpreted as "The incidence of lung cancer among heavy smokers is much higher than among nonsmokers."

An interesting observation that relates to this issue is that *property inheritance* — which is exploited extensively in knowledge-representation systems and high-level artificial intelligence languages — is a *brittle* property with respect to the replacement of the nonfuzzy quantifier *all* with the fuzzy quantifier *almost all*. What this means is that if, in the inference rule,

$$\frac{\begin{array}{l} p \triangleq \text{All } A\text{'s are } B\text{'s,} \\ q \triangleq \text{All } B\text{'s are } C\text{'s,} \end{array}}{r \triangleq \text{All } A\text{'s are } C\text{'s,}}$$

the quantifier *all* in p and q is replaced by *almost all*, then the quantifier *all* in r should be replaced by *none-to-all*. Thus, a slight change in the quantifier *all* in the premises may result in a large change in the quantifier *all* in the conclusion.

Another point that should be noted relates to the close connection between fuzzy quantifiers and fuzzy probabilities. Specifically, a proposition of the form $p \triangleq QA$'s are B's, where Q is a fuzzy quantifier, implies that the conditional probability of the event A given the event B is a fuzzy probability that is equal to Q. What can be shown, in fact, is that most statements involving fuzzy probabilities may be replaced by semantically equivalent statements involving fuzzy quantifiers. This connection between fuzzy quantifiers and fuzzy probabilities plays an important role in expert systems and fuzzy temporal logic.

The main idea underlying the approach to fuzzy quantifiers is that the natural way of dealing with such quantifiers is to treat them as fuzzy numbers. However, this does not imply that the concept of a fuzzy quantifier is coextensive with that of a fuzzy number.

More generally, we shall view a fuzzy quantifier as a fuzzy number that provides a fuzzy characterization of the absolute or relative cardinality of one or more fuzzy or nonfuzzy sets. There are propositions, however, in which the question of whether or not a constituent fuzzy number is a fuzzy quantifier does not have a clear-cut answer.

A simple example may be of help at this point in providing an idea of how fuzzy quantifiers may be treated as fuzzy numbers. Specifically, consider the propositions:

$$\frac{\begin{array}{l} p \triangleq 90\% \text{ of students are single,} \\ q \triangleq 55\% \text{ of single students are male,} \end{array}}{r \triangleq Q \text{ of students are single and male,}}$$

in which r represents the answer to the question "What percentage of students are single males?" given the premises expressed by p and q.

Clearly, the answer is: $90\% \times 55\% = 49.5\%$, and, more generally, we can assert that:

$$\frac{\begin{array}{l} p \triangleq Q_1 \text{ of } A\text{'s are } B\text{'s,} \\ q \triangleq Q_2 \text{ of } (A \text{ and } B)\text{'s are } C\text{'s,} \end{array}}{r \triangleq Q_1 Q_2 \text{ of } A\text{'s are } (B \text{ and } C)\text{'s,}}$$

where Q_1 and Q_2 are numerical percentages, and A, B and C are labels of nonfuzzy sets or, equivalently, names of their defining properties.

Now suppose that Q_1 and Q_2 are given by the following p and q statements:

$p \triangleq$ Most students are single,

$q \triangleq$ A little more than a half of single students are male

---

$r \triangleq$? Q of students are single and male,

where the question mark indicates that the value of Q is to be inferred from p and q.

By interpreting the fuzzy quantifiers *most*, *a little more than a half*, and Q as fuzzy numbers that characterize, respectively, the proportions of single students among students, males among single students, and single males among students, we can show that Q may be expressed as the product of the fuzzy numbers *most* and *a little more than a half*. Thus,

$$Q = \text{Most} \otimes \text{A little more than a half}$$

and, more generally, for fuzzy Q's, A's, B's, and C's, we can assert the syllogism:

$p \triangleq Q_1$ of A's are B's,

$q \triangleq Q_2$ of (A and B)'s are C's,

$r \triangleq Q_1 \otimes Q_2$ of A's are (B and C)'s,

which will be referred to as the *intersection/product* syllogism shown in Figure 2.5.1 below.

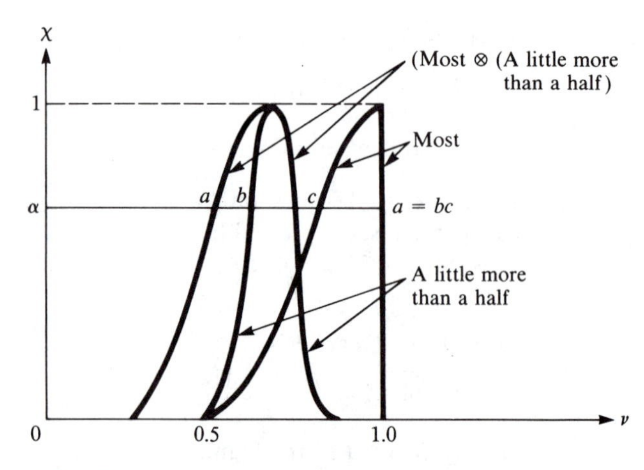

Figure 2.5.1. The intersection/product syllogism with fuzzy quantifiers.

Note that we have assumed that a fuzzy quantifier is a fuzzy number of type 1, that is, a fuzzy set whose membership function takes values in the unit interval. More generally, however, a fuzzy quantifier may be a fuzzy set of type 2 (or higher), in which case we shall refer to it as an *ultrafuzzy quantifier*. The membership functions of such quantifiers take values in the space of fuzzy sets of type 1, implying that the compatibility of an ultrafuzzy quantifier with a real number is a fuzzy number of type 1. For example, the fuzzy quantifier *not so many* would be regarded as an ultrafuzzy quantifier if the compatibility of *not so many* with 5, say, would be specified in a particular context as *rather high*, where *rather high* is interpreted as a fuzzy number in the unit interval.

Viewed from the perspective of test-score semantics, a semantic entity such as a proposition, predicate, predicate-modifier, quantifier, qualifier, command, question, etc., may be regarded as a system of elastic constraints whose domain is a collection of fuzzy relations in a database — a database that describes a state of affairs, a possible world, or a set of objects in a universe of discourse. The meaning of a semantic entity, then, is represented as a test which, when applied to the database, yields a collection of partial test scores. Upon aggregation, these test scores lead to an overall vector test score, τ, whose components are numbers in the unit interval, with τ serving as a measure of the compatibility of the semantic entity with the database. In this respect, test-score semantics subsumes both truth-conditional and possible-world semantics as limiting cases in which the partial and overall test scores are restricted to $\{$pass, fail$\}$ or, equivalently, $\{$true, false$\}$ or $\{1, 0\}$.

In more specific terms, the process of meaning representation in test-score semantics involves three distinct phases. In Phase 1, an *explanatory database frame* (EDF) is constructed. This is a collection of relational frames (names of relations, attributes, and attribute domain). For example, in the case of the proposition "$p \triangleq$ Over the past few years Jack earned far less than most of his close friends," the EDF might consist of the following relations: INCOME [*Name*; *Amount*; *Year*], which lists the income of each individual identified by his/her name as a function of the variable *Year*; FRIEND [*Name*; χ], where χ is the degree to which *Name* is a friend of Jack; FEW [*Number*; χ], where χ is the degree to which *Number* is compatible with the fuzzy quantifier FEW; MOST [*Proportion*; χ], in which χ is the degree to which *Proportion* is compatible with the fuzzy quantifier MOST; and FAR LESS [*Income* 1; *Income* 2; χ], where χ is the degree to which *Income* 1 fits the fuzzy predicate FAR LESS in relation to *Income* 2. Each of these relations is interpreted as an elastic constraint on the variables that are associated with it.

In Phase 2, a test procedure is constructed which acts on relations in the explanatory database and yields the test scores that represent the degrees to which the elastic constraints induced by the constituents of the semantic entity are satisfied. For example, in the case of p, the test procedure would yield the test scores for the constraints induced by the relations FRIEND, FEW, MOST, and FAR LESS.

In Phase 3, the partial test scores are aggregated into an overall test score, τ, which, in general, is a vector that serves as a measure of the compatibility of the semantic entity with an instantiation of EDF. As was stated earlier, the components of this vector are numbers in the unit interval, or, more generally, possibility/probability distributions over this interval. In particular, in the case of a proposition p for which the overall test score is a scalar, τ may be interpreted — in the spirit of truth-conditional semantics — as the degree of truth of the proposition with respect to the explanatory database ED (that is, an instantiation of EDF). Equivalently, τ may be interpreted as the possibility of ED given p, in which case we may say that p *induces a possibility distribution*. More concretely, we shall say that p *translates into a possibility assignment equation* where

$$p \to \Pi_{(X_1,\ldots,X_n)} = F,$$

where F is a fuzzy subset of a universe of discourse U, $X_1, \ldots, X_n$ are variables that are explicit or implicit in p, and $\Pi_{(X_1,\ldots,X_n)}$ is their joint possibility distribution. For example, in the case of the proposition "$p \triangleq$ Dan is tall," we have

$$\text{Dan is tall} \to \Pi_{Height(Dan)} = \text{TALL},$$

where TALL is a fuzzy subset of the real-line, $Height(Dan)$ is a variable that is implicit in p, and $\Pi_{Height(Dan)}$ is the possibility distribution of the variable $Height(Dan)$. Thus

$$\text{Poss}\{Height(Dan) = u\} = \chi_{\text{TALL}}(u),$$

where u is a specified value of the variable $Height(Dan)$, $\chi_{\text{TALL}}(u)$ is the grade of membership of u in the fuzzy set TALL, and $\text{Poss}\{X = u\}$ should be read as "the possibility that X is u." In effect, the proposition "Dan is tall" may be interpreted as an elastic constraint on the variable $Height(Dan)$, with the elasticity of the constraint characterized by the unary relation TALL.

We follow Zadeh [1982] with some examples from his outstanding paper.

■ ***Example 2.5.1***

$$SE \triangleq \text{Several balls, most of which are large.}$$

For this semantic entity, we shall assume that EDF comprises the following relations:

$$\begin{aligned} \text{EDF} \triangleq\ & \text{BALL}\,[Identifier;\ Size] \\ & +\text{LARGE}\,[Size;\ \chi] \\ & +\text{SEVERAL}\,[Number;\ \chi] \\ & +\text{MOST}\,[Proportion;\ \chi]. \end{aligned}$$

In this EDF, the first relation has n rows and is a list of the identifiers of balls and their respective sizes; in LARGE, χ is the degree to which a ball of size *Size* is large; in SEVERAL, χ is the degree to which *Number* fits the description *several*; and in MOST, χ is the degree to which *Proportion* fits the description *most*.

The test that yields the compatibility of SE with ED and thus defines the meaning of SE depends on the definition of fuzzy-set cardinality. In particular, using the sigma-count, the test procedure may be stated as follows:

1. Test the constraint induced by SEVERAL:

$$\tau_1 =_{\chi} \text{SEVERAL}\,[\,Number = n\,],$$

   which means that the value of *Number* is set to n and the value of χ is read, yielding the test score τ_1 for the constraint in question.

2. Find the size of each ball in BALL:

$$Size_i =_{Size} \text{BALL}\,[\,Identifier = Identifier_i\,],$$

$$i = 1, \ldots, n.$$

3. Test the constraint induced by LARGE for each ball in BALL:

$$\chi_{LB}(i) =_{\chi} \text{LARGE}\,[\,Size = Size_i\,].$$

4. Find the sigma-count of large balls in BALL:

$$\Sigma\,\text{Count}(\,LB\,) = \Sigma_{i\chi}\,LB(\,i\,).$$

5. Find the proportion of large balls in BALL:

$$PLB = \frac{1}{n}\Sigma_i \chi_{LB}(\,i\,).$$

6. Test the constraint induced by MOST:

$$\tau_2 =_{\chi} \text{MOST}\,[\,Proportion = PLB\,].$$

7. Aggregate the partial test scores:

$$\tau = \tau_1 \wedge \tau_2,$$

   where τ is the overall test score. The use of the min operator to aggregate τ_1 and τ_2 implies that we interpret the implicit conjunction in *SE* as the cartesian product of the conjuncts.

The use of fuzzy cardinality affects the way in which τ_2 is computed. Specifically, the employment of *FG* Count leads to:

$$\tau_2 = \sup_i(\,FG\,\text{Count}(\,LB\,) \cap n\,\text{MOST}),$$

which, expressed in terms of the membership functions of *FG* Count(*LB*) and MOST may be written as

$$\tau_2 = \sup_i\left(\chi_{FG\,\text{Count}(\,LB\,)}(\,i\,) \wedge \chi_{\text{MOST}}\left(\frac{i}{n}\right)\right).$$

The rest of the test procedure is unchanged. ■

■ ***Example 2.5.2***

Consider the proposition

$p \triangleq$ Over the past few years, Jack earned far less than most of his close friends.

In this case, we shall assume that EDF consists of the following relations:

$$\begin{aligned}\text{EDF} \triangleq\ & \text{INCOME}\,[\,Name;\ Amount;\ Year\,]\\ & +\text{FRIEND}\,[\,Name;\ \chi\,]\\ & +\text{FEW}\,[\,Number;\ \chi\,]\\ & +\text{FAR LESS}\,[\,Income\ 1;\ Income\ 2\!:\ \chi\,]\\ & +\text{MOST}\,[\,Proportion;\ \chi\,].\end{aligned}$$

Using the sigma-count, the test procedure may be described as follows:

1. Find Jack's income in $Year_i$, $i = 1, 2, \ldots,$ counting backward from present;

$$IN_i \triangleq {}_{Amount}\text{INCOME}\,[\,Name = Jack;\ Year = Year_i\,].$$

2. Test the constraint induced by FEW:

$$\chi_i \triangleq {}_{\chi}\text{FEW}\,[\,Year = Year_i\,].$$

3. Compute Jack's total income during the past few years:

$$TIN = \Sigma_i \chi_i IN_i,$$

in which the χ_i play the role of weighting coefficients.

4. Compute the total income of each $Name_j$ (other than Jack) during the past several years:

$$TIName_j = \Sigma_i \chi_i IName_{ji},$$

where $IName_{ji}$ is the income of $Name_j$ in $Year_i$.

5. Find the fuzzy set of individuals who earned far more than Jack. The grade of membership of $Name_j$ in this set is given by:

$$\chi_{FM}(Name_j) = {}_{\chi}\text{FAR LESS}\left[\,Income\ 1 = TIN;\ Income\ 2 = TIName_j\,\right].$$

6. Find the fuzzy set of close friends of Jack by intensifying the relation FRIEND:

$$CF = {}^{2}\text{FRIEND},$$

which implies that

$$\chi_{CF}(Name_j) = \left({}_{\chi}\text{FRIEND}\left[\,Name = Name_j\,\right]\right)^2.$$

7. Using the sigma-count, count the number of close friends of Jack:

$$\Sigma\,\text{Count}(CF) = \Sigma_j \chi\left[{}^{2}_{\text{FRIEND}}\right](Name_j).$$

8. Find the intersection of FM with CF. The grade of membership of $Name_j$ in the intersection is given by

$$\chi_{FM \cap CF}(Name_j) = \chi_{FM}(Name_j) \wedge \chi_{CF}(Name_j).$$

9. Compute the sigma-count of $FM \cap CF$:

$$\Sigma\,\text{Count}(FM \cap CF) = \Sigma_j \chi_{FM}(Name_j) \wedge \chi_{CF}(Name_j).$$

10. Compute the proportion of individuals in FM who are in CF:

$$\rho \triangleq \frac{\Sigma\,\text{Count}(FM \cap CF)}{\Sigma\,\text{Count}(CF)}.$$

11. Test the constraint induced by MOST:

$$\tau =_{\chi} \text{MOST}\,[\,Proportion = \rho\,],$$

which expresses the overall test score and thus represents the desired compatibility of ρ with the explanatory database.

For the proposition under consideration, the logical form has a more complex structure than in Example 2.5.1. Specifically, we have

$$\text{Prop}\big((\Sigma_j \chi_j / Name_j)/^2\text{FRIEND}\,[\,Name\ 2 = Jack\,]\big) \text{ is MOST},$$

where

$$\chi_j =_{\chi} \text{FAR LESS}\left[\,Income\,1 = TIN;\ Income\ 2 = TIName_j\,\right],$$

where $Name_j \neq Jack$ and

$$TIN = \Sigma_{i\chi}\text{FEW}(i)\,_{Amount}\text{INCOME}\,[\,Name = Jack;\ Year = Year_i\,]$$

and

$$TIName_j = \Sigma_{i\chi}\text{FEW}(i)\,_{Amount}\text{INCOME}\left[\,Name = Name_j;\ Year = Year_i\,\right]. \ ■$$

■ ***Example 2.5.3***

$$p \triangleq \text{They like each other.}$$

In this case there is an implicit fuzzy quantifier in p which reflects the understanding that not all members of the group referred to as *they* must necessarily like each other.

Since the fuzzy quantifier in p is implicit, it may be interpreted in many different ways. The test described below represents one such interpretation and involves, in effect, the use of an F Count.

Specifically, we associate with p the EDF

$$\begin{aligned} \text{EDF} \triangleq\ & \text{THEY}[\,Name\,] \\ & +\text{LIKE}[\,Name\,1;\ Name\ 2;\ \chi\,] \\ & +\text{ALMOST ALL}\,[\,Proportion;\ \chi\,], \end{aligned}$$

in which THEY is the list of names of members of the group to which p refers; LIKE is a fuzzy relation in which χ is the degree to which *Name* 1 likes *Name* 2; and ALMOST ALL is a fuzzy quantifier in which χ is the degree to which a numerical value of *Proportion* fits a subjective perception of the meaning of ALMOST ALL.

Let χ_{ij} be the degree to which $Name_i$ likes $Name_j$, $i \neq j$. If there are n names in THEY, then there are $(n^2 - n)$ χ_{ij}'s in LIKE with $i \neq j$. Denote the relation LIKE without its diagonal elements by LIKE*.

The test procedure which yields the overall test score τ may be described as follows:

1. Count the number of members in THEY:

$$n \triangleq \text{Count(THEY)}.$$

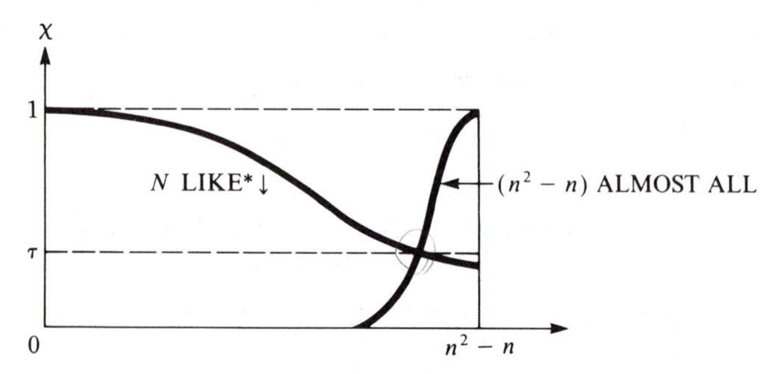

Figure 2.5.2. Computation of the test score for "They like each other."

2. Compute the *FG* Count of LIKE*:

$$C \triangleq FG\,\mathrm{Count}(\mathrm{LIKE}^*).$$

Note that C may be obtained by sorting the χ elements of LIKE* in descending order, which yields LIKE* ↓. Thus,

$$FG\,\mathrm{Count}(\mathrm{LIKE}^*) = N\,\mathrm{LIKE}^*\downarrow.$$

3. Compute the height (i.e., the maximum value) of the intersection of C and the fuzzy number $(n^2 - n)$ ALMOST ALL:

$$\tau = \sup(FG\,\mathrm{Count}(\mathrm{LIKE}^*) \cap (n^2 - n)\ \mathrm{ALMOST\ ALL}).$$

The result, as shown in Figure 2.5.2, is the overall test score. ■

CHAPTER THREE

Fuzzy Functions

3.1 FUZZY ALGEBRA AND UNCERTAINTY

We may view fuzzy logic as a special kind of many-valued logic. In fuzzy logic, the truth value of a formula, instead of assuming two values (0 and 1), can assume any value in the interval $[0, 1]$ and is used to indicate the degree of truth represented by the formula. For example, let $P(x)$ represent "x is a large number compared with unity"; then the truth value of $P(10^5)$ and $P(10^{-5})$ are certainly 1 and 0, respectively. As for $P(125)$, the truth value of it may be some value between 0 and 1, say 0.5.

We shall assume that well-formed formulas are defined to be exactly the same as those in two-valued logic. Letting $\chi(S)$ denote the truth value of a formula S, the evaluation procedure for a formula in fuzzy logic can be described as follows.

1. $\chi(S) = \chi(A)$ if $S = A$ and A is a ground atomic formula.
2. $\chi(S) = 1 - \chi(R)$ if $S = \bar{R}$.
3. $\chi(S) = \min[\chi(S_1), \chi(S_2)]$ if $S = S_1 \wedge S_2$.
4. $\chi(S) = \max[\chi(S_1), \chi(S_2)]$ if $S = S_1 \vee S_2$.
5. $\chi(S) = \inf[\chi(B(x)) \times D]$ if $S = (x)B$ and D is the domain of x.
6. $\chi(S) = \sup[\chi(B(x)) \times D]$ if $S = (\mathrm{Ex})B$ and D is the domain of x.

Note that if D is a finite set, then (5) and (6) become:

5′. $\chi(S) = \chi(B(a_1) \cdot \cdots \cdot B(a_n))$ if $S = (x)B$ and x assume values $a_1, \ldots, a_n$.
6′. $\chi(S) = \chi(B(a_1) + \cdots + B(a_n))$ if $S = (\mathrm{Ex})B$ and x assume values $a_1, \ldots, a_n$.

The reader should note that two-valued logic is a special case of fuzzy logic; all the rules stated above are applicable in two-valued logic.

■ ***Example 3.1.1***

Consider $S = (P \vee Q) \wedge (\bar{R})$. Assume $\chi(P) = 0.5$, $\chi(Q) = 0.7$, and $\chi(R) = 0.8$. Then

$$\begin{aligned}\chi(S) &= \min\{\max[\chi(P), \chi(Q)], 1 - \chi(R)\}\\ &= \min[\max(0.5, 0.7), 1 - 0.8]\\ &= \min(0.7, 0.2)\\ &= 0.2.\end{aligned}$$

Let $P(X)$ be the set of ordinary subsets of X. Here $P(X)$ is a Boolean lattice for $\cup$ and $\cap$, namely, L is a lattice if it is a partially ordered set such that

$$\forall a \in L, \quad \forall b \in L, \begin{cases} \exists! c \in L, c = \inf(a, b), \\ \exists! d \in L, d = \sup(a, b), \end{cases}$$

where inf and sup are, respectively, the greatest lower bound and the least upper bound.

L is complemented iff

$$\exists \mathbb{0} \in X, \qquad \exists \mathbb{1} \in X, \qquad \forall a \in L, \qquad \exists \bar{a} \in L,$$
$$\inf(a, \bar{a}) = \mathbb{0} \qquad \text{and} \qquad \sup(a, \bar{a}) = \mathbb{1}$$

and

$$\bar{a} \neq \mathbb{0} \ \text{ if } a \neq \mathbb{1}, \qquad \bar{a} \neq \mathbb{1} \ \text{ if } a \neq \mathbb{0}.$$

Here $\mathbb{0}$ and $\mathbb{1}$ are respectively the least and the greatest element of L. ($\forall a \in L, \inf(a, \mathbb{0}) = \mathbb{0}, \sup(a, \mathbb{1}) = \mathbb{1}$.) A lattice with a $\mathbb{0}$ and a $\mathbb{1}$ is a complete lattice. L is distributive iff sup and inf are mutually distributive. ■

Let $\tilde{P}(X)$ be the set of fuzzy subsets of X. Its structure can be induced from that of the real interval [0, 1]. Here [0, 1] is a pseudocomplemented distributive lattice where max and min play the role of sup and inf, respectively. The pseudocomplementation is complementation to 1, and $\tilde{P}(X)$, the set of mappings from X to [0, 1], is a pseudocomplemented distributive lattice.

We have the following properties for $\cup$, $\cap$, and $\bar{\ }$:

(a) Commutativity: $A \cup B = B \cup A$; $A \cap B = B \cap A$.
(b) Associativity: $A \cup (B \cup C) = (A \cup B) \cup C$,
$A \cap (B \cap C) = (A \cap B) \cap C$.
(c) Idempotency: $A \cup A = A$, $A \cap A = A$.
(d) Distributivity: $A \cup (B \cap C) = (A \cup B) \cap (A \cup C)$,
$A \cap (B \cup C) = (A \cap B) \cup (A \cap C)$.
(e) $A \cap \emptyset = \emptyset$, $A \cup X = X$
(f) Identity: $A \cup \emptyset = A$, $A \cap X = A$.
(g) Absorption: $A \cup (A \cap B) = A$, $A \cap (A \cup B) = A$.
(h) De Morgan's Laws: $\overline{(A \cap B)} = \bar{A} \cup \bar{B}$, $\overline{(A \cup B)} = \bar{A} \cap \bar{B}$.

(i) Involution: $\overline{\overline{A}} = A$.
(j) Equivalence formula: $(\overline{A} \cup B) \cap (A \cup \overline{B}) = (\overline{A} \cap \overline{B}) \cup (A \cap B)$.
(k) Symmetrical difference formula: $(\overline{A} \cap B) \cup (A \cap \overline{B}) = (\overline{A} \cup \overline{B}) \cap (A \cup B)$.

The only law of ordinary set theory that is no longer true is the excluded-middle law:

$$A \cap \overline{A} = \emptyset, \qquad A \cup \overline{A} = X.$$

However,

$$\forall A, \forall x \in X, \min[\chi_A(x), \chi_{\overline{A}}(x)] \leq 0.5$$

and

$$\max[\chi_A(x), \chi_{\overline{A}}(x)] \geq 0.5,$$

namely, A and $\overline{A}$ might overlap and $A \cup \overline{A}$ do not cover X exactly.

To check whether a formula A is valid or inconsistent in fuzzy logic, the simplest approach involves expanding A into conjunctive and disjunctive forms. Before defining these normal forms, we give some definitions:

1. A literal is a variable x_i or $\overline{x}_i$, the complement of x_i.
2. A clause is a disjunction of one or more than one literal.
3. A phrase is a conjunction of one or more than one literal.
4. A formula A is said to be in conjunctive normal form if

$$A = C_1 \wedge C_2 \wedge \cdots \wedge C_m, \qquad m \geq 1,$$

and every C_i, $1 \leq i \leq m$, is a clause.
5. A formula A is said to be in disjunctive normal form if

$$A = P_1 \vee P_2 \vee \cdots \vee P_m, \qquad m \geq 1,$$

and every P_i, $1 \leq i \leq m$, is a phrase.

In two-valued logic, it can be shown that every formula can be expressed in conjunctive and disjunctive normal forms owing to the existence of the distributive laws and De Morgan's laws. Since both of the laws mentioned above hold in fuzzy logic and since there is no syntactical difference between formulas in fuzzy logic and formulas in two-valued logic, we can easily see that formulas in fuzzy logic can also be expressed in conjunctive and disjunctive normal form.

Lemma 3.1.1

Let C be a clause. If C contains a complementary pair of literals, then C is fuzzily valid; namely, $\chi(C) \geq 0.5$.

Proof

Let $C = L_1 \vee L_2 \vee \cdots \vee L_n$. Assume L_i and L_j form such a complementary pair. Then $\chi(L_i) = 1 - \chi(L_j)$. For every possible assignment, either $\chi(L_i)$ or

$\chi(L_j)$ will be greater than or equal to 0.5. Therefore, $\max[\chi(L_i), \chi(L_j)] \geq 0.5$ for all possible assignments. Since

$$\chi(C) = \max[\chi(L_1), \chi(L_2), \ldots, \chi(L_n)],$$

$$\chi(C) = \max\left[\chi(L_1), \chi(L_2), \ldots, \chi(L_i), \ldots, \chi(L_j), \ldots, \chi(L_n)\right]$$
$$\geq \max\left[\chi(L_i), \chi(L_j)\right] \geq 0.5.$$

Thus C is fuzzily valid. Q.E.D.

Lemma 3.1.2

Let C be a clause. If C is fuzzily valid under any assignment of truth values, then C contains a complementary pair of literals.

Proof

Consider the assignment in which every literal of C is assigned a truth value smaller than 0.5; $\chi(C)$ will be smaller than 0.5 under this assignment, so C is not fuzzily valid. This is contradictory to the assumption that C is fuzzily valid. Q.E.D.

Combining Lemmas 3.1.1 and 3.1.2, we have the following theorem.

Theorem 3.1.1

Let C be a clause. C is fuzzily valid under any assignment of truth values iff C contains a complementary pair of literals.

Similarly, we can prove the following theorem concerning inconsistency in fuzzy logic.

Theorem 3.1.2

Let P be a phrase. P is fuzzily inconsistent iff P contains a complementary pair of literals.

Theorems 3.1.1 and 3.1.2 can be utilized to check the consistency of formulas in fuzzy logic. Suppose we want to see whether a formula A is valid. We can expand A into a conjunctive normal form:

$$A = C_1 \wedge C_2 \wedge \cdots \wedge C_n.$$

Then A is fuzzily valid iff every C_i is fuzzily valid. However, the fuzzy validity of a clause can be established through Theorem 3.1.1. Similarly, in case we want to check the fuzzy inconsistency of formula A, we can expand A into disjunctive normal form:

$$A = P_1 \vee P_2 \vee \cdots \vee P_n.$$

Then A is fuzzily inconsistent iff every phrase P_i is fuzzily inconsistent and the fuzzy inconsistency of a phrase can be established through Theorem 3.1.2.

It is quite clear that among the infinite number of distinct assignments of grade membership to the variables, there are a finite number of binary assignments (binary assignments of 0 or 1 to every variable).

It can be shown that the set of fuzzy functions compatible with a given Boolean function $F_B(x)$ is a sublattice of the lattice of fuzzy functions. It can also be shown that the "soft" algebra is a bounded, distributive, and symmetric lattice. Hence it is clear that a finite fuzzy lattice is isomorphic to a ring of sets. In the case of Boolean lattices, it is easy to prove that a finite lattice is Boolean if and only if it is isomorphic to the Boolean lattice of all subsets of a finite set. In any lattice we have the standard definition that an element ξ is an atom if $\xi \vdash 0$ (ξ covers 0) and a dual atom if $\xi \dashv 1$ (1 covers ξ), and thus all the elements that are immediate successors of the lower bound 0 are the atoms of the lattice.

Regarding Boolean algebra, it is obvious that the above definition of an atom and the classical definition, that a nonzero element ρ in a Boolean algebra B is an atom iff $\rho x = \rho$ or $\rho x = 0$ for every x in B, are identical.

Because the n-variable switching functions form a Boolean algebra of order 2^{2^n}, the algebra has 2^n atoms which are just the minterms. In fuzzy algebra, however, we prefer the original definition of an atom which implies that the atom is the minimal nonzero element of the lattice. Thus an element ρ is an atom of the fuzzy lattice L iff $\rho \neq 0$ and for every x, if $x \leq \rho$, then $x = \rho$ or $x = 0$.

In order for an element to be the minimal element of L it must be the conjunction of all fuzzy variables and their complements, and thus ensure the minimal grade membership required. Hence it is clear that the fuzzy lattice has only a single atom which has the form of $\Pi_j x_j \bar{x}_j$. The fuzzy algebra over n variables $(x_1, x_2, \ldots, x_n)$ has as its atom the element $x_1\bar{x}_1 x_2\bar{x}_2 \ldots x_j\bar{x}_j \ldots x_n\bar{x}_n$. This notion is a fuzzy analog of a one-point set. A fuzzy algebra Z is said to be atomic if for every nonzero element y in Z, there exists the atom $\Pi_j x_j \bar{x}_j \leq y$. In a similar way, it is clear that in a fuzzy algebra over n variables there exists a set of 2^n dual atoms which consist of the fuzzy variables and their complements.

The measure of uncertainty or the degree of fuzziness of a fuzzy set is assumed to express, on a global level, the difficulty of deciding the belonging grade of elements to the fuzzy set.

Let us consider a set I and a lattice L; any map from I to L is called an L-fuzzy set. Let us denote by $L(I)$ the class of all maps from I to L. It is possible to induce a lattice structure to $L(I)$ by the binary operations $\vee$ and $\wedge$, associating to any pair of elements f and g of $L(I)$ the elements $f \vee g$ and $f \wedge g$ of $L(I)$, defined point by point as

$$(f \vee g)(x) \equiv \text{l.u.b.}\{f(x), g(x)\},$$
$$(f \wedge g)(x) \equiv \text{g.l.b.}\{f(x), g(x)\},$$

where l.u.b. and g.l.b. denote respectively the least upper bound and the greatest lower bound of $f(x)$ and $g(x)$ in the lattice L.

For $L = [0, 1]$ we have

$$(f \vee g)(x) = \max\{f(x), g(x)\}$$
$$(f \wedge g)(x) = \min\{f(x), g(x)\}.$$

We try to introduce, for every element, or "fuzzy set" $f \in L(I)$, a measure of the degree of its "fuzziness." We require of this quantity, which we shall denote by $d(f)$, that it must depend only on the values assumed by f on I and satisfy at least the following properties.

P1: *$d(f)$ must be 0 if and only if f takes on I the values 0 or 1.*

P2: *$d(f)$ must assume the maximum value if and only if f assumes always the value $\frac{1}{2}$.*

P3: *$d(f)$ must be greater or equal to $d(f^*)$ where f^* is any "sharpened" version of f, that is, any fuzzy set such that $f^*(x) \geq f(x)$ if $f(x) \geq \frac{1}{2}$ and $f^*(x) \leq f(x)$ if $f(x) \leq \frac{1}{2}$.*

Let I be a finite set; this assumption and some others that we will make in the following simplify the mathematical formalism but may be suitably weakened in future generalizations. We note, however, that the finiteness of I corresponds to a large class of actual situations.

By defining $H(f)$, in a similar way to Shannon entropy, as

$$H(f) = -K \sum_{i=1}^{N} f(x_i) \log_e f(x_i),$$

where N is the number of elements in I and, K being a positive constant, we get

$$H(f \vee g) = -K \sum_{i=1}^{N} \max[f(x_i), g(x_i)] \log_e \max[f(x_i), g(x_i)]$$

$$H(f \wedge g) = -K \sum_{i=1}^{N} \min[f(x_i), g(x_i)] \log_e \min[f(x_i), g(x_i)].$$

Now let $d(f) = H(f) + H(\bar{f})$ be defined as the entropy of the fuzzy set f, where

$$\bar{f}(x) = 1 - f(x)$$

satisfies the following properties:

$$\bar{\bar{f}} = f \qquad \text{(Involution law)}$$

$$\left.\begin{aligned} \overline{f \vee g} &= \bar{f} \wedge \bar{g} \\ \overline{f \wedge g} &= \bar{f} \vee \bar{g}. \end{aligned}\right\} \qquad \text{(De Morgan laws).}$$

We explicitly note that $\bar{f}$, usually called the complement of f, is not the algebraic complement of f with respect to the lattice operations.

Clearly, $d(f) = d(\bar{f})$; moreover, $d(f)$ can be written using Shannon's function

$$S(x) = -x \log_e x - (1 - x) \log_e (1 - x)$$

as

$$d(f) = K \sum_{h=1}^{N} S(f(x_h)),$$

and $d(f)$ satisfies requirements P1 and P2. Requirement P3 is also satisfied. In fact, if f^* is a sharpened version of f, we have, by definition,

(a) $0 \leq f^*(x) \leq f(x) \leq \frac{1}{2}$, for $0 \leq f(x) \leq \frac{1}{2}$;
(b) $1 \geq f^*(x) \geq f(x) \geq \frac{1}{2}$, for $\frac{1}{2} \leq f(x) \leq 1$.

By the well-known property of Shannon's function $S(x)$, monotonically increasing in the interval $[0, \frac{1}{2}]$ and monotonically decreasing in $[\frac{1}{2}, 1]$ with a maximum at $x = \frac{1}{2}$, we immediately obtain from (a) and (b) that, for any value of $f(x)$,

$$S(f^*(x)) \leq S(f(x)), \qquad x \in I,$$

and

$$d(f^*) \leq d(f).$$

If we assume that $K = 1/N$, we obtain the functional

$$v(f) = \frac{1}{N} \sum_{h=1}^{N} S(f(x_h)),$$

which we shall call the "normalized entropy." This name is appropriate because, taking the logarithm in base 2, $\log_2$, one has

$$0 \leq v(f) \leq 1 \qquad \text{for all } f \in L(I).$$

We have been assuming that the functional $d(f)$ gives a measure of the fuzziness of f; this quantity may also be considered as measuring an amount of information even if its meaning is different from the standard one of Shannon's information theory.

We have an interesting case when the fuzzy set f is random; that is, f is a map

$$f\colon \quad \Omega \times I \to [0,1],$$

such that, for any fixed x, $f(\xi, x)$ is a random variable with respect to a given probability space (Ω, F, p), where Ω is the nonempty set of sample points, F a σ-field of subsets of Ω, and p a probability measure. For any fixed ξ, $f(\xi, x)$ is a fuzzy set. Let us consider the case where Ω has only a finite number M of elements $\xi_1, \ldots, \xi_M$, which may occur with probabilities $p(\xi_1), \ldots, p(\xi_M)$; we may introduce an average fuzzy set $\langle f \rangle$ as

$$\langle f(x) \rangle \equiv \sum_{i=1}^{M} f(x, \xi_i) p(\xi_i).$$

In such a case, the entropy of the fuzzy set is itself a random variable; in fact, if the event ξ_i happens, we have the fuzzy set $f(\xi_i, x)$ whose entropy is $d_i(f)$. In this case, it is meaningful to consider the average entropy given by

$$\sum_{i=1}^{M} p(\xi_i)d_i = \sum_{i=1}^{M} \sum_{j=1}^{N} p(\xi_i)S(f(\xi_i, x_j)).$$

Other proposed measures of fuzziness that satisfy $P1$–$P3$ are:

(i) $d(A) = F[\sum_{i=1}^{|X|} c_i f_i(\chi_A(x_i))]$ for finite X, where $c_i \in \mathbb{R}^+$, $\forall i$; f_i is a real-valued function satisfying (a) $f_i(0) = f_i(1) = 0$, (b) $f_i(x) = f_i(1 - x)$, $\forall x \in [0,1]$ and (c) f_i is strictly increasing on $[0, \frac{1}{2}]$; F is a positive increasing function. For linear F the following holds:

$$d(A_1) + d(A_2) = d(A_1 \cup A_2) + d(A_1 \cap A_2).$$

(ii) $d(A) = \sum_{i=1}^{|M|} |\chi_A(x_i) - \chi_{A_{0.5}}(x_i)|$, where $A_{0.5}$ is the 0.5-cut of A. Clearly (ii) is a special case of (i) when F is the identity and $\forall i$, $c_i = 1$, $f_i(x) = x$ when $x \in [0, 0.5]$.

It is interesting to note that the entropy is a special case of (i) when $F(x) = kx$, $k > 0$, and $\forall i$, $c_i = 1$ and

$$f_i(x) = -x \log_e(x) - (1 - x)\log_e(1 - x).$$

(iii) $d(A) = (1/p(X))\int_X F[\chi_A(x)]\, dp\,(x)$ where $F(y) = F(1 - y)$, $y \in [0,1]$; $F(0) = F(1) = 0$ and F is strictly increasing in $[0, 0.5]$.

3.2 FUZZY-VALUED SWITCHING FUNCTIONS

This section is concerned with the study of fuzzy-valued switching functions by means of the fuzzy algebra described in the previous section. Any fuzzy-valued switching function can be expressed in disjunctive and conjunctive normal forms, in a similar way to two-valued switching functions. As before, fuzzy-valued switching functions over n variables can be represented by the mapping

$$f\colon [0,1]^n \to [0,1].$$

Any fuzzy-valued switching function can be realized, using the basic fuzzy logic operations of max, min, and complement, by associating the multiplication factors to the variables at different stages.

In a disjunctive normal form, each phrase corresponds to a logic "gate" and each literal to an input line. The ratio between the cost of a logic gate and the cost of an input line will depend on the type of gates used in the realization. However, practically, the cost of an additional input line on an already existing gate will be several times less than the cost of an additional logic gate. On this basis, the elimination of gates will be the primary objective of the minimization process, leading to the following definition of a minimal expression.

Definition 3.2.1

A disjunctive normal form is regarded as a minimal-complexity form if there exists

(1) *no other equivalent form involving fewer phrases, and*
(2) *no other equivalent form involving the same number of phrases but a smaller total number of literals.*

Definition 3.2.2

A phrase f_j subsumes another phrase f_l iff f_j contains all the literals of f_l, and thus $\chi(f_j) \le \chi(f_l)$. A fuzzy phrase f_k is said to be a fuzzy implicant of F iff $\chi(f_k) \le \chi(F)$ under all possible assignments. A fuzzy implicant f_j is said to be a fuzzy prime implicant of F (F.P.I.) if it subsumes no other fuzzy implicant of F, that is, $\chi(f_j) \le \chi(f_k) \le \chi(F) \leftrightarrow k = j$. The minimal-complexity form must consist of a sum of phrases representing fuzzy prime implicants. In order to find the complete set of F.P.I.'s, the fuzzy consensus is defined and utilized as follows:

Theorem 3.2.1

Let P be a phrase of fuzzy literals from the set $\{x_i\}_{i=1}^n$. A disjunction D of any variable x_k and its complement $\bar{x}_k$, $1 \le k \le n$, can be appended to P without affecting the general value of the phrase iff there exists a variable x_i and its complement $\bar{x}_i$ in P, for some i, $1 \le i \le n$.

Theorem 3.2.2

Let C be a clause of fuzzy literals from the set $\{x_i\}_{i=1}^n$. A conjunction Q of any variable x_k and its complement $\bar{x}_k$, $1 \le k \le n$, can be appended to C without affecting the general value of the clause iff there exists a variable x_i and its complement $\bar{x}_i$ in C, for some i, $1 \le i \le n$.

In general if F is a conjunction of formulas, it can take a value ≤ 0.5 if certain conditions are satisfied. Clearly, if F is of the form

$$F = \left[\sum_j x_j \bar{x}_j \gamma_j\right]\beta, \qquad 1 \le j \le n,$$

when β and $\{\gamma_j\}_{j=1}^n$ are formulas in $\{x_i\}_{i=1}^n$, then $F \le 0.5$. This is a trivial case where one formula in the conjunction is ≤ 0.5 and thus F is ≤ 0.5. A more general case can be proved as follows.

Theorem 3.2.3

Let the set $\{F_i\}_{i=1}^\omega$ be a set of fuzzy formulas over $x_1, x_2, \ldots, x_n$, and let F be a conjunction of formulas from this set. A disjunction F_d, of any formula F_k, and its complement $\bar{F}_k$ can be appended to or deleted from the conjunction representing F, without affecting the value of F, if there exist functions F_s and F_t in the conjunction representing F such that $\bar{F}_s$ is subsumed by F_t.

Theorem 3.2.4

Let the set $\{F_j\}_{j=1}^\nu$ be a set of fuzzy formulas over $x_1, x_2, \ldots, x_n$, and let F be a disjunction of formulas from this set. A conjunction F_c of any formula F_r and its *complement* $\overline{F}_r$, *can be appended to or deleted from the disjunction representing* F, *without affecting the value of* F, *if there exist functions* F_i *and* F_m *in the disjunction representing* F *such that* F_i *is subsumed by* $\overline{F}_m$.

Definition 3.2.3

Let R *and* Q *be two phrases over the set of fuzzy variables* $x_1, x_2, \dots, x_n$. *The fuzzy consensus of* R *and* Q, *written* $R\psi Q$, *is defined to be the set of phrases* $\{R_iQ_i\}$, *where* $R = x_iR_i$ *and* $Q = \overline{x}_iQ_i$ (*or* $R = \overline{x}_iR_i$ *and* $Q = x_iQ_i$) *and* $x_i \in \{x_1, x_2, \dots, x_n\}$, *if the phrase* R_iQ_i *includes the conjunction* $x_j\overline{x}_j$ *for at least one* j, $j \in \{1, 2, \dots, n\}$. *If the phrase* R_iQ_i *does not include* $x_j\overline{x}_j$ *for any* j, $j \in \{1, 2, \dots, n\}$, *then*

$$R\psi Q = \{(R_iQ_ix_j\overline{x}_j | j = 1, 2, \dots, n), \qquad x_i \in (x_1, x_2, \dots, x_n)\}.$$

If none of the above occurs, then we say that

$$R\psi Q = 0.$$

The phrases added whenever $x_j\overline{x}_j \notin R_iQ_i$ are not needed *fuzzy prime implicants*. This can be seen from the following proposition.

Proposition 3.2.1

Let $R = x_iR_i$ *and* $Q = \overline{x}_iQ_i$ (*or* $R = \overline{x}_iR_i$ *and* $Q = x_iQ_i$), *and* R_iQ_i *does not include* $x_j\overline{x}_j$ *for any* j, $j \in \{1, 2, \dots, n\}$. *Then*

$$R + Q + (R\psi Q) = R + Q.$$

We shall define two kinds of fuzzy phrases. The first kind, to which we shall refer as a Type-1 phrase, are phrases that contain a conjunction of the form $x_j\overline{x}_j$ for at least one j, $j \in \{1, 2, \dots, n\}$. Otherwise we refer to the phrase as a Type-2 phrase. Clearly a Type-1 phrase cannot be subsumed by Type-2 phrases. However, they can subsume some of them. For the case where members of the set $\{R_iQ_i\}$ do not include a conjunction of the form $x_j\overline{x}_j$ for at least one j, $j \in \{1, 2, \dots, n\}$, two situations must be checked:

(a) R and Q are both Type-2 phrases. Since $\{R_iQ_ix_j\overline{x}_j | j = 1, 2, \dots, n\}$ is a set of Type-1 phrases covered by $R + Q$, this set is not needed.
(b) R is a Type-1 phrase and Q is a Type-2 phrase. In order for members of $\{R_iQ_i\}$ not to include a conjunction $x_j\overline{x}_j$ for any j, $j \in \{1, 2, \dots, n\}$, R must be of the form $\alpha x_i\overline{x}_i\beta$ and Q must be of the form $\gamma x_i\delta$ (or $\gamma'\overline{x}_i\delta'$) where the phrase $\alpha x_i\beta\gamma\delta$ (or $\alpha\overline{x}_i\beta\gamma'\delta'$) is a Type-2 phrase. Thus

$$R_i = \alpha x_i\beta \qquad (\text{or } \alpha\overline{x}_i\beta),$$

$$Q_i = \gamma\delta \qquad (\text{or } \gamma'\delta'),$$

and obviously the set $\{R_iQ_ix_j\overline{x}_j | j = 1, 2, \dots, n\}$ is covered by Q and is not needed.

We define a fundamental phrase as either a phrase of type-2 or a phrase of type-1 containing at least one literal of each variable. Clearly each phrase of type-2 in a fuzzy switching function is an essential prime implicant.

Theorem 3.2.5

Let R, Q, and W each represent a phrase. If $W \in R \psi Q$, then $R + Q \supseteq W$.

The set of axioms of fuzzy algebra and Theorems 3.2.1–3.2.4 form the basis for the method of fuzzy iterated consensus and minimization of fuzzy functions. It will be shown in the following that the successive addition of fuzzy consensus phrases to a sum-of-products expression and the removal of phrases that are included in other phrases ($x + xy = x$) will result in an expression that represents the function as the sum of all its F.P.I.'s.

Theorem 3.2.6

A sum-of-products expression $F = P_1 + P_2 + \cdots + P_r$ for the function $F(x_1, x_2, \ldots, x_n)$ is the sum of F.P.I.'s *of $F(x_1, x_2, \ldots, x_n)$ if and only if:*

1. *No phrase subsumes any other phrase, $P_j \not\subseteq P_i$ for any i and j, $i \neq j$, $i, j \in \{1, 2, \ldots, r\}$.*
2. *The fuzzy consensus of any two phrases, $P_i \psi P_j$, either does not exist ($P_i \psi P_j = 0$) or every phrase that belongs to the set describing $P_i \psi P_j$ subsumes some other phrase from the set $\{P_k\}_{k=1}^{r}$.*

The Minimization Algorithm

Input: *The set of phrases representing $F(x_1, x_2, \ldots, x_n)$.*

Output: *Set of* F.P.I.'s *of $F(x_1, x_2, \ldots, x_n)$.*

Step 1: *Compare each phrase with every other phrase in the expression, and remove any phrase that subsumes any other phrase.*

Step 2: *Add the fuzzy consensus phrases that do not subsume some other phrases.*

The process is iteratively repeated and it terminates when all possible fuzzy consensus operations have been performed. The remaining phrases include all the essential F.P.I.'s *of $F(x_1, x_2, \ldots, x_n)$.*

The minimization algorithm that is based on the above theorem may be broken down into three steps, namely:

1. Compute the fuzzy prime implicants of the function.
2. Remove the essential fuzzy prime implicants.
3. Find a minimal complexity cover of the remainder.

A fuzzy switching function that is expressed as the disjunction of some of its prime implicants, such that the deletion of any one would change the function represented, is said to be in *irredundant fuzzy disjunctive form* (IFDF). Fuzzy prime implicants that never appear in any of the IFDF's are called *absolutely dispensable*. Those that always appear are called *essential*, while those that appear in some but not all IFDF's are called *conditionally eliminable*.

Lemma 3.2.1

Let a fuzzy function F be represented by a disjunction of some of its prime implicants. If F contains A_i and some other prime implicants $A_{j1}, A_{j2}, \ldots, A_{jn}$ such that

$$A_i \leq A_{j1} + A_{j2} + \cdots + A_{jn},$$

then A_i is redundant in F.

Lemma 3.2.2

For a fuzzy function, F, and any one of its prime implicants, A_i, each IFDF, F', of F contains either A_i or some other prime implicants $A_{j1}, A_{j2}, \ldots, A_{jn}$ such that

$$A_i \leq A_{j1} + A_{j2} + \cdots + A_{jn} \text{ (irr)}.$$

From Lemma 3.2.2, all the IFDF's can be obtained by forming the disjunction of the prime implicants and considering the ways to replace each one by every disjunction of prime implicants it implies irredundantly.

The presence of a fuzzy prime implicant, A_i, or one of the disjunctions it implies irredundantly, $A_{j1} + A_{j2} + \cdots + A_{jn}$, $A_{k1} + A_{k2} + \cdots + A_{km}, \ldots$ can be represented algebraically by defining the *presence factor*, p_x, to represent the presence of A_x in an IFDF. From Lemma 3.2.2, the following Boolean relation must equal 1:

$$(p_i + p_{j1}p_{j2} \cdots p_{jn} + p_{k1}p_{k2} \cdots p_{km} + \ldots)$$

The above relation is called the *presence relation* of the fuzzy prime implicant A_i. Obviously, the presence relations of every prime implicant of a fuzzy function must equal one. This implies that the conjunction, P, of the presence relations for all the prime implicants of a fuzzy function, F, must equal one. P is called the *presence function* of F. When the presence function is expanded into its irredundant disjunctive form, each product term represents an IFDF. By Lemma 3.2.2, the presence function produces all the IFDF's. The next theorem follows directly from Lemma 3.2.2 and the definition of the presence function.

Theorem 3.2.7

A fuzzy prime implicant, A_i, is absolutely dispensable if and only if its presence factor, p_i, does not appear in the irredundant disjunctive form of the presence function.

The presence function can be reduced to irredundant disjunctive form using any of the algorithms from Boolean logic. Also, if a fuzzy prime implicant is known beforehand to be absolutely dispensable, then its presence factor can be assigned the value zero and its presence relation can be excluded from the presence function.

Before getting to absolute dispensability, the following theorem shows the necessary and sufficient conditions for a fuzzy prime implicant to be essential.

Theorem 3.2.8

A fuzzy prime implicant, A_i, is essential if and only if there is no irredundant implication relation among the prime implicants whose antecedent is A_i.

The essential prime implicants of a fuzzy function, F, are called the *core* of F. The simplest method for determining absolutely dispensable prime implicants is stated in the following theorem.

Theorem 3.2.9

A fuzzy prime implicant, A_i, is absolutely dispensable if there is at least one disjunction of essential prime implicants that it implies irredundantly.

Fuzzy prime implicants that are determined to be absolutely dispensable by Theorem 3.2.9 are said to be *Type 1* absolutely dispensable. The above theorems are illustrated by the following example. The following theorem provides for finding another type of absolutely dispensable fuzzy prime implicant, *Type 2*, which cannot be found by Theorem 3.2.9.

Theorem 3.2.10

A prime implicant, A_i, in a fuzzy function, F, is absolutely dispensable if it is the antecedent of at least one irredundant implication relation,

$$A_i \leq A_{j1} + A_{j2} + \cdots + A_{jn},$$

and no A_{jk} in its consequent is the antecedent of an irredundant implication relation whose consequent contains A_i.

Since any IFDF must contain all the essential fuzzy prime implicants, all essential prime implicants can be removed from the irredundant implication relations of the nonessential prime implicants to give a *modified* implication relation. However, some of these modified implication relations may now be redundant and can therefore be removed from further consideration. Theorem 3.2.10 also applies to modified irredundant implication relations, denoted by writing (mirr) beside the relation, as shown in Theorem 3.2.11 below.

Theorem 3.2.11

A fuzzy prime implicant, A_i, is absolutely dispensable if it is the antecedent of at least one modified irredundant implication relation,

$$A_i \leq A_{j1} + A_{j2} + \cdots + A_{jn}(\text{mirr}),$$

and no A_{jk} in its consequent is the antecedent of a modified irredundant implication relation whose consequent contains A_i. This theorem detects Type-3 *absolute dispensability.*

From combinatorial arguments it can be shown that there are 2^{4^n} *possible* fuzzy functions of n variables. However, most of these are simply nonminimized forms of a relatively few unique fuzzy functions. The problem can be appreciated best by considering the number of functions of two variables that exist: For Boolean functions there are 2^{2^2}, or 16; for fuzzy functions though, there are 2^{4^2}, or 65,536. It is obviously impossible to enumerate them by hand, which in the case of Boolean functions is quite reasonable for small n.

Since there are $\binom{n}{k}$ ways of selecting k out of n variables and each variable may be either complemented or left uncomplemented, the number of Type-2 phrases containing k literals is $\binom{n}{k}2^k$. It follows that the total number of Type-2 phrases is

$$\sum_{k=1}^{n} \binom{n}{k} 2^k = (1 + 2)^n - 1 = 3^n - 1.$$

As for the fundamental phrases of type-1, each such phrases contains, for each k, either x_k only, $\bar{x}_k$ only, or $x_k\bar{x}_k$. Thus there are 3^n such phrases. Since 2^n of them are simple, altogether there are $2 \cdot 3^n - 2^n - 1$ fundamental phrases. Clearly, any function can be defined by disjunction of a nonempty subset of the set of all fundamental phrases. Therefore there are at most $(2^{2 \cdot 3^n - 2^n - 1} - 1)$ distinct fuzzy switching functions. A set of phrases is called "independent" iff none of the phrases in the set implies any other in the same set.

We can use this definition in order to generate an independence relation R^* between the fundamental phrases. This relation can be graphically represented as an upper diagonal binary matrix showing the independence relation of the $(2 \cdot 3^n - 2^n - 1)$ elements.

Thus the number of distinct fuzzy switching functions will be given by M_n, where

$$M_n = 2 \cdot 3^n - 2^n - 1 + \sum_{i=1}^{\text{MAX}} tl_i$$

and tl_i is transitivity level i on the matrix, and MAX will denote the maximum level of transitivity for this particular n, given by the maximum cover fraction, such that

$$\text{MAX} = \binom{n}{n - \lfloor (n-2)/3 \rfloor - 1} 2^{(n - \lfloor (n-2)/3 \rfloor - 1)}.$$

By assuming that all fundamental phrases cannot combine, an upper bound on the number of minimized functions is given by

$$S_u = 2 + \sum_{i=1}^{\text{MAX}} \binom{2 \cdot 3^n - 2^n - 1}{i}, \qquad n \geq 1.$$

Recent results indicate that F_n, the exact number of fuzzy-valued switching functions of n variables, can be represented by the inequality

$$F_n^2 < F_{n+1} < 2^{3^n} * F_n^2,$$

which improves the upper bound results for large n.

Many important notions, fuzzy in nature, may be assigned a precise meaning using the concept of the membership function of a fuzzy set, thereby

Table 3.2.1

Number of Variables	Total Number of Functions	Exact Number of Minimized Functions	Lower Bound	Upper Bound
0	2	2	2	2
1	16	6	6	8
2	65,536	84	51	512
3	2^{64}	43,918	8,684	$2^{27} \simeq 10^8$
4	2^{256}	160,297,985,276	$2^{33} \simeq 8 * 10^9$	$2^{81} \simeq 2 * 10^{24}$

making them amenable to mathematical analysis as well as to engineering applications. One such example is the notion of a fuzzy switching function introduced to analyze hazard detection and transient phenomena in binary systems.

We feel that in many scientific models the amount of information is determined by the amount of the uncertainty — or, more exactly, it is determined by the amount by which the uncertainty has been reduced; that is, we can measure information as the decrease of uncertainty. The concept of information itself has been implicit in many areas of science, both as a substantive concept important in its own right and as a consonant concept that is ancillary to the entire structure of science. Thus, models using approximate information, especially in pattern recognition and classification, feature selection, branching questionnaires, optimal encoding, and optimal design of decision tables, and in fault/transient analysis of nonfuzzy systems, have many practical implications. The models may successfully utilize the concept of fuzzy-valued switching functions.

3.3 INTEGRATION AND DIFFERENTIATION OF FUZZY FUNCTIONS

Let f be a fuzzifying function from $[a, b] \subseteq \mathbb{R}$ to $\mathbb{R}$ such that $\forall x \in [a, b]$, $\tilde{f}(x)$ is a fuzzy number, i.e., a piecewise continuous convex normalized fuzzy set on $\mathbb{R}$. $\forall \alpha \in (0, 1]$, the equation $\chi_{\tilde{f}(x)}(y) = \alpha$ with x and α as parameters is assumed to have exactly two continuous solutions $y = f_\alpha^+(x)$ and $y = f_\alpha^-(x)$ for $\alpha \neq 1$ and only one, $y = f(x)$, for $\alpha = 1$, which is also continuous, f_α^+ and f_α^- are defined such that

$$f_{\alpha'}^+(x) \geq f_\alpha^+(x) \geq f(x) \geq f_\alpha^-(x) \geq f_{\alpha'}^-(x), \qquad \forall \alpha, \alpha',$$

such that $\alpha' \leq \alpha$. These functions are the α-level curves of $\tilde{f}$. The integral of any continuous α-level curve of $\tilde{f}$ over $[a, b]$ always exists. Following Dubois and Prade [1979b], it can be shown that an intuitive way of defining the integral $\tilde{I}(a, b)$ of $\tilde{f}$ over $[a, b]$ is to assign the membership value α to the integral of any α-level curve of $\tilde{f}$ over $[a, b]$. Using Zadeh's notation, $\tilde{I}(a, b)$ is the fuzzy set on $\mathbb{R}$

$$\tilde{I}(a, b) = \int_{\alpha \in (0,1]} \alpha \Big/ \int_a^b f_\alpha^-(x)\, dx + \int_{\alpha \in (0,1]} \alpha \Big/ \int_a^b f_\alpha^+(x)\, dx.$$

This definition is consistent with the extension principle. Clearly, $\forall T \in \mathbb{R}$, $\exists$ an α-level curve f_α, delimiting an area A whose surface is T, such that

$$\chi_{\tilde{I}(a,b)}(T) = \alpha \qquad \text{and} \qquad T = \int_a^b f_\alpha(x)\, dx.$$

When $\tilde{f}(x)$ has constant-membership intervals, the level curves may degenerate into "level areas."

Next, we discuss the relation to the integral of a nonfuzzy function over a fuzzy interval.

Let A and B be two fuzzy sets on $\mathbb{R}$. The extension principle allows us to define the integral of a real-valued ordinary function f over the fuzzy interval (A, B) bounded by A and B, say $I(A, B)$:

$$\chi_{I(A,B)}(Z) = \sup_{x,\,y:\ Z = \int_x^y f(u)\, du} \min[\chi_A(x), \chi_B(y)].$$

When our bound is fuzzy, let us consider the integral of f over (a, B):

$$\chi_{I(a,B)}(Z) = \sup_{y:\ Z = \int_a^y f(u)\, du} \chi_B(y) = \sup_{y:\ Z = F(y) - F(a)} \chi_B(y),$$

where F is an antiderivative of f. We see that $I(a, B) F(B) \ominus F(a)$ is the value of the extended $F(x) - F(a)$, when $x = B$. However, when both bounds are fuzzy, we can write

$$\begin{aligned}
\chi_{I(A,B)}(Z) &= \sup_{Z = F(y) - F(x)} \min[\chi_A(x), \chi_B(y)] \\
&= \sup_{x \in \mathbb{R}} \min \left[\chi_A(x), \sup_{y:\ Z = F(y) - F(x)} \chi_B(y)\right] \\
&= \sup_{x \in \mathbb{R}} \min \left[\chi_A(x), \chi_{I(x,B)}(Z)\right];
\end{aligned}$$

that is, $I(A, B) = A \circ I(\cdot, B) = A \circ (F(B) \ominus F(\cdot))$. Here $I(A, B)$ is the fuzzy value of the extended fuzzifying function $y = F(B) \ominus F(x)$ for $x = A$. Hence, $I(A, B) = F(B) \ominus F(A)$, which can be denoted $\int_A^B f(x)\, dx$. When both function and interval are fuzzy, the extension principle gives us

$$\chi_{I(A,B)}(Z) = \sup_{\substack{l \in L, (x,y) \in \mathbb{R}^2, x \le y \\ \int_x^y l(t)\, dt = Z}} \min[\chi_A(x), \chi_B(y), \chi_{\tilde{f}}(l)],$$

where $\tilde{f}$ is a fuzzifying function satisfying the assumptions of a and A, B are fuzzy numbers that delimit a fuzzy interval, $\tilde{I}(A, B)$ is the fuzzy integral. This is identical to

$$\chi_{\tilde{I}(A,B)}(Z) = \sup_{x \le y} \min \left[\chi_A(x), \chi_B(y), \sup_{\substack{l \in L \\ Z = \int_x^y l(t)\,dt}} \chi_{\tilde{f}}(l) \right],$$
$$= \sup_{x \le y} \min \left[\chi_A(x), \chi_B(y), \chi_{\tilde{I}(x,y)}(Z) \right],$$

and thus, $\tilde{I}(A, B)$ is the value of the extended $\tilde{I}(x, y)$ for $x = A$ and $y = B$.

It is easy to see that

$$\int_a^b [\tilde{f}(x) \oplus \tilde{g}(x)]\,dx = \int_a^b \tilde{f}(x)\,dx \oplus \int_a^b \tilde{g}(x)\,dx,$$

where $\int_a^b \tilde{f}(x)\,dx$, $\int_a^b \tilde{g}(x)\,dx$ denote the integrals of the fuzzifying functions $\tilde{f}$ and $\tilde{g}$, respectively. Also by using α-cuts we get

$$\int_{\tilde{a}}^{\tilde{b}} [f(x) + g(x)]\,dx = \int_{\tilde{a}}^{\tilde{b}} f(x)\,dx \oplus \int_{\tilde{a}}^{\tilde{b}} g(x)\,dx.$$

However,

$$\int_{\tilde{a}}^{\tilde{b}} f(x)\,dx \oplus \int_{\tilde{b}}^{\tilde{c}} f(x)\,dx \supseteq \int_{\tilde{a}}^{\tilde{c}} f(x)\,dx,$$

where the equality holds only if $\tilde{b}$ is a real number.

The treatment of differentiation follows the same line of reasoning. Let $\tilde{f}$ be a fuzzifying function from $\mathbb{R}$ to $\mathbb{R}$. The image of any $x \in D \subseteq \mathbb{R}$ is assumed to be a fuzzy number (i.e., a convex and normalized fuzzy set in R). Moreover, each α-level curve f_α of $\tilde{f}$ is assumed to have a derivative at any $x_0 \in D$. Then, the derivative of $\tilde{f}$ at x_0, denoted $(d\tilde{f}/dx)(x_0)$ is defined by its membership function

$$\chi_{(d\tilde{f}/dx)(x_0)}(P) = \sup_{f_\alpha:\,(df_\alpha/dx)(x_0) = P} \chi(f_\alpha),$$

where $\chi(f_\alpha) = \alpha$ by definition. ($\chi_{(d\tilde{f}/dx)(x)}(P) = 0$ if $\nexists\,\alpha$, $(df_\alpha/dx)(x_0) = P$).

Thus the membership value of P to $(d\tilde{f}/dx)(x_0)$ is the greatest level of all the α-level curves whose slope at x_0 is P. $(d\tilde{f}/dx)(x_0)$ is an estimate of the parallelism of the bundle of level curves at x_0. The less fuzzy $(d\tilde{f}/dx)(x_0)$, the more parallel the level curves.

S. S. L. Chang and Zadeh [1972] have defined the derivative and the integral of a function with a fuzzy parameter (viewed as a fuzzy bunch of ordinary functions) in the following way: Let $x \mapsto f(x, a)$ be a function from $\mathbb{R}$ to $\mathbb{R}$ depending on a real parameter a. Its extension $\tilde{f}$ when the parameter is a fuzzy set A on $\mathbb{R}$ is defined by

$$\chi_{\tilde{f}(x,A)}(y) = \sup_{a,\, y = f(x,a)} \chi_A(a),$$

This fuzzifying function $x \mapsto \tilde{f}(x, A)$ can also be viewed as a fuzzy bunch $F = \int \chi_A(a)/f(\cdot, a)$; in this latter approach the membership function of the derivative and the integral are respectively

$$\chi_{\frac{df}{dx}(x, A)}(y) = \sup_{a,\, y=(df/dx)(x,\, a)} \chi_A(a);$$

$$\chi_{\int_{x_0}^{x} f(t, A)\, dt}(y) = \sup_{a,\, y=\int_{x_0}^{x} f(t,\, a)\, dt} \chi_A(a).$$

In many cases the α-level curves are precisely the elements of the support of the associated fuzzy bunch; thus the two approaches give the same results. It should be noted that if we define a normalized fuzzy set $\beta(\tilde{f})$, we get that

$$\exists f\colon \chi_{\beta(\tilde{f})}(f) = 1$$

only if $\tilde{f}(u)$ is normalized for all u. However, if we try to recover $\tilde{f}$ from the knowledge of $\beta(\tilde{f})$, by a combination rule, it will not always be possible, as the following example indicates.

■ ***Example 3.3.1***

$$U = \{u_1, u_2\}, \qquad V = \{v_1, v_2\},$$

R	v_1	v_2
u_1	0.5	1
u_2	0.8	0.7

$$f_1\colon \begin{matrix} u_1 \mapsto v_1 \\ u_2 \mapsto v_1 \end{matrix} \qquad f_2\colon \begin{matrix} u_1 \mapsto v_1 \\ u_2 \mapsto v_2 \end{matrix} \qquad f_3\colon \begin{matrix} u_1 \mapsto v_2 \\ u_2 \mapsto v_1 \end{matrix} \qquad f_4\colon \begin{matrix} u_1 \mapsto v_2 \\ u_2 \mapsto v_2 \end{matrix}$$

$$\beta(\tilde{f}) = 0.5/f_1 + 0.5/f_2 + 0.8/f_3 + 0.7/f_4.$$

The membership value 1 is lost for $\beta(\tilde{f})$ and thus no combination rule enables us to recover $\chi_R(u_1, v_2)$. ■

In order to get a fuzzy mapping $\tilde{f}$ that can be recovered from $\beta(\tilde{f})$ by means of a combination value, we must define additional conditions. More specifically, a mapping f from U to V is defined for all elements in U; the definition of a fuzzy mapping could be adapted to get closer to the concept of a fuzzy set of mappings, that is to say, *genuine* mappings.

Other approaches for integration over fuzzy domains have been suggested in the literature. The above definition contrasts with them by the fact that the fuzzy domain is not a fuzzy set (of $\mathbb{R}$) but some area delimited by fuzzy bounds.

However, other approaches to integration exist, which enable us to integrate a mapping over a fuzzy set. Such extensions of the integration of a mapping over a set have been proposed in the literature by Ralescu [1976] and Yager [1980a], and will briefly be discussed here.

Given a measure space (X, A, P) where X is a nonempty set, A a σ-algebra on X, and P a measure on (X, A), given an integrable real function

f and a fuzzy set F on X, Ralescu [1976] defined the integral of f over F as

$$\int_F f\,dP = \int_X f \cdot \chi_F\,dP,$$

where $\forall \alpha \in (0,1]$, $F_\alpha \in A$.

This definition really reads $\int_F f = \int_{S(F)} f \cdot \chi_F$ where F is a fuzzy set of $\mathbb{R}$, and f is integrable on the support of F.

Ralescu's definition is closely related to the definition of a nonfuzzy probability of fuzzy events to be discussed later. Clearly, the above yields some number belonging to the interval $[\int_{F_1} f\,dP, \int_{S(F)} f\,dP]$ where F_1 is the one-cut of F. In other words, we get a weighted integral of f over X, F being viewed as a set of weights $\{\chi(x) | x \in X\}$. And indeed Klement [1980] proves that as soon as $\forall \alpha \in (0,1)$, $P(\{x \in X | \chi_F(x) = \alpha\}) = 0$, then $\int_F f\,dP$ equals the measure of some α-cut of F.

Dubois and Prade's [1979b] definition is more closely related to a concept introduced by Yager [1980a]. On the same measure space (X, A, P), and given an integrable real mapping f and a fuzzy set F such that $\forall \alpha \in (0,1]$, $F_\alpha \in A$, the fuzzy-valued integral of f over F can be defined as a fuzzy set $I_P(f, F) = \int_F f\,dP$ with membership function

$$\chi_{I_P(f,F)}(r) = \sup\left\{ \alpha | r = \int_{F_\alpha} f\,dP \right\}, \qquad \forall r \in \mathbb{R}.$$

Such a concept is expressed by Yager in a probabilistic framework, the fuzzy probability $P(F)$ of a fuzzy event F being defined as

$$\chi_{P(F)}(r) = \sup\{ \alpha | r = P(F_\alpha)\}, \qquad \forall r \in [0,1].$$

Clearly, the definition of the integral of a mapping over a real domain limited by fuzzy bounds has been proposed as an application of Zadeh's extension principle. This integral enjoys nice properties such as linearity, monotonicity, and others. Such a definition can be useful for laying the ground for a theory of fuzzy probabilities of fuzzy events, contrasting with Zadeh's approach [1968a]. Furthermore, differentiation can be extended to fuzzy-valued mappings, and thus lead to a study of fuzzy differential operations (FDEs), that is, equations where derivatives of fuzzy mappings are involved. This may seem appealing, especially to researchers interested in the analysis of imprecise systems and in the modelling of systems in a variety of fields (e.g., economics).

CHAPTER FOUR

Fuzzy Events and Fuzzy Statistics

4.1 FUZZY SETS VS. PROBABILITY

In this section we will discuss the question of the relation between the theory of fuzzy sets and the theory of probability. The main issues are:

(i) Is it true, as some claim, that the concept of a fuzzy set is merely a disguised form of subjective probability?
(ii) Are there problems that can be solved more effectively by the use of fuzzy-set techniques than by classical probability-based methods?

We use Zadeh's response to these questions [1978a]: In essence, the theory of fuzzy sets is aimed at the development of a body of concepts and techniques for dealing with sources of uncertainty or imprecision that are nonstatistical in nature. For example, the proposition "X is a small number," in which *small number* is a label of a fuzzy subset of nonnegative integers, defines the *possibility distribution* rather than the probability distribution of X. What this implies is that, if the degree to which an integer I fits one's subjective perception of *small number* is χ_I, then π_I, the possibility that X may take I as its value, is numerically equal to χ_I. Thus, the proposition "X is a small number," like the proposition "X is a number smaller than 6," conveys no information concerning the probability distribution of the values of X. In this sense, the uncertainty associated with the proposition "X is a small number" is nonstatistical in nature.

As mentioned earlier, fuzzy sets bear the same relation to the theory of possibility that measure theory does to the theory of probability. In this connection, it is important to recognize that, while some problems fall entirely within the province of probability theory and some entirely within that of possibility theory, in most cases of practical interest *both* theories must be used in combination to yield realistic solutions to problems in decision analysis under uncertainty. Thus, contrary to the belief expressed by some researchers, probability theory, by itself or in combination with the maximum-entropy principle, does not provide an adequate tool for the analysis of problems in which the available information is incomplete, imprecise, or unreliable.

These scientists may be more appreciative of what the theory of fuzzy sets has to offer after trying to solve the following simple problems by the use of conventional probability-based techniques. In each of these problems, the data are stated in the form of propositions expressed in English, and the answer, like the data, is expected to be in the form of a proposition:

X is a small number;
Y is much smaller than X.
How small is Y?

Most Swedes are tall;
Most tall Swedes are blond.
How many Swedes are blond?

It is unlikely that Joe is very young;
It is likely that Joe is young;
It is very unlikely that Joe is very old.
How likely is it that Joe is not old?

It is not quite true that Moti is very tall;
It is not true that Moti is short.
How tall is Moti?

Ted is much taller than most of his close friends.
How tall is Ted?

Although problems of this kind are not well-posed within the framework of classical mathematics and probability theory, they are illustrative of the types of problem that humans are capable of solving through the use of what might be called *approximate* or *fuzzy reasoning.* The theory of fuzzy sets and, in particular, fuzzy logic do provide a conceptual framework for the solution of imprecisely formulated problems. This is one of the reasons why so many investigators are currently exploring the usefulness of the theory of fuzzy sets in a wide variety of fields ranging from law and medicine to industrial process control and credibility analysis.

4.2 FUZZY STATISTICS

Ordinarily, imprecision and indeterminacy are considered to be statistical, random characteristics and are taken into account by the methods of probability theory. In real situations, a frequent source of imprecision is not only the presence of random variables, but the impossibility, in principle, of operating with exact data as a result of the complexity of the system, or the imprecision of the constraints and objectives. At the same time, classes of objects appear in the problems that do not have clear boundaries; the imprecision of such classes is expressed in the possibility that an element not only belongs or does not belong to a certain class, but that intermediate grades of membership are also possible.

Intuitively, a similarity is felt between the concepts of fuzziness and probability. The problems in which they are used are similar or even identical. These are problems in which indeterminacy is encountered due to random factors, inexact knowledge, or the theoretical impossibility (or lack of necessity) to obtain exact solutions. The similarity is also underscored by the fact that the intervals of variation of the membership grade of fuzzy sets and probability coincide. However, between the concepts of fuzziness and probability there are also essential differences.

Probability is an *objective* characteristic; the conclusions of probability theory can, in general, be *tested by experience*.

The membership grade is *subjective*, although it is natural to assign a lower membership grade to an event that, considered from the aspect of probability, would have a lower probability of occurrence. The fact that the assignment of a membership function of a fuzzy set is "nonstatistical" does not mean that we cannot use probability distribution functions in assigning membership functions. As a matter of fact, a careful examination of the variables of fuzzy sets reveals that they may be classified into two types: *statistical* and *nonstatistical*. The variable "magnitude of x" is an example of the former type. However, if one considers the "class of tall men," the "height of a man" can be considered to be a nonstatistical variable. In this case, for instance, if a man is under five feet tall, we would not call him a tall man by "everybody's" standard, and we would assign to him a low grade of membership in the class of tall men. Similarly, if a man is over seven feet, he certainly deserves a high grade of membership in the class of tall men.

The motivation for the development of fuzzy statistics is its philosophical and conceptual relation to subjective probability. In the subjectivistic view, probability represents the *degree of belief* that a given person has in a given event on the basis of given evidence. This view (called also *personalistic* or *judgmental* probability) can be best described by an early article by James Bernoulli,* who defines probability as a degree of confidence in a proposition of whose truth we cannot be certain. His "degree of confidence" is identified with the probability of an event and depends on the knowledge that the individual has at his disposal. Thus, it varies from individual to individual and can be best described as the art of guessing (*Ars conjectandi*).

---

*Bernoulli, J. (1713), *Ars Conjectandi*. Basle, Switzerland.

The difficulty in applying subjective probability stems from the vagueness associated with judgments made through subjective analysis. The postulates of subjective probability cannot be applied, as is widely recognized, to many interesting and useful theories of modern science that are inexact. Subjective probability can be regarded as a personal way of treating objective views, in that they are concerned with individual judgment. These judgments, however, are not additive since human behavior often contradicts the assumption of subjective probability, that an individual is using additive measures in his criteria for evaluation.

We shall remove the restrictive device known as additivity and formulate the basic structure of a fuzzy statistic to be applied as our analytical model of investigation of imprecise data.

In the classical approach, a *probability system* is a triple (Ω', S, P), where Ω' is an arbitrary set (the sample space which includes all possible outcomes), S is a set of events, and P is a real-valued function defined for each $A \in S$ such that:

1. $0 \leq P(A) \leq 1, \quad \forall A \in S$;
2. $P(\Omega') = 1$;
3. if $A_1, A_2, \ldots$ is any sequence of pairwise disjoint sets in S, then

$$P\left(\bigcup_n A_n\right) = \sum_n P(A_n).$$

Definition 4.2.1

Let B be a Borel field (σ-algebra) of subsets of the real line Ω. A set function $\mu(\cdot)$ defined on B is called a fuzzy measure if it has the following properties:

1. $\mu(\Phi) = 0 \qquad$ *(Φ is the empty set)*;
2. $\mu(\Omega) = 1$;
3. *If $\alpha, \beta \in B$ with $\alpha \subset \beta$, then $\mu(\alpha) \leq \mu(\beta)$*;
4. *If $\{\alpha_j | 1 \leq j < \infty\}$ is a monotone sequence, then*

$$\lim_{j \to \infty} \left[\mu(\alpha_j)\right] = \mu\left[\lim_{j \to \infty} (\alpha_j)\right].$$

Clearly, Φ, $\Omega \in B$; also, if $\alpha_j \in B$ and $\{\alpha_j | 1 \leq j < \infty\}$ is a monotonic sequence, then $\lim_{j \to \infty}(\alpha_j) \in B$. In the above definition, (1) and (2) mean that the fuzzy measure is bounded and nonnegative, (3) means monotonicity (in a similar way to finite additive measures used in probability), and (4) means continuity. It should be noted that if Ω is a finite set, then the continuity requirement can be deleted.

(Ω, B, μ) is called a fuzzy measure space; $\mu(\cdot)$ is the fuzzy measure of (Ω, B).

The fuzzy measure μ is defined on subsets of the real line. Clearly, $\mu[\chi_A \geq T]$ is a nonincreasing, real-valued function of T when χ_A is the membership function of set A. Throughout our discussion, we shall use ξ_T to

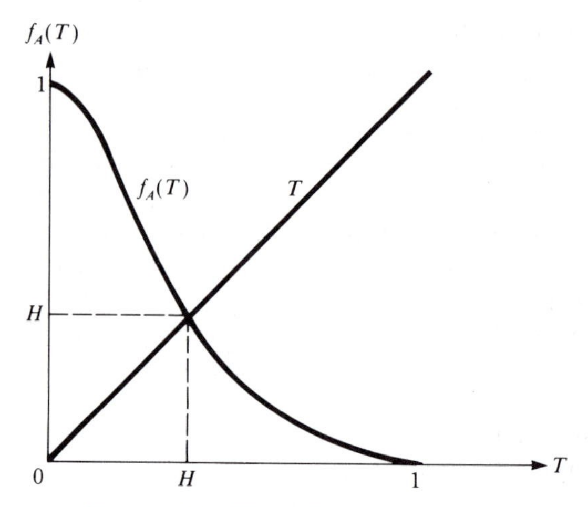

Figure 4.2.1. The evaluation of FEV(χ_A).

represent $\{x|\chi_A(x) \geq T\}$ and $\mu(\xi_T)$ to represent $\mu[\chi_A \geq T]$, assuming that our set A is well specified.

Let χ_A: $\Omega \to [0,1]$ and $\xi_T = \{x|\chi_A(x) \geq T\}$. The function χ_A is called a B-measurable function if $\xi_T \in B$, $\forall T \in [0,1]$. Definition 4.2.2 defines the *fuzzy expected value* (FEV) of χ_A when $\chi_A \in [0,1]$. Extension of this definition when $\chi_A \in [a, b]$, $a < b < \infty$, will be presented later.

Definition 4.2.2

Let χ_A be a B-measurable function such that $\chi_A \in [0,1]$. The fuzzy expected value (FEV) *of χ_A over a set A, with respect to the measure $\mu(\cdot)$, is defined as*

$$\operatorname*{Sup}_{T\in[0,1]} \{\min[T, \mu(\xi_T)]\},$$

where $\xi_T = \{x|\chi_A(x) \geq T\}$. Now, $\mu\{x|\chi_A(x) \geq T\} = f_A(T)$ is a function of the threshold T. The actual calculation of FEV(χ_A) *then consists of finding the intersection of the curves $T = f_A(T)$.*

The intersection of the two curves will be at a value $T = H$, so that FEV(χ_A) $= H \in [0,1]$. Figure 4.2.1 illustrates the above remarks.

It should be noted that when dealing with the FEV(η) where $\eta \notin [0,1]$, we should not use a fuzzy measure in the evaluation but rather a function of the fuzzy measure, μ^*, which transforms μ under the same transformation that χ and T undergo to η and T^*, respectively.

■ ***Example 4.2.1***

Using the base variable TALL, consider a population X which consists of z people and y people who have been assigned grades of membership χ_1 and χ_2, respectively, in the set of *tall* people, A. We require that $0 \leq \chi_1 < \chi_2 \leq 1$. The

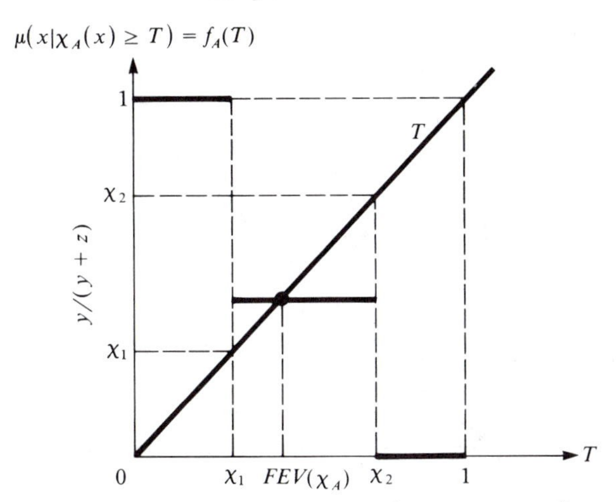

Figure 4.2.2. Graphical evaluation of FEV in the case $\chi_1 \leq y/(y+z) \leq \chi_2$.

fuzzy measure $\mu(\xi_T)$ is such that if $T \leq \chi_1$, $\mu(\xi_T) = 1$, while if $T > \chi_2$, $\mu(\xi_T) = 0$. For $\chi_1 < T \leq \chi_2$,

$$\mu(\xi_T) = y/(z+y).$$

We have therefore defined the measure as the proportion of the population whose grade of membership is larger than or equal to T over the entire population Ω. Thus

$$\mu(\xi_T) = \begin{cases} 0 & \text{if } \chi_2 < T \leq 1, \\ \dfrac{y}{z+y} & \text{if } \chi_1 < T \leq \chi_2, \\ 1 & \text{if } 0 \leq T \leq \chi_1. \end{cases}$$ ■

The graph for the determination of the FEV(χ_A) for this population is shown in Figure 4.2.2.

Figure 4.2.2 is not the only possible value of FEV(χ_A). We can also easily distinguish two other cases.

1. FEV(χ_A) = χ_1 if $\chi_1 \geq y/(y+z)$;
2. FEV(χ_A) = χ_2 if $\chi_2 \leq y/(y+z)$.

We shall expect that the calculation of the FEV(χ_A) for the discrete case will involve finding the intersection of the line $T = T$ with a function $\mu(x|\chi_A(x) \geq T) = f_A(T)$, as shown in Figure 4.2.3.

We shall now illustrate and obtain the relations between the FEV and the standard probabilistic expectation on one hand, and between the FEV and measures of central tendency, representing typical values of data sets, on the other.

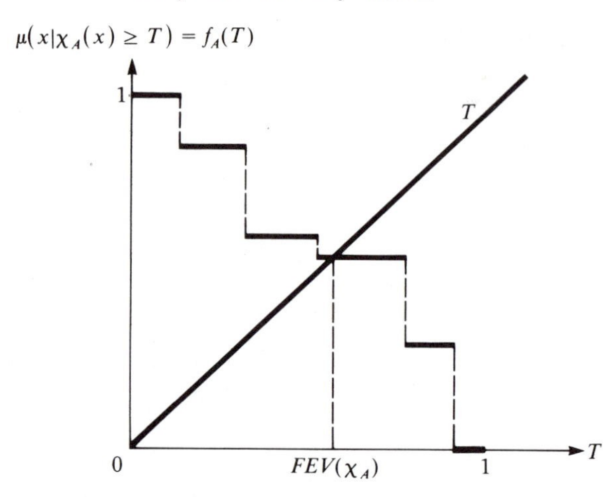

Figure 4.2.3. Graphical calculation of FEV(χ_A).

Definition 4.2.3

A value $T \in [0,1]$ is a subtypical value of a function ξ if $T \leq \mu(\xi_T)$, supertypical if $T \geq \mu(\xi_T)$, and typical if $T = \mu(\xi_T)$. Note that a typical value is both subtypical and supertypical.

Theorem 4.2.1

If a number T_1 is a subtypical value of two functions, ξ and η, and if, for every $T \geq T_1$, $\xi_T = \eta_T$, then $\mathrm{FEV}(\xi) = \mathrm{FEV}(\eta)$.

Theorem 4.2.1 implies that values of a function that are below a subtypical value may be ignored in calculating the FEV. Similarly, we show in Theorem 4.2.3, which is the dual of Theorem 4.2.1, that values of a function that are above a supertypical value may be ignored in calculating the FEV.

The B-measurable function χ is called the compatibility function. In the case of linguistic variables, the numerical variables whose values constitute what may be called the base variable describe a fuzzy restriction on the meaning of the linguistic variable. This fuzzy restriction (which is clearly subjective) on the values of the base variable is characterized by the compatibility function χ, which associates with each value of the base variable a number in the interval $[0,1]$, representing its compatibility with the fuzzy restriction.

If the composition established in Definition 4.2.2 under the operation "max-min" seems like a pessimistic evaluation of the FEV, does a similar composition using the operation "min-max" act as an optimistic model? The answer to this question lies in the following theorem.

Theorem 4.2.2
Let $\chi\colon \Omega \to [0,1]$ *and*

$$\xi_T = \{x | \chi(x) \geq T\} \in B.$$

Then

$$\sup_{T \in [0,1]} \{\min[T, \mu(\xi_T)]\} = \inf_{T \in [0,1]} \{\max[T, \mu(\xi_T)]\},$$

where μ *is a fuzzy measure.*

An average is a value that is typical of a set of data. Since such typical values tend to lie centrally within a set of data arranged according to magnitude, averages are also known as *measures of central tendency*. Several types of average can be defined, the most common being the arithmetic mean, the median, the mode, the geometric mean, and the harmonic mean.

Before discussing the relations of the FEV to the mean and to the median, we shall evaluate the use of the concept by constructing a combinatorial scheme to generalize Example 4.2.1.

Assume a finite set of data points where there are $(n + 1)$ distinct levels of compatibility, such that

$$0 \leq \chi_1 < \chi_2 < \cdots < \chi_{n+1} \leq 1,$$

which imply n distinct levels of fuzzy measure $\mu(\xi_T)$, excluding 0 and 1, represented by $\{\mu_j(\xi_T)\}_{j=1}^{n}$. Clearly the two sets representing

$$\{\chi_i\}_{i=1}^{n+1} \quad \text{and} \quad \{\mu_j(\xi_T)\}_{j=1}^{n}$$

are in increasing order, and we trivially exclude any permutation in each of the sets. Thus the set representing the union of these two sets has $(2n + 1)$ elements; in order to find all possible arrangements of these elements, we can view the problem of finding the number of arrangements as follows:

■ ***Example 4.2.1A***
Find the number of n-arrangements of $(2n + 1)$ objects, of which exactly $(n + 1)$ are alike of one kind and exactly n are alike of the second kind.

Solution
Let γ be the required number of arrangements. Let the $(2n + 1)$ objects be labeled $x_1, \ldots, x_{2n+1}$ with the $(n + 1)$ alike of one kind labeled $x_1, \ldots, x_{n+1}$, and the n alike of the second kind labeled $x_{n+2}, \ldots, x_{2n+1}$. Then there will be $(2n + 1)!$ n-arrangements of these (now distinct) labeled objects, such that we can set

$$(2n + 1)! = \gamma n!(n + 1)!,$$

since for each of the γ required arrangements, the first $(n + 1)$ alike, being labeled, can be interchanged in $(n + 1)!$ ways, and the n alike of the second kind, being labeled, can be interchanged in $n!$ ways. Thus

$$\gamma = [(2n + 1)!]/[n!(n + 1)!].$$ ■

We now present our claim.

Theorem 4.2.3

The median of the set of $2n + 1$ *numbers, representing* $\{\chi_i\}_{i=1}^{n+1}$ *and* $\{\mu_j(\xi_T)\}_{j=1}^{n}$ *as described above, where* $0 \le \chi_1 < \chi_2 < \cdots < \chi_{n+1} \le 1$ *for a finite* n, *arranged in order of magnitude (i.e., in an array left to right), is the* FEV.

Proof

In order to find the FEV we have to compare χ_i's, $i > 1$, with the proper $\mu_i(\xi_T)$, take the minimum of these two and then the maximum over all minimum values. Clearly, $\min(\chi_1, 1) = \chi_1$. It should be noted that χ_1 cannot appear to the right of the middle of the array since it has to be followed by at least all the members of the sequence $\{\chi_k\}_{k=2}^{n+1}$. It is important to note that the values of the two sequences are independent but not their place in the array. This is due to the fact that if $\chi_j \ge \chi_i$, then $\mu_j(\xi_T) \le \mu_i(\xi_T)$. Because of that and because the array is arranged in an increasing order, the first part of the array, including n numbers ($n + 1$ left to right, and excluding χ_1), is compared with the second part of the array, including the last n numbers (right to left). Since we take the minimum of every comparison, the results must lie in the first $(n + 1)$ numbers of the array. Out of these $(n + 1)$ numbers we are interested in the maximum, which is obviously the number in the $(n + 1)$th place since the array is increasing in order. This number is the *median* of the array. Q.E.D.

■ ***Example 4.2.2***

Let X be our tested population such that

x people are of age α (compatible value χ_1),

y people are of age β (compatible value χ_2),

and

z people are of age γ (compatible value χ_3),

where $x + y + z = X$, $1 \ge \chi_3 > \chi_2 > \chi_1 \ge 0$. ■

Since $n = 2$ there are 10 different arrangements as given below; the FEVs can easily be verified graphically and thus will be omitted here.

1. $\chi_1, \chi_2, \chi_3, z/X, (z+y)/X$; FEV $= \chi_3$.
2. $z/X, (z+y)/X, \chi_1, \chi_2, \chi_3$; FEV $= \chi_1$.
3. $\chi_1, z/X, \chi_2, \chi_3, (z+y)/X$;
4. $\chi_1, z/X, \chi_2, (z+y)/X, \chi_3$;
5. $z/X, \chi_1, \chi_2, (z+y)/X, \chi_3$;
6. $z/X, \chi_1, \chi_2, \chi_3, (z+y)/X$; } FEV $= \chi_2$.
7. $\chi_1, \chi_2, z/X, (z+y)/X, \chi_3$;
8. $\chi_1, \chi_2, z/X, \chi_3, (z+y)/X$; } FEV $= z/X$.
9. $\chi_1, z/X, (z+y)/X, \chi_2, \chi_3$;
10. $z/X, \chi_1, (z+y)/X, \chi_2, \chi_3$; } FEV $= (z+y)/X$.

■ ***Example 4.2.3***

Using the base variable "Hourly wages," let us assume a given population and a given subjective compatibility curve such that:

1 person is making \$3.00 → $\chi_1 = 0.40$;
3 persons are making \$4.00 → $\chi_2 = 0.50$;
4 persons are making \$4.20 → $\chi_3 = 0.55$;
2 persons are making \$4.50 → $\chi_4 = 0.60$;
2 persons are making \$10.00 → $\chi_5 = 1.00$.

Arranging the sequences

$$\{\chi_i\}_{i=1}^{5} \quad \text{and} \quad \{\mu_j(\xi_T)\}_{j=1}^{4},$$

where $\mu(\xi_T) = \mu\{w|\chi(w) \geq T\} = |\xi_T|/12$, in an array, we get the values

$$\left\{\tfrac{1}{6}, \tfrac{1}{3}, \tfrac{2}{5}, \tfrac{1}{2}, \tfrac{11}{20}, \tfrac{3}{5}, \tfrac{2}{3}, \tfrac{11}{12}, 1\right\}.$$

Both the FEV and the median are 11/20. The probabilistic expected value (mean) is 0.61 (using the same compatibility curve on the [0, 1] interval). ■

In this example, the FEV gives a better indication of the average hourly wage than the mean, since it is not affected by the extreme value of \$10.00.

It should be noted that a different population distribution might give a FEV that is different from the median.

■ ***Example 4.2.4***

The hourly wages of five people are \$2.20, \$2.50, \$2.70, \$3.50, and \$10.00; using the same compatibility data of Example 4.2.3, these wages give the following sequence of χ_i's, $1 \leq i \leq 5$:

$$0.25, \quad 0.3, \quad 0.35, \quad 0.45, \quad 1.$$

Namely, the median is 0.35 (\$2.70), the FEV is 0.4 (\$3.00), and the mean is 0.47 (\$3.80). ■

Since the median of a data set is a quantile that, like the expected value, acts as a "*center*" for a given distribution, we claim that the *first* $\mu_j(\xi_T)$ in the array (from left to right) which is greater than or equal to $\frac{1}{2}$ indicates the respective χ_j as the median of the data set. If no such $\mu_j(\xi_T)$ exists, then χ_1 is the median. We shall denote the above specified $\mu_j(\xi_T)$ and χ_j as μ_{median} and χ_{median}, respectively. Clearly, if $\mu_{\text{median}} \geq$ FEV, then $\chi_{\text{median}} \leq$ FEV; and if $\mu_{\text{median}} <$ FEV, then $\chi_{\text{median}} >$ FEV; since $\mu_{\text{median}} \geq 1/2$, the mean as obtained is closer to the FEV than to the median. That is,

$$|\text{Mean-FEV}| \leq |\text{Mean-Median}|,$$

or, using the empirical result for unimodal frequency curves that are moderately skewed (asymmetrical), we have

$$|\text{Mean-FEV}| \leq \tfrac{1}{3}|\text{Mean-Mode}|.$$

Similar results can be derived for grouped data. It should be pointed out that *different* compatibility curves with the same fuzzy measure and the same frequency distribution might yield a different FEV. This is not so when the standard median is used, since the FEV is also a function of the subjective evaluation.

Now let the domain of a function χ be the union of a finite number of subspaces $K = \{s_1, s_2, \ldots, s_n\}$ such that $\chi\colon K \to [0,1]$. If we consider a fuzzy measure space $(K, 2^K, \mu)$, we can write the FEV of χ as

$$\mathrm{FEV}(\chi) = \max_{K' \in 2^K} \left\{ \min\left[\min_{s \in K'} [\chi(s), \mu(K')] \right] \right\}.$$

We assume that $\chi(s_i) \leq \chi(s_{i+1})$ for $1 \leq i \leq n-1$. If this is not the case, then rearrangement of $\chi(s_i)$ is necessary. Then the following holds.

Theorem 4.2.4

The FEV *of* χ *in* $(K, 2^K, \mu)$ *can be written as*

$$\max_i \{ \min[\chi(s_i), \mu(K_i)] \},$$

where $K_i = \{s_i, s_{i+1}, \ldots, s_n\}$, $1 \leq i \leq n$.

Proof

Let $\chi(s_i) = \min_{s \in K'} \chi(s)$. Then

$$\max_{K' \in 2^K} \left\{ \min\left[\min_{s \in K'} [\chi(s), \mu(K')] \right] \right\} \leq \max_i \{ \min[\chi(s_i), \mu(K_i)] \},$$

and since $\{K_i | 1 \leq i \leq n < 2^K\}$, the reverse inequality holds, too. The equality follows. Q.E.D.

Even though the power set 2^K has 2^n members, Theorem 4.2.4 calls for a monotone sequence of subsets of K such that $K_1 > K_2 > \cdots > K_n$, so that we can find the FEV by calculating $\min[\chi(s_i), \mu(K_i)]$ at n points at most.

Theorem 4.2.5

The FEV *of* χ *in* $(K, 2^K, \mu)$ *can be written as*

$$\min[\chi(s_j), \mu(K_j)]$$

iff

$$\chi(s_{j-1}) \leq \mu(K_j) \leq \chi(s_j)$$

or

$$\mu(K_j) > \chi(s_j) \geq \mu(K_{j+1}).$$

Theorem 4.2.5 proves that there is no need to evaluate $\min[\chi(s_j), \mu(K_j)]$ for all i but only at point j fulfilling the requirements of the theorem. Moreover, since χ is a known function, there is a need to evaluate $\mu(K_i)$ for three different points only.

The relation between the FEV, when $\mu(\cdot)$ is taken to be additive (probability measure) and the probabilistic expected value can be established by means of the following theorem.

Theorem 4.2.6

Let (Ω, B, p) *be a probability space and let* $\chi: \Omega \to [0,1]$ *be a B-measurable function. Then*

$$|\Delta| = \left| \int_{\Omega} \chi(x)\, dp - \sup_{T \in [0,1]} \{\min[T, p(\xi_T)]\} \right| \leq \tfrac{1}{4},$$

where $\xi_T = \{x | \chi(x) \geq T\}$.

Some other interesting properties of the FEV are now discussed.

Let $\chi_1 \in [0,1]$ and $\chi_2 \in [0,1]$; clearly if χ_1 and χ_2 are B-measurable, then $\max(\chi_1, \chi_2)$, $\min(\chi_1, \chi_2)$ and $(1 - \chi_1)$ are also B-measurable. Thus it is trivial to show that the following hold, if computed under the same measure μ:

1. Let $K \in [0,1]$, then FEV $K = K$.
2. If $\chi_1 \leq \chi_2$, then FEV $\chi_1 \leq$ FEV χ_2.
3. FEV$\{\min(\chi_1, \chi_2)\} \leq \min\{\text{FEV } \chi_1, \text{FEV } \chi_2\}$.
4. FEV$\{\max(\chi_1, \chi_2)\} \geq \max\{\text{FEV } \chi_1, \text{FEV } \chi_2\}$.
5. If $A \subset B$, then FEV χ over set A is smaller than or equal to FEV χ over set B.
6. FEV χ over set C, where $C = A \cup B$, is larger than or equal to the maximum of the FEV χ over set A and FEV χ over set B.
7. FEV χ over set D, where $D = A \cap B$, is smaller than or equal to the minimum of the FEV χ over set A and the FEV χ over set B.
8. Let $a \in A$ and $x \in X$ such that

   $$\chi: X \times A \to [0,1],$$

   and $\chi(x, a)$ be a B-measurable function of x for an arbitrary a. Then

   (i) FEV$\{\text{Sup}_{a \in A} \chi(x, a)\} \geq \text{Sup}_{a \in A}\{\text{FEV}\, \chi(x, a)\}$
   (ii) FEV$\{\text{Inf}_{a \in A} \chi(a, x)\} \leq \text{Inf}_{a \in A}\{\text{FEV}\, \chi(x, a)\}$.

9. Let $K \in [0,1]$. Then FEV$\{\min(K, \chi)\} = \min\{K, \text{FEV } \chi\}$.
10. Let $K \in [0,1]$. Then FEV$\{\max(K, \chi)\} = \max\{K, \text{FEV } \chi\}$.

4.3 PROBABILITY MEASURES OF FUZZY EVENTS

In Chapter 2, discussion centered on the algebra of fuzzy sets and on some of the properties of grades of membership in such fuzzy sets. In this chapter, we examine fuzzy probability and probabilistic processes and illustrate several of the points made with examples. In ordinary probability theory, probabilities are defined in a given sample space. In the discrete case, the collection of points in the sample space where an event A occurs describes the event. Thus, an event is the same as an aggregate of sample points; that is, an event A consists of, or contains, certain points representing an experiment where A occurs. Then, by definition, for the discrete case it follows that the probability of any event A is the sum of the probabilities of all sample points in the sample space.

We now turn to the question of grades of membership again. In the case wherein we deal with ordinary probability theory, grades of membership in a set can take on only the values unity or zero, corresponding to certain (unity) membership and no (zero) membership. Thus, for example, the grade of membership of an integer in the set of integers is unity, while for noninteger numbers, the grade of membership is zero.

There are, however, concepts for which the notion of either belonging, or not belonging, to a given set makes little sense. We give a few examples. Heights are distributed probabilistically. Since, in principle, all heights are possible, we can assign, on the basis of exhaustive experiments, a probability density of heights $P(h)\,dh$, which gives the probability of having height in the interval $(h, h + dh)$. We want the distribution to be normalized so that $\int_0^\infty P(h)\,dh = 1$. But if we now ask the question "What is the probability of the event 'tall' among people whose heights are distributed as above?", then the set of tall people must be defined. We choose to make this definition by incorporating a grade of membership $\chi_T(h) \in [0,1]$ in the set of tall people, with

$$\lim_{h \to 0} \chi_T(h) \to 0 \qquad \text{and} \qquad \lim_{h \to \infty} \chi_T(h) \to 1.$$

There are other questions that come to mind: We can ask "What constitutes the set of crowded streets?" We ask "What constitutes a 'suitable' response of a circuit to an input signal?" To answer each of these questions, we need a suitable mathematics into which is built a notion of *grade of membership* in a set.

Zadeh [1968b] has provided a framework such that inexact, or fuzzy, concepts can be discussed rigorously within the confines of an extension of probability theory. In ordinary probability theory, given a random variable X in one dimension, we define

$$P\{X = x\} = f(x),$$

where $f(x)$ is the probability density of the random variable X. Then we may define the (cumulative) distribution function $P(x)$, which gives $P\{X \le x\}$, as

$$P(x) = \int_{-\infty}^{x} f(x')\,dx',$$

for $x' \in [-\infty, \infty]$. Then with $dP(x) \equiv f(x)\,dx$, the above equation can be written as

$$P(x) = \int_{-\infty}^{x} dp(x').$$

It is held that

$$\lim_{x \to \infty} P(x) \equiv \int_{-\infty}^{\infty} dp(x) = 1;$$

that is, the distribution function is normalizable. For a fuzzy random variable X_A, by analogy with the above equation, we write

$$P\{X_A = x\} \equiv \chi_A(x) f(x),$$

thereby associating with each x a grade of membership in the set A. Then we define a quantity $P(A; x)$ as

$$P(A; x) \equiv P\{\chi_A \leq x\} = \int_{-\infty}^{x} \chi_A(x')\, dp(x').$$

Corresponding to the normalization condition of ordinary probability theory, we write that

$$\lim_{x \to \infty} P(A, x) \equiv P(A),$$

where

$$P(A) = \int_{-\infty}^{\infty} \chi_A(x)\, dp(x).$$

This equation is Zadeh's definition of a fuzzy event A in a one-dimensional space. We now seek to formalize the above argument to multi-dimensional spaces by referring directly to Zadeh's [1968b] paper.

A probability space is assumed to be a triplet $(\mathbb{R}^n, B, P)$, where B is the σ-field of Borel sets in $\mathbb{R}^n$ and P is a probability measure over $\mathbb{R}^n$. A point in $\mathbb{R}^n$ is denoted by $\vec{x}$.

Let a set $A \in B$. Then on defining a characteristic function of, or grade of membership in, the set A by $\chi_A(\vec{x})$: $\mathbb{R}^n \to [0, 1]$, we define the probability of A as

$$P(A) = \int_{\mathbb{R}^n} \chi_A(\vec{x})\, dp(\vec{x}),$$

where

$$\int_{\mathbb{R}^n} dp(\vec{x}) = 1,$$

or

$$P(A) = \langle \chi_A \rangle = E\{\chi_A\},$$

where we use the notation $\langle\ \rangle$ to denote the expected value of a function, and where $E\{\chi_A\}$ is the expected value of χ_A.

■ ***Example 4.3.1***

We examine a two-dimensional example. Let two probability distributions be independent and given as

$$dP(x_1) = e^{-x_1}\, dx_1 \qquad \text{and} \qquad dP(x_2) = x_2 e^{-x_2}\, dx_2.$$

Since both $dP(x_1)$ and $dP(x_2)$ must be positive and normalizable, both x_1 and $x_2 \in [0, \infty]$. We ask: Are x_1 and x_2 similar? To answer this question, we must define a set A which includes the notion of similarity. We write

$$A = \{x_1, x_2 \mid |x_1 - x_2| \approx 1\}.$$

A suitable grade of membership in such a set is given by

$$\chi_A(x_1, x_2) = e^{-|x_1 - x_2|}.$$

If $x_1 = x_2$, $\chi_A(x_1, x_2) = 1$, in which case x_1 and x_2 are identical. If $x_1 \approx x_2$, then they are similar. We now ask for the occurrence of such a set. We have

$$P(A) = \int_0^\infty \int_0^\infty e^{-|x_1 - x_2|} e^{-(x_1 + x_2)} x_2 \, dx_1 \, dx_2.$$

The integral can be evaluated, with the result that

$$P(A) = \tfrac{3}{8}.$$

Thus, the set A is realized three eighths of the "time." ■

As an application of such an idea to a more practical problem, consider two frequency standards (or calibrating devices). One is specified by a probability density function

$$dP(w_1) = \delta(w_1 - w_0) \, dw_1;$$

that is, its frequency output is "exact." The other is less precise. It is, say, crystal-controlled, and as the crystal heats or cools, its output is characterized by a distribution

$$dP(w) = \frac{1}{\sqrt{2\pi}\,\sigma} e^{-(w - w_0)^2/2\sigma^2} \, dw,$$

where we shall require that $w_0 > 0$. We ask: How similar are the two frequency standards? To put it another way, we can ask whether the crystal-controlled device has a suitable output if we want all w's in some range Δw. To answer the question, we can define a set $\Omega = \{w, w_1 \mid |w_1 - w| < \Delta w\}$, and choose a grade of membership

$$\chi_\Omega(w, w_1) = e^{-|w - w_1|/\Delta w}.$$

Then, for $P(\Omega)$, we have, by virtue of the density function

$$P(\Omega) = \frac{1}{\sqrt{2\pi}\,\sigma} \int_{-\infty}^{\infty} e^{-|w - w_0|/\Delta w} e^{-(w - w_0)^2/2\sigma^2} \, dw.$$

Recalling that $w_0 > 0$, we have:

$$P(\Omega) = \frac{1}{\sqrt{2\pi}\,\sigma} \left\{ \int_{-\infty}^{w_0} e^{-(w_0 - w)/\Delta w} e^{-(w - w_0)^2/2\sigma^2} \, dw \right.$$

$$\left. + \int_{w_0}^{\infty} e^{-(w - w_0)/\Delta w} e^{-(w - w_0)^2/2\sigma^2} \, dw \right\}.$$

The details of evaluating the above integrals will be omitted. The result is that

$$P(\Omega) = e^{\sigma^2/(2(\Delta w)^2)} \frac{2}{\sqrt{\pi}} \int_{\sigma/(\sqrt{2}\,\Delta w)}^{\infty} e^{-x^2} \, dx,$$

or, on defining the complementary error function by

$$\operatorname{Erfc}\left(\frac{\sigma}{\sqrt{2}\,\Delta w}\right) = \frac{2}{\sqrt{\pi}} \int_{\sigma/(\sqrt{2}\,\Delta w)}^{\infty} e^{-x^2\,dx},$$

we have

$$P(\Omega) = e^{\sigma^2/(2(\Delta w)^2)} \operatorname{Erfc}\left(\frac{\sigma}{\sqrt{2}\,\Delta w}\right).$$

We examine the expression for $P(\Omega)$. If $\Delta w \to \infty$, then all w and w_1 are similar. This is indeed the case, since

$$\lim_{\Delta w \to \infty} e^{\sigma^2/(2(\Delta w)^2)} \to 1,$$

and

$$\lim_{\Delta w \to \infty} \operatorname{Erfc}\left(\frac{\sigma}{\sqrt{2}\,\Delta w}\right) \to 1,$$

which implies that $P(\Omega) = 1$. As $\Delta w \to 0$,

$$\lim_{\Delta w \to 0} P(\Omega) \to \lim_{\Delta w \to 0} e^{\sigma^2/(2(\Delta w)^2)} \int_{\sigma/(\sqrt{2}\,\Delta w)}^{\infty} e^{-u^2}\,du.$$

We need the asymptotic expansion of $\operatorname{Erfc}(x) = (2/\sqrt{\pi})\int_x^{\infty} e^{-u^2}\,du$. It is

$$\lim_{x \to \infty} \frac{2}{\sqrt{\pi}} \int_x^{\infty} e^{-u^2}\,du = \sqrt{2}\,\frac{e^{-(1/2)x^2}}{x}(1 + o(x^{-3})).$$

Thus,

$$\lim_{\Delta x \to 0} P(\Omega) = \frac{\Delta w}{\sigma} \to 0 \qquad \text{for } \sigma > 0.$$

If, however, $\sigma = \sqrt{\langle w^2 \rangle - \langle w_0 \rangle^2}$ is of the order of Δw,

$$\lim_{\Delta w \to 0} P(\Omega) \to \frac{w}{o(\Delta w)} \approx 1.$$

If frequency calibration need not be done with extreme precision, it may be that, for $P(\Omega) \geq 0.9$, say, the crystal-controlled standard will be adequate, and, in all probability, less expensive. To determine a $P(\Omega) \geq 0.9$ becomes a problem of placing various ratios of $(\sigma/\Delta w)$ into the expression of $P(\Omega)$.

■ ***Example 4.3.2***

We now generalize Zadeh's definition of $P(A)$ in a seemingly trivial way. (However, there are interesting implications to the generalization that will form the basis for much of the work in the remainder of this chapter.) Consider, rather than chronological age, "biological" age defined by

$$x_B = \alpha t,$$

when the quantity α, an aging rate, is distributed according to a distribution

$dP(\alpha)$ such that $\int_{\mathbb{R}} dP(\alpha) = 1$. Let membership in the set of aged people at time t be denoted by

$$A(t) = \{\alpha | \alpha > \alpha_m; t\}.$$

Set $\chi_{A(t)}(\alpha) \equiv \chi_A(\alpha, t) = 1 - e^{-\alpha t/\alpha_m t_m}$ and let, for $\alpha \in [0, \infty]$,

$$dP(\alpha) = \frac{1}{\alpha_0^2} \alpha e^{-\alpha/\alpha_0}\, d\alpha.$$

Then

$$P(A(t)) \equiv P_A(t) = \frac{1}{\alpha_0^2} \int_0^{\infty} (1 - e^{-\alpha t/\alpha_m t_m}) \alpha e^{-\alpha/\alpha_0}\, d\alpha.$$

We find that

$$P_A(t) = 1 - (1 + \lambda t)^{-2},$$

where $\lambda = \alpha_0/\alpha_m t_m$. For positive, finite α_0, α_m, and t_m,

$$\lim_{t \to 0} P_A(t) \to 0 \qquad \text{and} \qquad \lim_{t \to \infty} P_A(t) \to 1.$$

For any finite $t > 0$,

$$\lim_{\lambda \to 0} P_A(t) \to 0 \qquad \text{and} \qquad \lim_{\lambda \to \infty} P_A(t) \to 1.$$

■

This time-dependent extension of Zadeh's work leads, in succeeding sections, to differential equations satisfied by functions such as $\chi_A(\alpha, t)$, or to differential and integro-differential equations satisfied by $P_A(t)$. Thus, by means of a time-dependent generalization of Zadeh's work, there can be constructed a theory of the probability of time-dependent fuzzy events.

■ ***Example 4.3.3***

We can generalize Zadeh's results in another way by demanding that functions themselves contain, or be dependent upon, grades of membership in a set. This is now illustrated by means of an example. Let a group of capacitors, each of which initially has a value C_0, be constructed of various dielectrics whose permittivities, ε, have different (distributed) changes in value when subjected to radiation. Then we can define a "fuzzy damaged capacitor" by the statement that

$$C_D(\varepsilon) = C_0 \left(1 - \chi_D(\varepsilon) \frac{C_1}{C_0}\right)$$

for $C_1 \le C_0$. Here $\chi_D(\varepsilon)$ is the grade of membership in the set of damaged dielectrics. Then, if the density function of ε's after irradiation is $dP(\varepsilon)$, we have

$$P(D) = \frac{C_1}{C_0} \int_0^{\varepsilon_0} \chi_D(\varepsilon)\, dP(\varepsilon),$$

where ε_0 is the maximum of all the values of the dielectric constant prior to irradiation. We can also form

$$\int_0^{\varepsilon_0} C_D(\varepsilon)\, dP(\varepsilon) = C_0(1 - P(D)).$$

■

The above example, then shows that there may be functions of physical and engineering interest that can depend on grades of membership. We shall next make use of the time-dependent generalization of Example 4.3.2, combined with the idea of a function depending on a grade of membership, to discuss functions $f(\chi_A(x, t), t)$ when there exists a probability density $dP(x)$ to calculate quantities of the form

$$\int_R f(\chi_A(x,t),t)\,dP(x).$$

Time-Dependent Processes

In Example 4.3.2, we found that the quantity $P_A(t)$, the probability of being aged at time t, was given by

$$P_A(t) = 1 - (1 + \lambda t)^{-2} \qquad \text{with } \lambda = \alpha_0/\alpha_m t_m.$$

The time-dependent grade of membership involved in the example is

$$\chi_A(\alpha, t) = 1 - e^{-\alpha t/\alpha_m t_m},$$

so that $\chi_A(\alpha, t)$ satisfies the differential equation

$$\frac{\partial \chi_A}{\partial t} + \frac{\alpha}{\alpha_m t_m}\chi_A = \frac{\alpha}{\alpha_m t_m}.$$

Corresponding to this relation there is a differential equation satisfied by $P_A(t)$. It is

$$\frac{dP_A(t)}{dt} + \left(\frac{2\lambda}{1+\lambda t}\right)P_A(t) = \frac{2\lambda}{1+\lambda t}.$$

In this section, we shall derive equations satisfied by $P_A(t)$, when either $\chi_A(\cdot)$, or $dP(\cdot)$ vary with time. These will then be applied to various problems of some practical interest.

Let

$$P_A(t) = \int_R \chi_A(x)\,p(x,t)\,dx,$$

and form

$$\frac{dP_A(t)}{dt} = \int_R \chi_A(x)\frac{\partial p}{\partial t}\,dx.$$

Assume that a normalizable $p(x, t)$ satisfies the differential equation

$$\frac{\partial p}{\partial t} + \lambda(x)p = q(x,t),$$

and thus

$$\frac{dP_A(t)}{dt} = \int_R \chi_A(x)\{q(x,t) - \lambda(x)p(x,t)\}\,dx$$

or

$$\frac{dP_A(t)}{dt} = \langle q(t)\rangle_z - \int_R \chi_A(x)\lambda(x)p(x,t)\,dx,$$

by defining the Zadeh average of a function $q(x,t)$ as

$$\langle q(t)\rangle_z = \int_R \chi_A(x)q(x,t)\,dx.$$

Equivalently, $\langle q(t)\rangle_z \equiv \langle \chi_A q(t)\rangle$, and $P_A(t)$ satisfies the differential equation

$$\frac{dP_A(t)}{dt} + g_A(t)P_A(t) = \langle q(t)\rangle_z,$$

where

$$g_A(t)P_A(t) = \int_R \chi(x)\lambda(x)p(x,t)\,dx.$$

Consider now the case for which, again,

$$P_A(t) = \int_{R(x)} \chi_A(x)p(x,t)\,dx,$$

and let $p(x,t)$ satisfy an integro-differential equation of the form

$$\frac{\partial p}{\partial t} + \lambda p = \int_{R(x')} q(x',x)p(x',t)\,dx',$$

where λ is a constant. It is assumed that the function $p(x',t)$ is such that

$$\int_{R(x')} p(x',t)\,dx' = 1, \qquad \forall t.$$

We have

$$\frac{dP_A}{dt} + \lambda P_A = \int_{R(x)} \chi_A(x)\int_{R(x')} q(x,x')p(x',t)\,dx'\,dx.$$

Assuming validity of the interchange of the order of integration, we write:

$$\int_{R(x)} \chi_A(x)\int_{R(x')} q(x,x')p(x',t)\,dx'\,dx$$
$$= \int_{R_1(x')} p(x',t)\int_{R_1(x)} \chi_A(x)q(x',x)\,dx\,dx',$$

where the interchange in the x–x' plane has carried the ranges $R(x)$ and $R(x')$ into $R_1(x')$ and $R_1(x)$, as shown.

Let $f_A(x') = \int_{R_1(x)} \chi_A(x)q(x',x)\,dx'$, whence

$$\frac{dP_A}{dt} + \lambda P_A = \langle f_A(t)\rangle,$$

where

$$\langle f_A(t) \rangle \equiv \int_{R_1(x')} f_A(x') p(x', t)\, dx'$$

is an ordinary probabilistic average. We see that if $p(x,t)$ satisfies an integro-differential equation such as that above, $P_A(t)$ satisfies an ordinary inhomogeneous differential equation.

We examine next the case for which the density function $p(x,t)$ satisfies an integro-differential equation of the convolution type; namely,

$$\frac{\partial p}{\partial t} + \lambda p = \int_0^t q(t-\tau) p(x,\tau)\, d\tau,$$

where again λ is a constant and $p(x,t)$ is assumed normalizable. Then

$$\frac{dP_A}{dt} = \int_{R(x)} \left(-\lambda p + \int_0^t q(t-\tau) p(x,\tau)\, d\tau \right) \chi_A(x)\, dP(x),$$

or

$$\frac{dP_A}{dt} + \lambda P_A = \int_0^t q(t-\tau) P_A(\tau)\, d\tau,$$

on assuming that interchange of the order of integration is valid.

■ ***Example 4.3.4***

Consider a set of capacitors that have been irradiated. Each of them originally had a capacitance C_0, and each was constructed of a different dielectric. Dielectrics have different sensitivities to radiation and, after being irradiated, the dielectric constants will be, in general, at values $\varepsilon \leq \varepsilon_m$, where ε_m is the largest of the dielectric constants prior to irradiation. There will be a distribution of ε's that will be found acceptable, in the sense that a capacitance $C_D(\varepsilon)$ will be in an acceptable range of values for a given circuit application. Let $C_D(\varepsilon)$, the capacitance after radiation, be denoted by

$$C_D(\varepsilon) = \frac{C_0}{1 + \eta \chi_D(\varepsilon)},$$

where $\eta \equiv C_1/C_0$ with C_1 an arbitrarily chosen value of capacitance and where $\chi_D(\varepsilon) \in [0,1]$ as ε ranges from some ε_1 to ε_m. We assume that

$$\chi_D(\varepsilon) = \frac{\varepsilon - \varepsilon_1}{\varepsilon_m - \varepsilon_1}.$$

Let the probability density function of dielectric constants after radiation be

$$dP(\varepsilon) = \frac{d\varepsilon}{\varepsilon_m - \varepsilon_1}, \qquad \text{for } \varepsilon \in [\varepsilon_1, \varepsilon_m],$$

for which, obviously,

$$\int_{\varepsilon_1}^{\varepsilon_m} dP(\varepsilon) = 1.$$

Let the capacitors $C_D(\varepsilon)$ be incorporated into R-C circuits, and charged to a value $q(0)$; then let a switch, S_1, be closed. A discharge of a given capacitor will then occur such that $q_D(t,\varepsilon)$, the charge at time t on the capacitor $C_D(\varepsilon)$, will be

$$q_D(t) = q(0)\,e^{-t/\tau_D(\varepsilon)},$$

where $\tau_D(\varepsilon) = \tau_0/(1+\eta\chi_D(\varepsilon))$ and $\tau_0 = RC_0$, the time constant of the pre-irradiated RC_0 circuit.

We want to average $q_D(t)$ over the probability distribution of the ε's. We have

$$\langle q_D(t)\rangle = \frac{q(0)}{\varepsilon_m - \varepsilon_1}\int_{\varepsilon_1}^{\varepsilon_m} e^{-(t/\tau_0)(1+\eta(\varepsilon-\varepsilon_1)/(\varepsilon_m-\varepsilon_1))}\,d\varepsilon,$$

or

$$\langle q_D(t)\rangle = \frac{q(0)}{\varepsilon_m - \varepsilon_1}e^{-(t/\tau_0)(1-[\eta\varepsilon_1/(\varepsilon_m-\varepsilon_1)])}\int_{\varepsilon_1}^{\varepsilon_m} e^{-(t/\tau_0)\eta\varepsilon/(\varepsilon_m-\varepsilon_1)}\,d\varepsilon.$$

The integral is easily evaluated, with the result that

$$\langle q_D(t)\rangle = q(t)\frac{\tau_0}{\eta t}(1 - e^{-\eta t/\tau_0}),$$

where

$$q(t) = q(0)\,e^{-t/\tau_0}$$

is the charge at time t on the capacitance C_0. Thus, the influence of changes in capacitance value, when averaged over the given grade of membership and the given probability distribution, is contained in the factor

$$F_\eta(t) = \frac{\tau_0}{\eta t}(1 - e^{-\eta t/\tau_0}).$$

As $t \to 0$, $F_\eta(t) \to 1$, so that, initially, $\langle q_D(0)\rangle = q(0)$. As $t \to \infty$, $F_\eta(t) \to \tau_0/\eta t \to 0$. Now recall that $\eta = C_1/C_0$, is variable. Although C_0 is fixed, we are free to choose C_1 as any value $C_1 \geq 0$. If $C_1 = 0$, $F_0(t) = 1$ and the circuit is again unperturbed. It is just this concept of η being variable that allows flexibility in decision-making with regard to use of radiation-changed capacitors. Suppose, for example, that the value of $\langle q_D(t)\rangle$ at $t = \tau_0$ is some q_0. If we want faster decay times, then we simply choose that portion of the group of capacitors which will guarantee this, while if we want slower decay times, we choose capacitors $C'_D(\varepsilon) < C_0/(1+\eta)$. ■

There is another way of looking at the example we have discussed. Assume, for example, that capacitances after irradiation are characterized by a value

$$C_D(\varepsilon) = C_0(1 - \chi_D(\varepsilon)).$$

Let "acceptable damaged capacitors" be denoted by

$$C_D(\varepsilon) = C_0(1 - \eta\chi_D(\varepsilon)),$$

where, again, $\eta = C_1/C_0 < 1$. Then

$$C_0 - C_1 \leq C_D(\varepsilon) \leq C_0$$

as $\chi_D(\varepsilon)$ varies over its range.

The discharge of a capacitor, charged initially to a value $q(0)$, will now take place according to

$$q_D(t) = q(0)e^{-t/\tau_D(\varepsilon)},$$

where

$$\tau_D(\varepsilon) = \tau_0(1 - \eta\chi_D(\varepsilon)).$$

If $\eta\chi_D(\varepsilon) \ll 1$, we can find the following average:

$$\langle q_D(t)\rangle = \frac{q(0)}{\varepsilon_m - \varepsilon_1}\int_{\varepsilon_1}^{\varepsilon_m} \exp\left(-t/\tau_0\left(1 - \eta\frac{\varepsilon - \varepsilon_1}{\varepsilon_m - \varepsilon_1}\right)\right) d\varepsilon.$$

This can be written, on letting $\Delta\varepsilon_R = \varepsilon_m - \varepsilon_1$, as

$$\langle q_D(t)\rangle = \frac{q(0)}{\Delta\varepsilon_R}\int_{\varepsilon_1}^{\varepsilon_m} e^{-(t/\tau_0)[\Delta\varepsilon_R/(\Delta\varepsilon_R + \eta\varepsilon_1 - \eta\varepsilon)]}\, d\varepsilon.$$

Let $\Delta\varepsilon_R + \eta\varepsilon_1 - \eta\varepsilon = u$. Then

$$\langle q_D(t)\rangle = -\frac{q(0)}{\Delta\varepsilon_R\eta}\int_{\Delta\varepsilon_R}^{\Delta\varepsilon_R(1-\eta)} e^{-(t/\tau_0)(\Delta\varepsilon_R/u)}\, du.$$

Let

$$\frac{t}{\tau_0}\frac{\Delta\varepsilon_R}{u} = v,$$

so that

$$dv = -\frac{t\,\Delta\varepsilon_R}{\tau_0 u^2}\, du.$$

Then

$$\langle q_D(t)\rangle = \frac{q(0)t}{\tau_0\eta}\int_{t/\tau_0}^{(t/\tau_0)(1/[1-\eta])} e^{-v}\frac{dv}{v^2}.$$

The nth-order exponential integral is defined as

$$E_n(x) \equiv \int_x^{\infty} e^{-v}\frac{dv}{v^n}.$$

We therefore have

$$\langle q_D(t)\rangle = \frac{q(0)t}{\tau_0\eta}\left\{E_2\left(\frac{t}{\tau_0}\right) - E_2\left(\frac{t}{[\tau_0(1-\eta)]}\right)\right\}.$$

We look for the connection between this solution and the previous one. The asymptotic form of the function $E_2(x)$ is e^{-x}/x^2. Thus,

$$\lim_{t\to\infty}\langle q_D(t)\rangle = \frac{q(0)\tau_0}{\eta t}\{e^{-t/\tau_0} - e^{-t/\tau_0(1-\eta)}\}.$$

If η is small, we have

$$\lim_{t\to\infty} \langle q_D(t)\rangle = \frac{q(t)\tau_0}{\eta t}(1 - e^{-(t\eta/\tau_0)}),$$

where, again, $q(t)$ is the unperturbed charge at time t.

4.4 GENERAL FUZZY SYSTEMS

In our extension of Zadeh's work, we have formulated problems in which a scalar obeys an equation of the form

$$\frac{df}{dt} + aq(\alpha)f = bq(\alpha,t); \qquad f = f(\alpha,t),$$

and where the quantities α are distributed with density $dP(\alpha)$.

These equations can be taken as a single component of a matrix equation of the more general form

$$\frac{d\vec{f}}{dt} + \boldsymbol{A}(\alpha)\vec{f} = \boldsymbol{B}(\alpha,t)\vec{q}(\alpha,t),$$

where, if $\vec{f}$ and $\vec{q}$ are vectors of n components, then $\boldsymbol{A}(\alpha)$ and $\boldsymbol{B}(\alpha)$ are $(n \times n)$ matrices. We are avoiding the case for which $\boldsymbol{A} = \boldsymbol{A}(\alpha,t)$ since there will be, in that case, a state transition matrix $\boldsymbol{\Phi}(t, t_0; \alpha)$ and solutions can be found analytically only in special cases for which $\boldsymbol{A}(\alpha,t)$ is diagonal (or independent of time).

A solution to this equation is written as

$$\vec{f}(\alpha,t) = e^{At}\vec{f}(\alpha,0) + \int_0^t e^{A(t-\tau)}\boldsymbol{B}(\alpha,\tau)\vec{q}(\tau)\,d\tau.$$

The eigenvalues of $\boldsymbol{A}(\alpha)$ are found by calculating the roots S_i of the determinant

$$|S\boldsymbol{I} - \boldsymbol{A}| = 0.$$

The resulting characteristic equation will yield n roots, some of which may be repeated. If none is repeated, then $\boldsymbol{A}$ can be diagonalized by forming

$$\Lambda(\alpha) = \boldsymbol{P}^{-1}(\alpha)\boldsymbol{A}(\alpha)\boldsymbol{P}(\alpha),$$

where $\boldsymbol{P}(\alpha)$ is the matrix whose columns are the n independent eigenvectors. If repetition of roots occurs, then the matrix can be reduced to Jordan canonical form.

Given the above, we can formally write down the Zadeh average of $\vec{f}$ as

$$\langle \vec{f}_A(t)\rangle_z = \langle e^{At}\vec{f}(\alpha,0)\rangle_z + \int_0^t \langle e^{A(t-\tau)}\boldsymbol{B}(\alpha,\tau)\rangle\vec{q}(\tau)\,d\tau,$$

where each averaged quantity is the average of a vector.

We have defined

$$\langle \vec{f}_A(t) \rangle_z \equiv \int_{R(\alpha)} \chi_A(\alpha) \vec{f}(\alpha, t)\, dP(\alpha).$$

Clearly,

$$\langle \vec{f}_A(t) \rangle_z = \langle \chi_A \vec{f} \rangle,$$

with $\langle \chi_A \vec{f} \rangle$ an ordinary probabilistic average. Since the integral of a vector is the integral of each of its components, the integrations will proceed as in earlier sections. The quantity $\langle e^{At} \vec{f}(\alpha, 0) \rangle_z$ is again a vector. Let $\vec{f}(\alpha, 0)$ be independent of α so that

$$\langle e^{At} \rangle_z = \int_{R(\alpha)} \chi_A(\alpha) e^{A(\alpha)t}\, dP(\alpha).$$

The integral of the matrix is the matrix of the integrals of each term. Again, it is clear that, with $e^{A(\alpha)t}$ expressed in terms of its eigenvalues through the prescription

$$e^{A(\alpha)t} = \mathscr{L}^{-1}\left[(s\boldsymbol{I} - \boldsymbol{A})^{-1}\right],$$

with $\mathscr{L}^{-1}$ meaning the inverse Laplace transform, the quantity $\langle e^{At} \rangle_z$ can be calculated, at least in principle. We now present several simple examples of dealing with such systems.

■ ***Example 4.4.1***

Let $\dot{\vec{X}} = \alpha \boldsymbol{A} \vec{X}(\alpha, t)$ be a system such that $\boldsymbol{A}$ is a nonfuzzy time-invariant $n \times n$ matrix, and α is a fuzzy parameter, distributed such that

$$\int_0^\infty dp(\alpha) = \int_0^1 d\alpha = 1.$$

Thus,

$$\vec{X}(\alpha, t) = e^{\alpha A t} \vec{X}(\alpha, 0)$$

and $\langle \vec{X}(\alpha, t) \rangle$ can be calculated. For example, consider $\boldsymbol{A}$ to be diagonal, in the form

$$\boldsymbol{A} = \begin{bmatrix} 1 & 0 \\ 0 & 2 \end{bmatrix}.$$

Clearly,

$$e^{\alpha A t} = \begin{bmatrix} e^{\alpha t} & 0 \\ 0 & e^{2\alpha t} \end{bmatrix}.$$

let $\vec{X}(\alpha, 0) = \begin{bmatrix} 1 \\ 2 \end{bmatrix}$. Thus,

$$\langle \vec{X}(\alpha, t) \rangle = \begin{bmatrix} \int_0^1 e^{\alpha t}\, d\alpha \\ \int_0^1 2e^{2\alpha t}\, d\alpha \end{bmatrix} = \begin{bmatrix} \frac{1}{t}(e^t - 1) \\ \frac{1}{t}(e^{2t} - 1) \end{bmatrix}.$$

It is interesting to note that if we have no fuzziness in this example, that is, $\alpha \equiv 1$, then

$$\vec{X}(t) = \begin{bmatrix} e^t \\ 2e^{2t} \end{bmatrix}. \qquad \blacksquare$$

■ ***Example 4.4.2***

Consider a deterministic linear dynamical system with two sets of input, $\vec{u}$ and $\vec{\alpha}$. The elements of input set $\vec{u}$ can be controlled where the other input vector $\vec{\alpha}$ consists of disturbance inputs that are elements of uncontrollable fuzzy disturbances. Thus, the system is represented by

$$\dot{\vec{X}}(\alpha, t) = \boldsymbol{A}\vec{X}(\alpha, t) + \boldsymbol{B}[\vec{u}(t) + \vec{\alpha}], \qquad \vec{\alpha} = \begin{bmatrix} \alpha \\ \vdots \\ \alpha \end{bmatrix},$$

where $0 \leq \alpha \leq 1$ and they are distributed such that

$$\int_0^\infty dp(\alpha) = \int_0^1 d\alpha = 1.$$

Thus,

$$\vec{X}(\alpha, t) = e^{At}\vec{X}_0 + \int_0^t e^{A(t-\tau)}\boldsymbol{B}[\vec{u}(t) + \vec{\alpha}]\, d\tau$$

and

$$\begin{aligned} \langle \vec{X}(\alpha, t) \rangle &= e^{At}\vec{X}_0 + \int_0^t e^{A(t-\tau)}\boldsymbol{B}\vec{u}(\tau)\, d\tau \\ &\quad + \int_0^1 \int_0^t e^{A(t-\tau)}\boldsymbol{B}\vec{\alpha}\, d\tau\, d\alpha \\ &= \underbrace{e^{At}\vec{X}_0 + \int_0^t e^{A(t-\tau)}\boldsymbol{B}\vec{u}(\tau)\, d\tau}_{\text{Nonfuzzy part}} + \underbrace{\tfrac{1}{2}\int_0^t e^{A(t-\tau)}\boldsymbol{B}\, d\tau}_{\text{Fuzzy part}}. \end{aligned}$$

In essence it looks like an added input to $\vec{u}(t)$ in the form of the fuzzy part. Consider, for example, a system where

$$\boldsymbol{A} = \begin{bmatrix} 0 & 1 \\ -2 & -3 \end{bmatrix} \quad \text{and} \quad \boldsymbol{B} = \begin{bmatrix} 2 \\ 0 \end{bmatrix}.$$

Let

$$u(t) = \begin{cases} 0, & t < 0 \\ e^{-t}, & t \geq 0. \end{cases}$$

Let $\vec{X} = \boldsymbol{Q}\vec{y}$, where

$$\boldsymbol{Q} = \begin{bmatrix} 1 & 1 \\ -1 & -2 \end{bmatrix}.$$

Thus

$$\dot{\vec{y}} = \begin{bmatrix} -1 & 0 \\ 0 & -2 \end{bmatrix}\vec{y} + \begin{bmatrix} 4 \\ -2 \end{bmatrix}(u + \alpha)$$

or

$$\dot{\vec{y}} = \boldsymbol{A}_1\vec{y} + \boldsymbol{B}_1(u + \alpha),$$

such that

$$\vec{y}(\alpha,t) = e^{A_1 t}\left[\vec{y}(\alpha,0) + \int_0^t e^{-A\tau}\boldsymbol{B}[u(\tau)+\alpha]\,d\tau\right],$$

where

$$e^{A_1 t} = \begin{bmatrix} e^{-t} & 0 \\ 0 & e^{-2t} \end{bmatrix}.$$

Let $\vec{y}(\alpha,0)=0$, so that

$$\begin{aligned}\vec{y}(\alpha,t) &= \begin{bmatrix} e^{-t} & 0 \\ 0 & e^{-2t} \end{bmatrix}\int_0^t \begin{bmatrix} e^{\tau} & 0 \\ 0 & e^{2\tau} \end{bmatrix}\begin{bmatrix} 4 \\ -2 \end{bmatrix}(e^{-\tau}+\alpha)\,d\tau \\ &= \begin{bmatrix} 4te^{-t} \\ -2e^{-t}+2e^{-2t} \end{bmatrix} + \begin{bmatrix} 4\alpha(1-e^{-t}) \\ \alpha(e^{-2t}-1) \end{bmatrix}.\end{aligned}$$

However,

$$\begin{aligned}\vec{X} = \boldsymbol{Q}\vec{y}(\alpha,t) &= \begin{bmatrix} 1 & 1 \\ -1 & -2 \end{bmatrix}\vec{y}(\alpha,t) \\ &= \begin{bmatrix} 4te^{-t}-2e^{-t}+2e^{-2t} \\ -4te^{-t}+4e^{-t}-4e^{-2t} \end{bmatrix} + \begin{bmatrix} \alpha(3-4e^{-t}+e^{-2t}) \\ \alpha(-2+4e^{-t}-2e^{-2t}) \end{bmatrix}.\end{aligned}$$

Hence,

$$\begin{aligned}\langle \vec{X}(\alpha,t)\rangle &= \begin{bmatrix} 4te^{-t}-2e^{-t}+2e^{-2t} \\ -4te^{-t}+4e^{-t}-4e^{-2} \end{bmatrix} + \int_0^1 \begin{bmatrix} \alpha(3-4e^{-t}+e^{-2t}) \\ \alpha(-2+4e^{-t}-2e^{-2t}) \end{bmatrix} d\alpha \\ &= \begin{bmatrix} 4te^{-t}-4e^{-t}+\left(\frac{5}{2}\right)e^{-t}+\frac{3}{2} \\ -4te^{-t}+6e^{-t}-5e^{-2t}-1 \end{bmatrix}.\end{aligned}$$

In the general time-varying case, where

$$\dot{\vec{X}}(\alpha,t) = \boldsymbol{\Psi}(\alpha,t)\vec{X}(\alpha,t),$$

a solution for $\vec{X}(\alpha,t)$ can, in principle, be found via a fuzzy-state transition matrix $\boldsymbol{\Phi}(t,t_0)$, in the form of

$$\vec{X}(\alpha,t) = \boldsymbol{\Phi}_f(t,t_0)\vec{X}(t,t_0),$$

such that

1) $\boldsymbol{\Phi}_f(\alpha,t_0,t_0) = I$, the identity matrix;
2) $(\partial/\partial t)\boldsymbol{\Phi}_f(\alpha,t,t_0) = \boldsymbol{\Psi}(\alpha,t)\boldsymbol{\Phi}_f(\alpha,t,t_0)$.

Thus

$$\langle \vec{X}(\alpha,t)\rangle = \int_{R(\alpha)} \boldsymbol{\Phi}_f(\alpha,t,t_0)\vec{X}(\alpha,t_0)\,dp(\alpha).$$ ■

4.5 POSSIBILITY MEASURES

Zadeh [1977a] proposed, in studying the representation of meaning in natural language, the concept of a possibility measure.

Let f be a mapping from Ω to $[0,1]$, then a measure μ defined by

$$\forall A \subset \Omega: \quad \mu(A) = \begin{cases} \sup_{\omega \in A} f(\omega), & \text{if } A \neq \emptyset, \\ 0, & \text{if not,} \end{cases}$$

is called a *possibility measure.*

We follow Puri and Ralescu [1982] in their presentations of possibility measures.

Lemma 4.5.1

Any possibility measure is a fuzzy measure if, and only if, the corresponding distribution function f is such that $\sup_{\omega \in \Omega} f(\omega) = 1$.

Proof

Let μ be a possibility measure; then under the assumption of the lemma we have $\mu(\Omega) = 1$. Conversely, if $\mu(\Omega) = 1$, then $\exists \omega \in \Omega$: $f(\omega) = 1$ and the assumption is satisfied. The monotonicity property is verified since

$$\forall A, \quad B \subset \Omega, \quad A \subset B: \sup_{\omega \in A} f(\omega) \leq \sup_{\omega \in B} f(\omega).$$

It is easy to verify that a possibility measure μ satisfies the F-additivity property

$$\forall A, B \subset \Omega: \quad \mu(A \cup B) = \max(\mu(A), \mu(B)).$$ Q.E.D.

A measure μ satisfying the following properties is called a g_λ-*measure* ($\lambda \in]-1, \infty[$):

$$\mu(\Omega) = 1,$$

$$\forall A, B \subset \Omega, \quad A \cap B = \emptyset: \quad \mu(A \cup B) = \mu(A) + \mu(B) + \lambda\mu(A)\mu(B).$$

As for probability measures, it can be established that the g_λ-measures satisfy

$$\forall A, B \subset \Omega: \quad \mu(A \cup B) = \frac{\mu(A) + \mu(B) - \mu(A \cap B) + \lambda\mu(A)\mu(B)}{1 + \lambda\mu(A \cap B)}$$

and

$$\forall A \subset \Omega: \quad \mu(A) + \mu(\bar{A}) = 1 - \lambda\mu(A)\mu(\bar{A}) \in \begin{cases} [1,2[, & \text{if } \lambda \in]-1,0], \\ ]0,1], & \text{if } \lambda \in [0,\infty[. \end{cases}$$

Lemma 4.5.2

Any possibility measure μ is a g_λ-measure($\lambda \in]-1, \infty[$), if, and only if, μ is a Dirac measure.

Proof

We notice that any Dirac measure is a possibility measure. Let μ be a possibility measure and f its possibility distribution; then

$$\forall \omega, \omega' \in \Omega: \quad \mu(\{\omega, \omega'\}) = \max(f(\omega), f(\omega')).$$

If μ is also a g_λ-measure ($\lambda \in]-1, \infty[$), then

$$\forall \omega, \omega' \in \Omega: \quad \mu(\{\omega, \omega'\})$$
$$= \max(f(\omega), f(\omega')) + \min(f(\omega), f(\omega'))(1 + \lambda \max(f(\omega), f(\omega'))).$$

Since $\lambda \in]-1, \infty[$ and $\forall \omega$: $f(\omega) \in [0,1]$, we must have

$$\forall \omega, \omega' \in \Omega: \quad \min(f(\omega), f(\omega')) = 0,$$

which means that μ is a Dirac measure. In other words, if μ is to be a g_λ-measure, μ must be a Dirac measure. It is also a sufficient condition, since a Dirac measure is also a g_λ-measure. Q.E.D.

Lemma 4.5.3

Let μ and v be two measures such that $\forall A \subset \Omega$, $\mu(A) + v(\bar{A}) = 1$; then μ is a consonant belief function if and only if v is a possibility measure.

Proof

Let μ be any consonant belief function; then

$$\forall A, B \subset \Omega: \quad \mu(A \cap B) = \min(\mu(A), \mu(B))$$

and

$$\begin{aligned}\forall A, B \subset \Omega: \quad v(A \cup B) &= 1 - \mu(\overline{A \cup B}) = 1 - \mu(\bar{A} \cap \bar{B})\\ &= 1 - \min(\mu(\bar{A}), \mu(\bar{B}))\\ &= 1 - \min(1 - v(A), 1 - v(B))\\ &= \max(v(A), v(B)),\end{aligned}$$

which implies that v is a possibility measure. Conversely, let v be a possibility measure, then

$$\forall A, B \subset \Omega: \quad v(A \cup B) = \max(v(A), v(B))$$

and

$$\begin{aligned}\forall A, B \subset \Omega: \quad \mu(A \cap B) &= 1 - v(\overline{A \cap B}) = 1 - v(\bar{A} \cup \bar{B})\\ &= 1 - \max(v(\bar{A}), v(\bar{B}))\\ &= 1 - \max(1 - \mu(A), 1 - \mu(B))\\ &= \min(\mu(A), \mu(B)),\end{aligned}$$

which implies that μ is a consonant belief function. Q.E.D.

The above one-to-one correspondence suggests new definitions of subsets of possibility measures from already defined subsets of consonant belief functions. Hence, to a simple support function μ there corresponds a measure v defined by

$$\forall B \subset \Omega: \quad v(B) = \begin{cases} 1, & \text{if } B \cap A \neq \emptyset, \\ t, & \text{if } B \cap A = \emptyset, \quad B \neq \emptyset, \\ 0, & \text{if } B = \emptyset, \end{cases}$$

(with $t = 1 - s \in [0, 1]$), which could be called a *t-one-possibility measure* focused on A. To a certainty measure ($s = 1$) there corresponds what could be called a zero–one possibility measure. Finally, to the vacuous belief function there corresponds the one–one possibility measure or the zero–one possibility measure focused on Ω; this could be called more suggestively the *maximum possibility measure* because, in such a case, any proposition A ($A \neq \emptyset$) has the maximum degree of possibility to be true.

Clearly, any possibility measure is uniquely determined by a function $f\colon X \to [0, 1]$, via the formula

$$\pi(A) = \sup_{x \in A} f(x), \qquad A \subset X.$$

Indeed, it suffices to take $f(x) = \pi(\{x\})$, $x \in X$.

The next question arises, then, naturally: Is any possibility measure a fuzzy measure?

Puri and Ralescu [1982] give two counterexamples which show that, even in "nice" cases, a possibility measure is *not* a fuzzy measure.

■ ***Example 4.5.1***

Take $X = \mathbb{R}$, $f(x) = 1$, and π defined as $\sup_{x \in A} f(x)$, $A \subset X$. Consider $A_n = (n, \infty)$; observe that $\pi(\cap_{n=1}^{\infty} A_n) = 0$, while $\lim_{n \to \infty} \pi(A_n) = 1$; thus, it is not true that

$$A_1 \supset A_2 \supset A_3 \supset \cdots \Rightarrow \mu\left(\bigcap_{n=1}^{\infty} A_n\right) = \lim_{n \to \infty} \mu(A_n). \qquad ■$$

■ ***Example 4.5.2***

Take $X = [0, 1]$, $f(x) = 1$ for $x \in [0, 1)$, and $f(1) = 0$. Consider $A_n = [1 - 1/n, 1]$; thus $\pi(\cap_{n=1}^{\infty} A_n) = 0$, while $\lim_{n \to \infty} \pi(A_n) = 1$. ■

In the rest of this section, we suppose that $X = \mathbb{R}^k$ (the k-dimensional euclidean space). The reader who feels that this assumption is too restrictive may take X to be any metric space without isolated points.

Theorem 4.5.1

Let $f\colon \mathbb{R}^k \to [0,1]$ and $\pi(A) = \sup_{x \in A} f(x)$, $A \subset \mathbb{R}^k$ be the associated possibility measure. If π is a fuzzy measure, then $f(x) = 0$ at every point of continuity of f.

Proof

Take a point $x_0 \in \mathbb{R}^k$ such that f is continuous at x_0. Define $A_n = \{x \in \mathbb{R}^k \mid \|x - x_0\| < 1/n, x \neq x_0\}$ (here $\| \; \|$ denotes the euclidean norm in $\mathbb{R}^k$). Obviously $A_n \neq \emptyset$; also $A_1 \supset A_2 \supset \cdots$ and $\cap_{n=1}^{\infty} A_n = \emptyset$. Since π satisfies (FM4), it follows that

$$\lim_{n \to \infty} \pi(A_n) = 0.$$

Let $(x_j)_j$ be a sequence such that $x_0 = \lim_{j \to \infty} x_j$, $x_j \neq x_0$. If $n \geq 1$ is fixed, then $x_j \in A_n$ for all $j \geq j_n$. Thus $f(x_j) \leq \sup_{x \in A_n} f(x)$ for $j \geq j_n$. It follows that

$0 \leq \limsup_{j \to \infty} f(x_j) \leq \sup_{x \in A_n} f(x) = \pi(A_n)$. Since this is true for any $n \geq 1$, we conclude that $\limsup_{j \to \infty} f(x_j) = 0$, thus $\lim_{j \to \infty} f(x_j) = 0$. Finally, since f is continuous at x_0, it follows that $f(x_0) = \lim_{y \to x_0} f(y) = 0$, which ends the proof.
Q.E.D.

Corollary 4.5.1
Let π be a possibility measure with a continuous "density" f. If π is a fuzzy measure, then $\pi = 0$.

The proof of this corollary is trivial.

The next question is: when is a possibility measure a fuzzy measure? The implicator is always true if the set X is finite. In general, if f is upper semicontinuous, any possibility measure π as given above is a capacity; that is, it satisfies

$$A_1 \supset A_2 \supset \cdots \Rightarrow \mu\left(\bigcap_{n=1}^{\infty} A_n \right) = \lim_{n \to \infty} \mu(A_n),$$

where the A_n's are compact subsets of the topological space X.

CHAPTER FIVE

Fuzzy Relations

5.1 BINARY AND n-ARY FUZZY RELATIONS

If X is the Cartesian product of n universes of discourse $X_1, \ldots, X_n$, then an n-ary *fuzzy relation* R in X is a fuzzy subset of X. Then R may be expressed as the union of its consistent fuzzy singletons $\chi_R(x_1, \ldots, x_n)/(x_1, \ldots, x_n)$; that is,

$$R = \int_{X_1 \times \cdots \times X_n} \chi_R(x_1, \ldots, x_n)/(x_1, \ldots, x_n),$$

where χ_R is the membership function of R.

Common examples of (binary) fuzzy relations are: *much greater than*, *resembles*, *is relevant to*, and *is close to*. For example, if $X_1 = X_2 = (-\infty, \infty)$, the relation *is close to* may be defined by

$$\text{is close to} \triangleq \int_{X_1 \times X_2} e^{-\alpha|x_1 - x_2|}/(x_1, x_2),$$

where α is a scale factor. Similarly, if $X_1 = X_2 = 1 + 2 + 3 + 4$, then the relation *much greater than* may be defined by the relation matrix

R	1	2	3	4
1	0	0.4	0.8	1
2	0	0	0.4	0.8
3	0	0	0	0.4
4	0	0	0	0

in which the (i, j)th element is the value of $\chi_R(x_1, x_2)$ for the ith value of x_1 and the jth value of x_2.

The fuzzy relation "x is much greater than y" in N may be defined subjectively by a membership function such as

$$\chi(x, y) = \begin{cases} 0, & \text{if } x - y \le 0, \\ \left[1 + 10(x-y)^{-1}\right]^{-1} & \text{if } x - y > 0. \end{cases}$$

■ ***Example 5.1.1***

Let $x = (\alpha_1, \alpha_2, \ldots, \alpha_n)$ and $y = (\beta_1, \beta_2, \ldots, \beta_n)$ be two points in the n-dimensional euclidean space R^n. The fuzzy relation "y is in the neighborhood of x" is a fuzzy set in R^n which may be defined subjectively by a membership function such as

$$\chi(x, y) = \frac{1}{\exp\|x - y\|},$$

where $\|x - y\| = [(\alpha_1 - \beta_1)^2 + (\alpha_2 - \beta_2)^2 + \cdots + (\alpha_n - \beta_n)^2]^{1/2}$. ■

Definition 5.1.1

The composition of two fuzzy relations A and B, denoted by $B \circ A$, is defined as a fuzzy relation in X whose membership function is related to those of A and B by

$$\chi_{B \circ A}(x, y) = \operatorname*{Sup}_{v}\left[\min[\chi_A(x, v), \chi_B(v, y)]\right], \qquad v, x, y \in X.$$

composition

Definition 5.1.2

Let X and Y be two spaces of objects, and h be a mapping from X to Y. Let B be a fuzzy set in Y with membership function $\chi_B(y)$. The fuzzy set A in X induced by the inverse mapping h^{-1} is defined by

$$A = \left\{(x, \chi_A(x)) | x = h^{-1}(y) \quad \text{and} \quad \chi_A(x) = \chi_B(y), \quad y \in B\right\}.$$

Now consider the converse problem. Suppose A is a given fuzzy set in X and h is a mapping from X to Y. What is the membership function for fuzzy set B in Y which is induced by this mapping? To answer this question, we consider the following two cases separately.

Case A

The mapping h is one-to-one. The fuzzy set B in Y induced by the mapping h is defined as

$$B = \left\{(y, \chi_B(y)) | y = h(x) \quad \text{and} \quad \chi_B(y) = \chi_A(x), \quad x \in A\right\}.$$

Note that since h is one-to-one, for any $y \in Y$, where y is an image of some x in A, that is, $y = h(x)$, there does not exist x' other than x such that $y = h(x')$. Hence $\chi_B(y)$ is uniquely defined by $\chi_A(x)$.

Case B

The mapping is many-to-one. In this case, the following ambiguity arises. Suppose x_1 and x_2 are two distinct points with different grades of membership in A and are mapped to the same point y in Y. Then what grade of membership in B should be assigned to y? To resolve this ambiguity, we agree to assign the larger of the two grades of membership to y. If there are n distinct points $x_1, x_2, \ldots, x_n$ in A with different grades of membership that are mapped to the same point y in Y, the grade of membership in Y will be the largest of these grades of membership.

To combine the above two cases, we give the following general definition:

Definition 5.1.3

Let X and Y be two spaces of objects, and h be a mapping from X to Y. Let A be a fuzzy set in X with membership function $\chi_A(x)$. The fuzzy set B in Y induced by the mapping h is defined by

$$B = \left\{ (y, \chi_B(y)) \mid y = h(x) \quad \textit{and} \quad \chi_B(y) = \max_{x \in T^{-1}(y)} \chi_A(x), \quad x \in A \right\},$$

where $T^{-1}(y)$ is the set of points in X that are mapped into Y by h.

If R is a relation from U to V (or, equivalently, a relation in $U \times V$), and S is a relation from V to W, then the *composition* of R and S is a fuzzy relation from U to W denoted by $R \circ S$ and defined by

$$R \circ S = \int_{U \times W} \operatorname{Sup}_{v} \left[(\chi_R(u, v) \wedge \chi_S(v, w)) \right] / (u, w).$$

If U, V, and W are finite sets, then the relation matrix for $R \circ S$ is the max–min product of the relation matrices for R and S.

If R is an n-ary fuzzy relation in $X_1 \times \cdots \times X_n$, then its *projection* (*shadow*) on $X_{i_1} \times \cdots \times X_{i_k}$ is a k-ary fuzzy relation R_q in X which is defined by

$$R_q \triangleq \operatorname{Proj} R \text{ on } X_{i_1} \times \cdots \times X_{i_k} \triangleq P_q R$$

$$\triangleq \int_{X_{i_1} \times \cdots \times X_{i_k}} \left(V_{x_{(\bar{q})}} \chi_R(x_1, \ldots, x_n) \right) / (x_{i_1}, \ldots, x_{i_k}),$$

where q is the index sequence $(i_1, \ldots, i_k)$; $x_{(q)} \triangleq (x_{i_1}, \ldots, x_{i_k})$; $\bar{q}$ is the complement of q; and $V_{x_{(\bar{q})}}$ is the supremum of $\chi_R(x_1, \ldots, x_n)$ over the x's that are in $x_{(\bar{q})}$.

The domain of a fuzzy relation R is denoted by dom R and is a fuzzy set defined by

$$\chi_{\operatorname{dom} R}(x) = \operatorname{Sup}_{y \in Y} \left[\chi_R(x, y) \right], \qquad x \in X.$$

Similarly, the range of R is denoted by ran R and is defined by

$$\chi_{\text{ran } R}(y) = \operatorname*{Sup}_{x \in X}[\chi_R(x, y)], \qquad y \in Y.$$

The height of R is denoted by $h(R)$ and is defined by

$$h(R) = \operatorname*{Sup}_{x}\left\{\operatorname*{Sup}_{y}[\chi_R(x, y)]\right\}.$$

A fuzzy relation is subnormal if $h(R) < 1$ and normal if $h(R) = 1$.

The support of R is denoted by $S(R)$ and is defined to be the exact subset of $X \times Y$ over which $\chi_R(x, y) > 0$.

Specifically, a *similarity relation*, S, in X is a fuzzy relation that is:

1. reflexive; that is,

$$\chi_s(x, y) = 1 \qquad \text{iff } x \equiv y;$$

2. symmetric; that is,

$$\chi_s(x, y) = \chi_s(y, x), \qquad \forall x, y \in X; \text{ and}$$

3. transitive; that is,

$$\chi_s(x, z) \geq \max_{y}\{\min[\chi_s(x, y), \chi_s(y, z)]\}.$$

Similarly, a dissimilarity relation D can be defined as the complement of S with

$$\chi_d(x, y) = 1 - \chi_s(x, y); \quad x, y \in X.$$

If $\chi_d(x, y)$ is interpreted as a distance function $d(x, y)$, then transitivity implies that

$$1 - d(x, z) \geq \max_{y}\{\min[1 - d(x, y), \;\; 1 - d(y, z)]\},$$

and since

$$\min[1 - d(x, y), \;\; 1 - d(y, z)] = 1 - \max[d(x, y), \;\; d(y, z)],$$

we can conclude that

$$d(x, z) \leq \max[d(x, y), \;\; d(y, z)], \quad \forall x, y, z \in X,$$

which implies the triangle inequality.

A *fuzzy restriction* is a fuzzy relation that acts as an elastic constraint on the values that may be assigned to a variable. More specifically, if Y is a variable that takes values in a universe of discourse X, then a fuzzy restriction $R(Y)$ on the values that may be assigned to Y is a fuzzy relation in X such that the assignment of a value y to Y requires a stretch of the restriction expressed by

$$\text{Degree of stretch} = 1 - \chi_{R(Y)}(y),$$

where $\chi_{R(Y)}(y)$ is the grade of membership of y in $R(Y)$; namely,

$$g = y: \quad \chi_{R(Y)}(y),$$

where g denotes a generic value of Y and $\chi_{R(Y)}(y)$ is the "degree of ease" with which y may be assigned to Y.

As a simple illustration, suppose that $X = 0 + 1 + 2 + \cdots$ and that Y is a variable labeled "small integer." Assume that the fuzzy set *small integer* is defined by

$$\text{Small integer} = 1/0 + 1/1 + 0.8/2 + 0.6/3 + 0.4/4 + 0.2/5.$$

Then if g is a generic value of the variable "small integer" and we assign the value 2 to this variable, we have

$$g = 2: \quad 0.8,$$

which implies that the fuzzy restriction labeled *small integer* must be stretched to the degree 0.2 to allow the assignment of the value 2 to the variable "small integer."

More generally, if $Y = (Y_1, \ldots, Y_n)$ is an n-ary variable taking values in the Cartesian product space

$$X = X_1 \times \cdots \times X_n,$$

then an n-ary fuzzy relation $R(Y_1, \ldots, Y_n)$ in X is a *fuzzy restriction* if it acts as an elastic constraint on the values that may be assigned to Y. An n-ary variable that is associated with a fuzzy restriction on the values that may be assigned to it is said to be an *n-ary fuzzy variable*.

Until now we have focused our attention upon fuzzy relations in the sense of fuzzy sets on a cartesian product of universes. Other kinds of relations involving fuzziness include, for example, *tableaus* of fuzzy sets whose columns refer to the universes and whose rows contain $(k + 1)$-tuples of labels of fuzzy sets.

Other extensions are interval-valued fuzzy relations discussed by Ponsard [1977] and fuzzy relations between fuzzy sets studied by E. Sanchez [1976].

5.2 FUZZY GRAPHS

The concept of a fuzzy graph is a natural generalization of crisp graphs, using fuzzy sets, and in many cases the extension principle. Our exposition of fuzzy graphs follows the outstanding extension of crisp graphs, given by Rosenfeld in [Zadeh et al., 1975].

A fuzzy graph $\tilde{G}$ is a pair $(\tilde{V}, \tilde{E})$ where $\tilde{V}$ is a fuzzy set on V and $\tilde{E}$ is a fuzzy relation on $V \times V$ such that

$$\chi_{\tilde{E}}(v, v') \leq \min(\chi_{\tilde{V}}(v), \chi_{\tilde{V}}(v')).$$

The above inequality states that the strength of the link between two vertices cannot exceed the degree of "importance" or of "existence" of the vertices. In other words, $\tilde{E}$ is a fuzzy relation on $\tilde{V} \times \tilde{V}$ in the sense that dom($\tilde{E}$) and ran($\tilde{E}$) are contained in $\tilde{V}$. However, in some situations it may be desirable to relax this inequality.

Classical concepts and definitions pertaining to graphs have been extended to fuzzy graphs by Rosenfeld in [Zadeh et al., 1975]:

A path whose length is n in a fuzzy graph is a sequence of distinct vertices $v_0, v_1, \ldots, v_n$ such that $\chi_{\tilde{E}}(v_{i-1}, v_i) > 0, \forall i = 1, n$. The strength of the path is

$$\min_i \chi_{\tilde{E}}(v_{i-1}, v_i), \quad 1 \le i \le n, \qquad \text{and} \qquad \chi_{\tilde{V}}(v_0) \quad \text{if } n = 0.$$

A strongest path joining two vertices v_0 and v_n has a strength $\chi_{\tilde{\tilde{E}}}(v_0, v_n)$, where $\tilde{\tilde{E}}$ is the *transitive closure* of $\tilde{E}$.

In an ordinary graph, the distance between two vertices is the length of the shortest path linking them. A set U of vertices is called a cluster of order k iff:

(i) $\forall v, v' \in U, \quad d(v, v') \le k,$
(ii) $\forall v \notin U, \exists v' \in U, \quad d(v, v') > k,$

where $d(v, v')$ denotes the distance between v and v'. When $k = 1$, a k-cluster is called a clique, i.e., a *maximum complete subgraph*. In a fuzzy graph, a nonfuzzy subset U of V is called a fuzzy cluster of order k if

$$\min_{\substack{v \in U \\ v' \in U}} \chi_{\tilde{E}^k}(v, v') > \max_{v \notin U} \left[\min_{v' \in U} \chi_{\tilde{E}^k}(v, v') \right],$$

where $\tilde{E}^k$ is the kth power of $\tilde{E}$.

The following definitions assume that the set of vertices is not fuzzy and that $\tilde{E}$ is symmetrical ($\chi_{\tilde{E}}(v, v') = \chi_{\tilde{E}}(v', v)$). The degree of a vertex v is

$$\text{dg}(v) = \sum_{v' \neq v} \chi_{\tilde{E}}(v, v').$$

The minimum degree of G is $\delta(G) = \min_v \text{dg}(v)$. Here G is said to be λ-degree connected iff

(i) $\forall v, v' \in V, \quad \chi_{\tilde{E}}(v, v') \neq 0 \quad (\text{if } v \neq v')$, and
(ii) $\delta(G) \ge \lambda$.

A λ-degree component of G is a maximal λ-degree connected subgraph of G. For any $\lambda > 0$, the λ-degree components of a fuzzy graph are disjoint.

As in the theory of nonfuzzy sets, the notion of fuzzy graphs may be explained in terms of fuzzy relations.

A nonzero fuzzy relation Q on X is weakly reflexive and symmetric iff there is a universe Y and a fuzzy relation R on $X \times Y$ such that $Q = R \circ R^{-1}$.

Clearly, $\chi_{R \circ S}(x, z)$ is the strength of a set of chains linking x to z, each chain of form $x - y - z$, such that the strength of such a chain is that of the weakest link. The strength of the relation between x and z is that of the strongest chain between x and z.

When the related universes X and Y are finite, a fuzzy relation R on $X \times Y$ can be represented as a matrix $[R]$ whose generic term $[R]_{ij}$ is

$$\chi_R(x_i, y_j) = r_{ij}, \qquad i = 1, \quad n, \qquad j = 1, \quad m,$$

where $|X| = n$ and $|Y| = m$.

The composition of finite fuzzy relations can thus be viewed as a matrix product. With $[S]_{jk} = s_{jk}$, $k = 1,\ p,\ p = |Z|$,

$$[R \circ S]_{ik} = \sum_j r_{ij} s_{jk},$$

where Σ is in fact the operation max and *product* the operation min.

Since $R \circ S$ can be written proj$[c(R) \cap c(S); X \times Z]$, where R and S are respectively on $X \times Y$ and $Y \times Z$, other compositions can be defined. For example, changing min to $*$, we define $R \boxed{*} S$ as

$$\chi_{R \boxed{*} S}(x, z) = \sup_y (\chi_R(x, y) * \chi_S(y, z)).$$

If $*$ is associative, and nondecreasing with respect to each of its arguments, the sup-$*$ composition satisfies associativity, distributivity over union, and monotonicity.

Let R be a fuzzy relation on $X \times X$, R is max–min transitive iff $R \circ R \subseteq R$, or, more explicitly,

$$\forall (x, y, z) \in X^3, \quad \chi_R(x, z) \geq \min(\chi_R(x, y), \quad \chi_R(y, z)).$$

Other transitivities, associated with other kinds of compositions of fuzzy relations, can be defined. Generally R is said to be max-$*$ transitive iff $R \boxed{*} R \subseteq R$. A specific case of max-$*$ transitive is where $a * b$ is given by:

1. $a \mathbin{\dot{\cap}} b = \max(0, a + b - 1)$ (bold intersection);
2. $a \mathbin{\square} b = \frac{1}{2}(a + b)$ (arithmetic mean);
3. $a \vee b = \max(a, b)$ (union);
4. $a \mathbin{\dot{+}} b = a + b - ab$ (probabilistic sum).

Bandler and Kohout [1980b] have defined the triangle product of two fuzzy relations, R and S, as

$$(R \triangleleft S)_{ik} = \frac{1}{N_j} \sum_j (R_{ij} \to S_{jk}),$$

when N_j is the number of elements and $\to$ is a fuzzy implication operator. The triangle product is nonsymmetric and in essence represents the (mean) degree to which the fuzzy afterset $a_i R$, defined by

$$a_i R = \{b \in B | a_i R b, a_i \in A, R \in \mathcal{B}(A \to B)\},$$

is contained in the fuzzy foreset Sc_j, defined by

$$Sc_j = \{a \in A | aSc_j, c_j \in C, S \in \mathscr{B}(A \to C)\}.$$

Namely, it is the mean degree to which being related by R to a_i implies relationship by S to c_k. Clearly, the afterset aR of $a \in X_1$ is the fuzzy subset of X_2 consisting of those $y \in X_2$ to which a is related, each with its degree; thus producing

$$\chi_{aR}(y) = \chi_R(a, y), \quad R \subseteq X_1 \times X_2.$$

Similarly, the foreset Sc of $c \in X_3$ is a fuzzy subset of X_2 consisting of those $y \in X_2$ which are related to c, each with its degree of intensity given by

$$\chi_{Sc}(y) = \chi_S(y, c), \quad S \subseteq X_2 \times X_3.$$

An example of the application of this concept as a tool for analysis and synthesis of the behavior of complex natural and artificial systems has been illustrated by Bandler and Kohout in [1980b].

A special kind of fuzzy relation, based on a new composition, is called the α-composite fuzzy relation, and has been introduced by Sanchez [1976]–[1984]. Formally, let $Q \subseteq X \times Y$ and $R \subseteq Y \times Z$ be two fuzzy relations. We define $T = Q@R$, $T \subseteq X \times Z$, the @-composite fuzzy relation of Q and R, by

$$(Q@R)(x, z) = \min_y [Q(x, y) \alpha R(y, z)],$$

where $y \in Y$, for all $(x, z) \in X \times Z$.

The following results are quite trivial:

(i) If $b, d \in L$, where L is a fixed complete Brouwerian lattice* and if $b \leq d$, then $a\alpha b \leq a\alpha d, \forall a \in L$.

(ii) If $R_1, R_2 \subseteq Y \times Z$ and if $R_1 \subseteq R_2$, then

$$Q@R_1 \subseteq Q@R_2, \quad \forall Q \subseteq X \times Y,$$

where

$$(Q@R_i)(x, z) = \min_y [Q(x, y) \alpha R_i(y, z))$$

and

$$y \in Y, \quad \forall (x, z) \in X \times Z.$$

As an example of this concept, consider a set of medical patients P, having a set of symptoms S, with a set of diagnoses D. Given Q, $Q \subseteq P \times S$, a fuzzy relation between patients and symptoms, and T, $T \subseteq P \times D$, a fuzzy relation between patients and diagnoses, under the assumption that the range of these fuzzy relations is a Brouwerian lattice, we can investigate the concept of medical knowledge observed from Q and T. The expression of this medical

*A Brouwerian lattice L is a lattice in which, for any given elements x and y, the set of all Z in L such that $x \wedge z \leq y$, where $x \wedge z$ denotes the greatest lower bound of x and z, contains a greatest element denoted $x \alpha y$.

knowledge can be computed by determining the greatest fuzzy relation $R \subseteq S \times D$ such that

$$R(s,d) \to [Q(p,s) \to T(p,d)]$$

for all $s \in S$, $d \in D$, and $p \in P$.

Since a Brouwerian logic is a propositional calculus, that is, a lattice with 0 and I, in which

$$P \to Q = I \quad \text{iff } P \leq Q,$$
$$P \to (Q \to R) = (P \wedge Q) \to R, \quad \forall P, Q, R,$$

we can conclude that $R(s,d)$ can be determined as above if and only if

$$[R(s,d) \wedge Q(p,s)] \to T(p,d)$$

or

$$R(s,d) \wedge Q(p,s) \leq T(p,d).$$

Hence, $\forall (p,d) \in P \times D$,

$$\bigvee_s [R(s,d) \wedge Q(p,s)] \leq T(p,d)$$

or simply, $R \circ Q \subseteq T$.

However, it is clear that the greatest R such that $R \circ Q \subseteq T$ is equal to

$$R_{\max} = Q^{-1} \text{ⓐ} T,$$

which gives us the medical knowledge associated with Q and T.

Given, now, a patient p_i with a known fuzzy set of symptoms $Q_{p_i}(s)$, $\forall s \in S$, the composition $R_{\max} \circ Q_{p_i} = T_{p_i}$ expresses the "closest" or the "most reliable" fuzzy set of diagnoses presented by p_i.

In general, the correspondence from $f(p)$, p being a natural object, to $\chi_C(p)$ is relational. In other words, C is assumed to be characterized by a relational tableau R with linguistic entries that allow us to set translation rules of fuzzy propositions related to inexact, ill-defined, and vague dependencies. The translation of R into a fuzzy relation R_r is quite simple. Thus, given a fuzzy set A related to the attributes, the composition $R_r \circ A$ corresponds to the linguistic interpolation yielding new grades of membership for C.

5.3 SIMILARITY RELATIONS

Especially important in science are the relations that are reflexive, symmetric, and transitive; these relations are known in classical set theory as *equivalence relations*. The concept of a *similarity* relation is essentially a generalization of the concept of an equivalence relation.

Definition 5.3.1

A similarity relation, S, is a fuzzy relation in X that is reflexive, symmetric, and transitive. Thus, let x_i and x_j be elements of X, and let $\chi_S(x_i, x_j)$ denote the *grade-membership of the ordered pair (x_i, x_j) in S. Then S is a similarity relation in* X iff, $\forall x, y, z \in X$,

$$\chi_s(x, x) = 1 \quad (\textit{reflexive}), \tag{5.3.1}$$

$$\chi_s(x, y) = \chi_s(y, x) \quad (\textit{symmetric}), \tag{5.3.2}$$

$$\chi_s(x, z) \geq \max_y [\min(\chi_s(x, y), \chi_s(y, z))] \quad (\textit{transitive}). \tag{5.3.3}$$

Clearly, this definition of transitivity is the explicit expression of general transitivity under the max–min composition denoted by $\circ$; namely,

$$S \supset S \circ S.$$

It should be noted that if one uses a different definition of composition, for example, max-product, denoted by $*$, then

$$S \supset S * S,$$

or, more explicitly,

$$\chi_s(x, z) \geq \max_y [\chi_s(x, y) * \chi_s(y, z)].$$

Clearly, the *n-purlieus relation* defined as

$$\chi_s^n(x, y) \triangleq \operatorname*{Sup}_{\vec{x} \in \Omega} \{\min[\chi_s(x, x_1), \ldots, \chi_s(x_{n-1}, y)]\},$$

where $\vec{x} = (x_1, x_2, \ldots, x_{n-1}) \in \hat{X}$ ($\hat{X}$ is the $(n-1)$-fold Cartesian product of X with itself, where $x, y \in X$) implies that, for all $x, y \in X$ and all $n \geq 1$,

$$0 \leq \chi_s^n(x, y) \leq \chi_s^{n+1}(x, y) \leq 1.$$

Consequently, $\lim_{n \to \infty} \chi_s^n(x, y) \triangleq \bar{\chi}_s(x, y)$ exists by the monotone convergence principle; namely, for every $\varepsilon > 0$, there is an integer N such that

$$|\chi_s^n(x, y) - \bar{\chi}_s(x, y)| < \varepsilon \quad \text{for} \quad n > N.$$

Since the sequence is nondecreasing and is bounded from above and from below, we can conclude that existence of the limit is due to the following theorem:

Theorem 5.3.1

A bounded nondecreasing sequence $\{a_i\}_i$ has a limit; the limit is the smallest number that is not less than any a_j.

Definition 5.3.2

Let x and y be two elements of X, and let $\chi_s^n(x, y)$ be the n-purlieus relation as defined above. Then we define the propinquity $\bar{\chi}(x, y)$ in $[0, 1]$ such that

$$\bar{\chi}(x, y) = \lim_{n \to \infty} \chi_s^n(x, y).$$

Definition 5.3.3

Let $x, y \in X$. Then x and y are said to have a threshold relation (xR_Ty) iff $\bar{\chi}_s(x, y) \geq T$.

Theorem 5.3.2

$\forall x, y, z \in X\ [\bar{\chi}_s(x, z) \geq \min[\bar{\chi}_s(x, y), \bar{\chi}_s(y, z)]]$.

Proof

(i) $xR_T x$, $\forall T \in [0, 1]$, since $1 = \chi_s(x, x) \leq \bar{\chi}_s(x, x) \leq 1, \quad x \in X$.
(ii) $xR_T y$ iff $yR_T x$ since

$$\lim_{n \to \infty} \chi_s^n(x, y) = \bar{\chi}_s(x, y) = \bar{\chi}_s(y, x) = \lim_{n \to \infty} \chi_s^n(y, x).$$

(iii) $xR_T y \wedge yR_T z \rightarrow xR_T z$ since

$$\bar{\chi}_s(x, z) \geq \min[\bar{\chi}_s(x, y), \bar{\chi}_s(y, z)].$$

Q.E.D.

Clearly, we can associate with every relation an appropriate matrix to represent the relation, and hence we can classify the patterns using the partition induced by the threshold relation. This relation is similar to the idea of α-level-sets of a fuzzy relation R, denoted by R_α (and is a nonfuzzy set in $X \times Y$), which give rise to the notion of relation matrices.

Before proceeding, we shall investigate some of the important properties of fuzzy or *inexact matrices*. The results are a generalization of Boolean matrix theory and some known results on lattice matrices.

When the related universes X and Y are finite, a fuzzy relation R on $X \times Y$ can be represented as a matrix $[R]$ whose generic term $[R]_{ij}$ is

$$\chi_R(x_i, y_j) = r_{ij}, \quad i = 1, n, \quad j = 1, m,$$

where $|X| = n$ and $|Y| = m$.

The composition of finite fuzzy relations can thus be viewed as a matrix product, with $[S]_{jk} = s_{jk}$, $k = 1,\ p$, and $p = |Z|$, such that

$$[R \circ S]_{ik} = \sum_j r_{ij} s_{jk},$$

where Σ is in fact the operation max and *product* the operation min.

Let R be a fuzzy relation on $X \times X$ where $|X| = n$. The mth power of a fuzzy relation is defined as $R^m = R \circ R^{m-1}$, $m > 1$, and $R^1 = R$. The following propositions will be discussed:

(i) The power of R either converges to idempotent R^c for a finite c or oscillates with finite period (if R^m does not converge, then it must oscillate with a finite period since $|X|$ is finite and the composition is deterministic and cannot introduce numbers not in R originally);
(ii) If $\forall i, j, \exists k$ such that $r_{ij} \leq \min(r_{ik}, r_{kj})$, then R converges to R^c where $c \leq n - 1$.

Other results in more particular cases can be found in the next section.

Since $R \circ S$ can be written proj$[c(R) \cap c(S); \quad X \times Z]$ where R and S are respectively on $X \times Y$ and $Y \times Z$, other compositions may be introduced by modifying the operator used for the intersection.

Changing min to $*$, we define $R \boxed{*} S$ through

$$\chi_{R \boxed{*} S}(x, z) = \sup_y (\chi_R(x, y) * \chi_s(y, z)).$$

Zadeh [1971a] proved that when $*$ is associative and nondecreasing with respect to each of its arguments, the sup-$*$ composition satisfies associativity, distributivity over union, and monotonicity. Examples of such operators are *product* and *bold intersection*.

We may encounter another kind of alternative compositions, inf-max compositions. The following property holds: $\overline{R \circ S} = \overline{R} \bar{\circ} \overline{S}$ where $\bar{\circ}$ denotes inf-max composition.

In the following section we present a more detailed discussion of fuzzy relations presented by fuzzy matrices.

5.4 FUZZY MATRICES

In order to use the notion of a proximity relation for cluster analysis, we must develop a re-evaluation procedure that will produce the transitive closure of the proximity matrix (the matrix describing all proximity relations between its elements).

Motivated by this idea, we shall investigate the properties of fuzzy matrices under the operations of fuzzy logic. It should be noted that some of the results can be obtained in many other mathematical configurations since the following discussion is basically a generalization of Boolean matrix theory.

It is important to note that there are several aspects of inexact matrix theory that are interesting in their own right. However, our discussion will be centered on those properties that can be applied toward the solution of the classification problem of static patterns under the subjective measure of similarity.

Let Z be the fuzzy algebra that is a distributive lattice with unique identities under $+$ and $*$, e_+ and e_*, respectively. Consider Z_{pq}, the complete set of $p \times q$ matrices with elements in Z. Two elements,

$$S = [s_{ij}] \qquad \text{and} \qquad T = [t_{ij}] \quad \text{in } Z_{pq}$$

are regarded as equal iff $s_{ij} = t_{ij}$ for all i and j. We now define the following compositions in Z_{pq}.

Definition 5.4.1

(i) SUM: $S + T = W$ iff $w_{ij} = s_{ij} + t_{ij}$ *for all i and j.*
(ii) LATTICE PRODUCT: $S * T = W$ iff $w_{ij} = s_{ij}t_{ij}$ *for all i and j.*
(iii) $S \geq T$ iff $s_{ij} \geq t_{ij}$ *for all i and j.*
(iv) $S = \bar{T}$ iff $s_{ij} = \bar{t}_{ij}$ *for all i and j.*

Evidently, the system $[Z_{pq}, +, *]$ is a lattice with the universal element E_* (all entries e_*) and the zero element E_+ (all entries e_+).

Scalar multiplication:
$W = rS = Sr$ iff $w_{ij} = rs_{ij}$, $S \in Z_{pq}$, $r \in Z$.

Matrix product:
$W = ST$ iff $w_{ij} = \Sigma_{k=1}^{q} s_{ik} t_{kj}$, $S \in Z_{pq}$, $T \in Z_{qm}$, *and* $W \in Z_{pm}$, *where the symbol* Σ *is used to denote the sum with respect to the operation* + (max) *of* Z.

Transpose:
$W = S^t$ iff $w_{ij} = s_{ji}$, $S \in Z_{pq}$, $W \in Z_{qp}$.

Permanent:
$\text{per}(S) = |S| = \Sigma_i [\Pi_j^p (s_{ji_j})]$, *where* $S \in Z_{pp}$, $|S| \in Z$, *and the summation is taken over all permutations* $(i_1, i_2, \ldots, i_p)$.

Adjoint:
$W = \text{Adj}\, S$ iff $w_{ij} = S_{ij}$ *where* $S \in Z_{pp}$, $W \in Z_{pp}$, *and* S_{ij} *is the cofactor of* s_{ji} *in* $|S|$.

It is interesting to note that Z_{pp} forms a lattice-ordered semi-group with the matrix product as third binary composition, and the matrix I (all diagonal entries e_* and all others e_+) serves as unity.

A vector is called *fuzzy* iff all its entries are elements of Z. A matrix is called *fuzzy* iff all its rows are fuzzy vectors. Since our interest is mainly in *fuzzy symmetric matrices* we will refer to them just as fuzzy or inexact matrices. Thus for the set of all $p \times p$ fuzzy symmetric matrices Z_{pp}, we have the following definitions and propositions.

Definition 5.4.2
Let $A = [a_{ij}]$ *and* $B = [b_{ij}]$ *be two* $p \times p$ *fuzzy matrices.*

(i) $A + B = [\max(a_{ij}, b_{ij})]$;
(ii) $AB = [\max_k\{\min(a_{ik}, b_{kj})\}]$;
(iii) $A \leq B$ *exists* iff $a_{ij} \leq b_{ij}$, $\forall i, j$.

It is easy to prove that $A \leq B$ implies

$$SA \leq SB \text{ and } AT \leq BT, \quad \forall S, \quad T \in Z_{pp}.$$

$A \leq B$ evidently implies $A + B = B$ and conversely. One immediately verifies that the set of $p \times p$ fuzzy matrices is a monoid under matrix multiplication. Namely, the set of $p \times p$ fuzzy matrices is closed under fuzzy matrix multiplication and the unit $p \times p$ matrix (all diagonal entries 1 and all other 0) is a fuzzy matrix.

Definition 5.4.3
A fuzzy matrix A is called constant if all its rows are equal.

The following propositions are immediate consequences of the above definitions.

Proposition 5.4.1
If A is a constant fuzzy matrix and $B \geq I$ is a fuzzy matrix of the same order, then AB and BA are constant fuzzy matrices.

Proof

(i) By definition 5.4.3, a_{ik} is the same for all i. By definition,

$$AB = \left[\max_k \left\{ \min(a_{ik}, b_{kj}) \right\} \right]$$

and thus the elements in AB are independent of i; and therefore AB is constant. This is true regardless of whether A and B are symmetric or not.

(ii) $BA = [\max_k \{\min(b_{ik}, a_{kj})\}]$. $B \geq I$ implies $b_{ii} = 1, \forall i, i = 1, 2, \ldots, n$.

Thus since a_{kj} is independent of k, denote a_{kj} by a_j and

$$\max_k \left\{ \min(b_{ik}, a_{kj}) \right\} = \min\left\{ a_j, \max_k (b_{ik}) \right\}.$$

However, $\max_k (b_{ik}) = 1$ and thus a_j is the minimum term regardless of k. Q.E.D.

It should be noted that if $B \not\geq I$, the second part is not valid.

■ ***Example 5.4.1***
Let

$$A = \begin{bmatrix} 0.5 & 0.3 \\ 0.5 & 0.3 \end{bmatrix} \quad \text{and} \quad B = \begin{bmatrix} 0.1 & 0.4 \\ 0.7 & 0.1 \end{bmatrix};$$

then

$$BA = \begin{bmatrix} 0.4 & 0.3 \\ 0.5 & 0.3 \end{bmatrix},$$

which is clearly not constant. ■

Proposition 5.4.2
Let $A = [a_{ij}]$ be a constant fuzzy matrix and $B = [b_{ij}]$ a fuzzy matrix. If for all i,

$$\max_j \left[b_{ij} \right] \geq [a_{ij}],$$

then $BA = A$.

Proof
For all i we have the following:

$$\max_j \left[b_{ij} \right] \geq [a_{ij}],$$

then

$$\max_k \left\{ \min(b_{ik}, a_{kj}) \right\} = \min\left\{ a_j, \max_k (b_{ik}) \right\} = a_j. \qquad \text{Q.E.D.}$$

Corollary 5.4.1

If A is a constant fuzzy matrix, then $A^2 = A$ (that is, fuzzy constant matrices are idempotent).

Theorem 5.4.1

Let A be a $p \times p$ fuzzy matrix. Then the sequence $A, A^2, A^3, \ldots$ is ultimately periodic.

Proof

Let $p = \{x_1, x_2, \ldots, x_m\}$ be the set of all the fuzzy elements that occur in matrix A. Then the number of different matrices that can be obtained by multiplying A is at most m^{p^2}, which is clearly finite. Q.E.D.

Proposition 5.4.3

Let A be a $p \times p$ matrix and let $B = A + I$, where $+$ denotes the operation max. *Then*

$$B \leq B^2 \leq \cdots \leq B^{p-1} = B^p = B^{p+1} = \cdots$$

Let

$$\tilde{A} = \operatorname*{Sup}_k A^k$$

where A is a $p \times p$ fuzzy matrix.

From Proposition 5.4.3, it is clear that if $B = A + I$, then $\tilde{B} = B^{p-1}$.

Proof

We shall prove a stronger result which claims that there exists an integer $q \leq p - 1$ such that

$$B \leq B^2 \leq \cdots \leq B^q = B^{q+1} = \cdots$$

For simplicity of the proof we shall denote the operation min by concatenation. Let $B = (a_{ij})$ and $B^h = (b_{ij})$. Then

$$B^{h+1} = \left(\sum_{k=1}^{p} b_{ik} a_{kj} \right) = \left(b_{ij} + \sum_{k=1}^{p} b_{ik} a_{kj} \right),$$

since $a_{jj} = 1$. Thus

$$b_{ij} \leq \sum_{k=1}^{p} b_{ik} a_{kj} \qquad \text{and} \qquad B^h \leq B^{h+1}.$$

This shows that $B \leq B^2 \leq \cdots$. We must show that this chain does not strictly increase indefinitely. It will be sufficient to show that $B^{p-1} = B^p$. Since we already know $B^{p-1} \leq B^p$, we need only show that $B^p \leq B^{p-1}$. We are using the fact that $\leq$ is a partial ordering, and hence is antisymmetric.

Consider an off-diagonal element of B^p. It is of the form

$$\sum_{k_{n-1}=1}^{p} \sum_{k_{n-2}=1}^{p} \cdots \sum_{k_1=1}^{p} a_{ik_1} a_{k_1 k_2} \cdots a_{k_{p-2} k_{p-1}} a_{k_{p-1} j}.$$

Clearly there are $p - 1 + 2 = p + 1$ subscripts, so not all of them can be distinct

(Dirichlet's principle). Consider a term of the above form and suppose there exists an integer s such that $j = k$. The term is of the form

$$a_{ik_1} a_{k_1 k_2} \cdots a_{k_{s-1} j} a_{j k_{s+1}} \cdots a_{k_{p-1} j},$$

but since $ab \leq a$, this is contained in the term

$$a_{ik_1} \cdots a_{k_{s-1} j},$$

which is the i–j entry of B^s and is contained in the i–j entry of B^{p-1}. The previous argument is symmetric if $i = k_s$, for some integer s.

The remaining case occurs if there exist integers r and s such that $k_s = k_r$. Assuming $s < r$, we get

$$a_{ik_1} \cdots a_{k_{s-1} k_r} a_{k_r k_{s+1}} \cdots a_{k_{r-1} k_r} a_{k_r k_{r+1}} \cdots a_{k_{p-1} j},$$

which is contained in

$$a_{ik_1} \cdots a_{k_{s-1} k_r} a_{k_r k_{r+1}} a_{k_r k_{r+1}} \cdots a_{k_{p-1} j}.$$

This term is contained in the i–j entry of B^{p-1}. Therefore $B^p \leq B^{p-1}$ and $B^p = B^{p-1}$. Q.E.D.

Proposition 5.4.4

Let A be a $p \times p$ fuzzy matrix and let $B = A + I$ ($+ \equiv$ max), where I is of the same order. Then $\tilde{B} = \operatorname{adj} B$.

Before proving Proposition 5.4.4, the following discussion is quite helpful. Let B be a symmetric matrix with diagonal of 1's. Then we can consider a fuzzy undirected finite graph G having a *primitive connection matrix* B corresponding to the grade memberships of the edges of G, where the columns and rows of B represent the vertices of G.

For any primitive connection matrix, we define the *characteristic fuzzy matrix* or *fuzzy transmission matrix* $\psi(B) = [x_{ij}]$ such that x_{ij} is the fuzzy transmission function of the two-terminal system connecting vertex i to j. It is clear that $\psi(B)$ is a symmetric matrix, since the graph is an undirected one, and thus

$$x_{ij} = x_{ji}, \forall i, j \qquad \text{and} \qquad x_{ii} = 1, \forall i.$$

During the process of computing the characteristic fuzzy matrix, simplification of the fuzzy structures are possible.

Lemma 5.4.1

Let B be a square fuzzy matrix of order p. Then there exists an integer $q \leq p - 1$ such that

$$B^q = B^{q+1} = \cdots = \psi(B).$$

Proof of Lemma 5.4.1

Let $B = [b_{ij}]$. The ij entry of B^2 is

$$\sum_{k=1}^{p} b_{ik} b_{kj},$$

and this term has the grade of membership of

$$\max_k \left[\min(b_{ik}, b_{kj})\right]$$

iff there is a direct path between vertices i and j or there is a path from i to j through one intermediate vertex. Extending this argument to B^i, it is clear that no path requires more than $(p-2)$ intermediate vertices, since there are only p vertices and internal loops are excluded. Hence, the entry of B^{p-1} has the grade of membership of $\max_{\text{subterms}}\{ij \text{ terms of } B^{p-1}\}$ iff i and j are connected, namely $B^{p-1} = \psi(B)$. Q.E.D.

Proof of Proposition 5.4.4

Let G be the fuzzy undirected finite graph having a primitive connection matrix B. By using network-analysis techniques we can prove that the transmission matrix $T_G = [t_{ij}]$ of graph G is given by

$$T_G = \operatorname{adj} B,$$

that is $B_{ii} = 1$, and the transmission t_{ij} from v_i to v_j, $i \neq j$, equals B_{ij}. The fact that $B_{ii} = 1$ follows immediately from $b_{ii} = 1$. Now we consider any pair (i, j), $i, j = 1, 2, \ldots, p$ with $i \neq j$.

Let S be the set of permutations

$$s = \begin{pmatrix} 1 & \cdots & p \\ s_1 & \cdots & s_p \end{pmatrix},$$

with $s_j = i$. Let $s = u \cdot v \cdots$ be the decomposition of s into cyclic permutations, such that u is of the form

$$u = (j, i, \ldots, j),$$

and let U be the set of all such permutations u. Also let

$$B_s = b_{1s_1}, \ldots, b_{ps_p};$$

then we have $B_s = B_u \cdot B_v \cdots$ and thus $B_u \geq B_s$. Since $U \subseteq S$, we get

$$\sum_{s \in S} B_s = \sum_{u \in U} B_u.$$

Evidently,

$$\sum_{u \in U} B_u = b_{ji} t_{ij},$$

whereas $\sum_{s \in S} B_s$ consists of all terms of $|B|$ containing b_{ji}.

Hence t_{ij} is the cofactor of b_{ji} in $|B|$; that is, $t_{ij} = B_{ij}$, which implies $T_G = \operatorname{adj} B$.

Furthermore, this proof shows that if B and $\psi(B)$ are the *fuzzy primitive connection* and *transmission* matrices of some graph G, respectively, then $\psi(B) = \operatorname{adj} B$. Q.E.D.

Proposition 5.4.5

Let B be a $p \times p$ fuzzy matrix such that $B = A + I$. Then

(i) $\psi(\psi(B)) = \psi(B)$ $[\tilde{B} = \tilde{\tilde{B}}]$;
(ii) $B \leq \psi(B)$ $[\tilde{B} \geq B]$;
(iii) $\psi(B^2) = \psi(B)$ $[\tilde{B}^2 = \tilde{B}]$.

Proof

(i) $\psi(B) = B^{p-1} = B^p = \cdots = B^{p^2}$, $\psi(\psi(B)) = \psi(B^p) = (B^p)^p = B^{p^2} = B^p = \psi(B)$.
(ii) $B \leq B^{p-1} = \psi(B)$.
(iii) $\psi(B^2) = (B^2)^{p-1} = B^{2p} = \cdots = B^p = B^{p-p} = \psi(B)$. Q.E.D.

Theorem 5.4.2
Let A and B be $p \times p$ fuzzy matrices and let I be a $p \times p$ unit matrix. If $C = A + I$ and $D = B + I$, then

$$\widetilde{C + D} = \widetilde{\tilde{C}\tilde{D}} = \widetilde{\tilde{D}\tilde{C}}.$$

Proof
$\tilde{C} \geq C$ and $\tilde{D} \geq I$ imply $\tilde{C}\tilde{D} \geq C$. Similarly $\tilde{C}\tilde{D} \geq D$ and therefore $\tilde{C}\tilde{D} \geq C + D$, which implies $\widetilde{\tilde{C}\tilde{D}} \geq \widetilde{C + D}$. However, $\tilde{C} \leq \widetilde{C + D}$ and $\tilde{D} \leq \widetilde{C + D}$ imply $\tilde{C}\tilde{D} \leq \widetilde{C + D}$. The result follows immediately since $\widetilde{\tilde{C}\tilde{D}} \leq \widetilde{\widetilde{C + D}} = \widetilde{C + D}$. Q.E.D.

Theorem 5.4.3
Let A and B be $p \times p$ fuzzy matrices and let $C = A + I$ and $D = B + I$, when I is the $p \times p$ unit matrix. Then

$$C \leq \tilde{D} \quad \text{iff } \tilde{C} \leq \tilde{D}.$$

Proof
$C \leq \tilde{C}$ implies that if $\tilde{C} \leq \tilde{D}$, then $C \leq \tilde{D}$. $C \leq \tilde{D}$ implies that $\tilde{C} \leq \tilde{\tilde{D}} = \tilde{D}$. Q.E.D.

For $n \times n$ fuzzy matrices $R = [r_{ij}]$ and $P = [p_{ij}]$ with their elements having values in the unit interval [0, 1], we define

$$R \ominus P = \left[r_{ij} \ominus p_{ij}\right],$$

where

$$r_{ij} \ominus p_{ij} \triangleq \begin{cases} r_{ij} & \text{if } r_{ij} > p_{ij}, \\ 0 & \text{otherwise.} \end{cases}$$

We deal only with $n \times n$ fuzzy matrices. A matrix R is called *transitive* if $R^2 \leq R$. This matrix represents a fuzzy transitive relation. The above definition of transitivity is equivalent to what is called max-min transitivity, that is, matrix $R = [r_{ij}]$ is transitive if and only if

$$\min(r_{ik}, r_{kj}) \leq r_{ij} \quad \text{for all } k.$$

This definition is most basic and seems to be convenient when fuzzy matrices are generalized to certain matrices over other algebras.

We have already considered convergence of powers of a matrix R such that $R \leq R^2$, that is,

$$\min(r_{ik}, r_{kj}) \geq r_{ij} \qquad \text{for some } k.$$

If $R \leq R^2$, then we have $R \leq R^2 \leq R^3 \leq \cdots$. The matrix R such that $R \leq R^2$ is called *compact*. A fuzzy transitive relation is important, as well as a compact relation. We examine convergence of powers of a transitive matrix, that is, a matrix R such that $R \geq R^2$. If R is a transitive matrix, then we have $R \geq R^2 \geq R^3 \geq \cdots$. In the sequence of powers of a matrix R, if $R^k = R^{k+1}$ for some positive integer k, then R is called *convergent*.

Clearly, if R is transitive or compact, then R is convergent because of its monotonicity.

The definition of the operation $\ominus$ differs from ordinary definitions, but there are some cases where this operation is convenient. When we reduce fuzzy matrices, fuzzy graphs, or systems represented by them, the operation $\ominus$ is essential. For example, if R is an $n \times n$ matrix such that $R^n = 0$ (0 is the zero matrix), then

$$R \ominus (R^2 + R^3 + \cdots + R^{n-1}) = R \ominus (R \times R^+)$$

is considered to be a fuzzy transitive reduction of R, where

$$R^+ = R + R^2 + \cdots + R^n$$

is the transitive closure. This fact shows the importance of the operation $\ominus$. Other examples of applications of $\ominus$ are fuzzy models of information retrieval. It is suitable to employ the operation $\ominus$ when we extend the reduction operation of Boolean systems to that of fuzzy systems.

Theorem 5.4.4

If R is an $n \times n$ transitive matrix, then

$$(R \ominus (R \times P))^n = (R \ominus (R \times P))^{n+1}$$

for any $n \times n$ matrix P.

The proof of the above theorem is quite simple and will be omitted here.

■ ***Example 5.4.2***

Let

$$R = \begin{bmatrix} 0.7 & 0.4 & 0.5 \\ 0 & 0.2 & 0.3 \\ 0 & 0 & 0 \end{bmatrix} \quad \text{and} \quad P = \begin{bmatrix} 0.4 & 0 & 0.7 \\ 0.5 & 0.3 & 0 \\ 0 & 0.2 & 0.2 \end{bmatrix}.$$

Then

$$R^2 = \begin{bmatrix} 0.7 & 0.4 & 0.5 \\ 0 & 0.2 & 0.3 \\ 0 & 0 & 0 \end{bmatrix} \times \begin{bmatrix} 0.7 & 0.4 & 0.5 \\ 0 & 0.2 & 0.3 \\ 0 & 0 & 0 \end{bmatrix} = \begin{bmatrix} 0.7 & 0.4 & 0.5 \\ 0 & 0.2 & 0.2 \\ 0 & 0 & 0 \end{bmatrix} \leq R,$$

which means that R is transitive.

Next we compute $S = R \ominus (R \times P)$, S^2, and S^3. We have:

$$R \times P = \begin{bmatrix} 0.7 & 0.4 & 0.5 \\ 0 & 0.2 & 0.3 \\ 0 & 0 & 0 \end{bmatrix} \times \begin{bmatrix} 0.4 & 0 & 0.7 \\ 0.5 & 0.3 & 0 \\ 0 & 0.2 & 0.2 \end{bmatrix} = \begin{bmatrix} 0.4 & 0.3 & 0.7 \\ 0.2 & 0.2 & 0.2 \\ 0 & 0 & 0 \end{bmatrix},$$

$$S = R \ominus (R \times P) = \begin{bmatrix} 0.7 & 0.4 & 0 \\ 0 & 0 & 0.3 \\ 0 & 0 & 0 \end{bmatrix},$$

$$S^2 = \begin{bmatrix} 0.7 & 0.4 & 0 \\ 0 & 0 & 0.3 \\ 0 & 0 & 0 \end{bmatrix} \times \begin{bmatrix} 0.7 & 0.4 & 0 \\ 0 & 0 & 0.3 \\ 0 & 0 & 0 \end{bmatrix} = \begin{bmatrix} 0.7 & 0.4 & 0.3 \\ 0 & 0 & 0 \\ 0 & 0 & 0 \end{bmatrix},$$

$$S^3 = \begin{bmatrix} 0.7 & 0.4 & 0.3 \\ 0 & 0 & 0 \\ 0 & 0 & 0 \end{bmatrix} \times \begin{bmatrix} 0.7 & 0.4 & 0 \\ 0 & 0 & 0.3 \\ 0 & 0 & 0 \end{bmatrix} = \begin{bmatrix} 0.7 & 0.4 & 0.3 \\ 0 & 0 & 0 \\ 0 & 0 & 0 \end{bmatrix} = S^2.$$

Thus we have $S^3 = S^4$. ■

Clearly from the above theorem we can obtain the following two corollaries.

Corollary 5.4.2
If R is an $n \times n$ transitive matrix, then

$$(R \ominus (P \times R))^n = (R \ominus (P \times R))^{n+1}$$

for any $n \times n$ matrix P.

Corollary 5.4.3
If R is an $n \times n$ transitive matrix, then $R^n = R^{n+1}$.

We now consider conditions under which an $n \times n$ transitive matrix R fulfills the relationship $R^{n-1} = R^n$, where $n \geq 2$.

Theorem 5.4.5
Let R be an $n \times n$ transitive matrix. If

$$\min(R, I) \leq P \leq R$$

and

$$\sum_{i=1}^{n} (r_{ij} + r_{ji}) \leq r_{jj} \text{ for some } j,$$

where + indicates max, *then $P^{n-1} = P^n$.*

■ ***Example 5.4.3***
Let

$$R = \begin{bmatrix} 0 & 0.6 & 0.7 \\ 0 & 0.6 & 0.5 \\ 0 & 0.4 & 0.5 \end{bmatrix} \quad \text{and} \quad P = \begin{bmatrix} 0 & 0.6 & 0.3 \\ 0 & 0.6 & 0.4 \\ 0 & 0 & 0.5 \end{bmatrix}.$$

Then we have

$$R^2 = \begin{bmatrix} 0 & 0.6 & 0.7 \\ 0 & 0.6 & 0.5 \\ 0 & 0.4 & 0.5 \end{bmatrix} \times \begin{bmatrix} 0 & 0.6 & 0.7 \\ 0 & 0.6 & 0.5 \\ 0 & 0.4 & 0.5 \end{bmatrix} = \begin{bmatrix} 0 & 0.6 & 0.5 \\ 0 & 0.6 & 0.5 \\ 0 & 0.4 & 0.5 \end{bmatrix} \leq R,$$

$$R^3 = \begin{bmatrix} 0 & 0.6 & 0.5 \\ 0 & 0.6 & 0.5 \\ 0 & 0.4 & 0.5 \end{bmatrix} \times \begin{bmatrix} 0 & 0.6 & 0.7 \\ 0 & 0.6 & 0.5 \\ 0 & 0.4 & 0.5 \end{bmatrix} = \begin{bmatrix} 0 & 0.6 & 0.5 \\ 0 & 0.6 & 0.5 \\ 0 & 0.4 & 0.5 \end{bmatrix} = R^2,$$

$$P^2 = \begin{bmatrix} 0 & 0.6 & 0.3 \\ 0 & 0.6 & 0.4 \\ 0 & 0 & 0.5 \end{bmatrix} \times \begin{bmatrix} 0 & 0.6 & 0.3 \\ 0 & 0.6 & 0.4 \\ 0 & 0 & 0.5 \end{bmatrix} = \begin{bmatrix} 0 & 0.6 & 0.4 \\ 0 & 0.6 & 0.4 \\ 0 & 0 & 0.5 \end{bmatrix},$$

$$P^3 = \begin{bmatrix} 0 & 0.6 & 0.4 \\ 0 & 0.6 & 0.4 \\ 0 & 0 & 0.5 \end{bmatrix} \times \begin{bmatrix} 0 & 0.6 & 0.3 \\ 0 & 0.6 & 0.4 \\ 0 & 0 & 0.5 \end{bmatrix} = \begin{bmatrix} 0 & 0.6 & 0.4 \\ 0 & 0.6 & 0.4 \\ 0 & 0 & 0.5 \end{bmatrix} = P^2. \quad \blacksquare$$

Similarly we have the following theorem.

Theorem 5.4.6

Let R be an $n \times n$ transitive matrix. If $\min(R, I) \leq P \leq R$ *and*

$$\sum_{i=1}^{n} (p_{ij} + p_{ji}) \leq p_{jj} \quad \textit{for some } j,$$

where $+ = \max$, *then* $P^{n-1} = P^n$.

As a special case of Theorems 5.4.5 and 5.4.6, we obtain the following corollary.

Corollary 5.4.4

If R is an $n \times n$ transitive matrix and

$$\sum_{i=1}^{r} (r_{ij} + r_{ji}) \leq r_{jj} \quad \textit{for some } j,$$

where Σ, $+$ *are* max *operations, then* $R^{n-1} = R^n$.

Specific applications of these results to clustering analysis will be illustrated in Chapter 7.

CHAPTER SIX

Fuzzy Logics

6.1 MULTIVALENT LOGICS

Multivalent logics considered here are indenumerably-valued logics whose truth space is the real interval $[0, 1]$. We are not concerned here with the "fuzzification" of binary and finite multivalued logical calculi. We follow excellent investigations on this subject by both Rescher [1968] and Dubois and Prade [1979b]. Let Φ be a propositional variable containing elementary propositions $P, Q, R, \ldots$ joined with logical connectives. Φ is symbolically written as $\Phi = f(P, Q, R, \ldots)$.

Four transformations can be defined on Φ:

1. *Identity*: $I(\Phi) = \Phi$;
2. *Negation*: $N(\Phi) = \neg\Phi$;
3. *Reciprocity*: $R(\Phi) = f(\neg P, \neg Q, \neg R, \ldots)$;
4. *Correlativity*: $C(\Phi) = \neg R(\Phi)$;

where $\neg$ denotes the unary connective for negation of a proposition. These transformations, for a function compositional law, have a Klein group structure whose table is given below.

	I	N	R	C
I	I	N	R	C
N	N	I	C	R
R	R	C	I	N
C	C	R	N	I

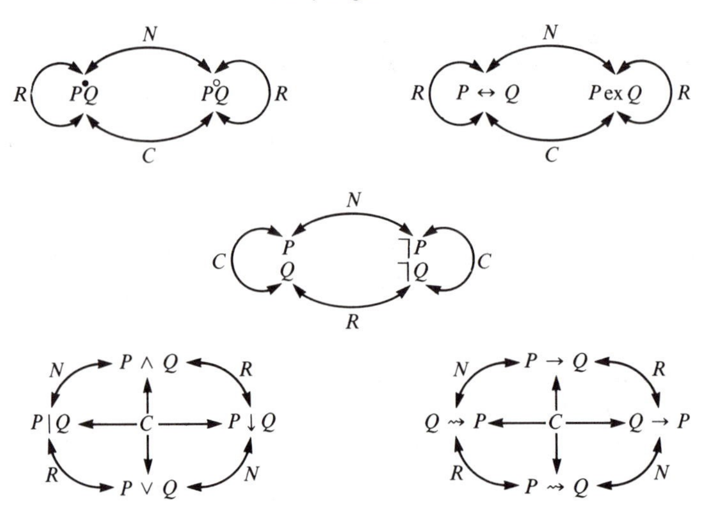

Figure 6.1.1.

The explicit connection of these transformations and binary connectives in the case of binary propositional calculus is shown in Table 6.1.1. The truth tables of the 16 standard binary connectives are given where $\cdot$ denotes tautology, $\vee$ disjunction, $\to$ implication, $\leftrightarrow$ equivalence, $\wedge$ conjunction, $\mid$ is Sheffer's connective, ex denotes exclusive disjunction, $\downarrow$ is Peirce's connective, and $\circ$ denotes contradiction. ($\rightsquigarrow$ has no common name.)

These 16 connectives are exchanged through I, R, N, and C as shown in Dubois and Prade [1979b] and in Figure 6.1.1.

In the multivalent logics underlying fuzzy-set theories we always denote by $v(P)$ the truth value of a proposition P, $v(P) \in [0, 1]$. Also, the valuation of the negation is $v(\neg P) = 1 - v(P)$. Hence, $v(\neg\neg P) = v(P)$.

Table 6.1.1

P	Q	$P \dot{} Q$	$P \vee Q$	$Q \to P$	P	$P \to Q$	Q	$P \leftrightarrow Q$	$P \wedge Q$
1	1	1	1	1	1	1	1	1	1
1	0	1	1	1	1	0	0	0	0
0	1	1	1	0	0	1	1	0	0
0	0	1	0	1	0	1	0	1	0

P	Q	$P \mid Q$	P ex Q	$\neg Q$	$Q \rightsquigarrow P$	$\neg P$	$P \rightsquigarrow Q$	$P \downarrow Q$	$P \circ Q$
1	1	0	0	0	0	0	0	0	0
1	0	1	1	1	1	0	0	0	0
0	1	1	1	0	0	1	1	0	0
0	0	1	0	1	0	1	0	1	0

The implication connective $\rightarrow$ is always defined as $v(P \rightarrow Q) = v(\neg P \vee Q)$ and the equivalence as $v(P \leftrightarrow Q) = v[(P \rightarrow Q) \wedge (Q \rightarrow P)]$; ex, $|$, $\downarrow$, and $\rightsquigarrow$ are expressed as the negation of $\leftrightarrow$, $\wedge$, $\vee$, and $\leftarrow$, respectively; the tautology and the contradiction are defined respectively as:

$$v(\dot{P}) = v(P \vee \neg P); \qquad v(\mathring{P}) = v(P \wedge \neg P).$$

More generally,

$$v(P\dot{}Q) = v((P \vee \neg P) \vee (Q \vee \neg Q));$$
$$v(P^{\circ}Q) = v((P \wedge \neg P) \wedge (Q \wedge \neg Q)).$$

Three cases should be identified:

(i) Logic Associated with ($\tilde{P}(X), \cup, \cap, {}^{-}$)

The disjunction and the conjunction underlying $\cup$ and $\cap$ are, respectively,

$$v(P \vee Q) = \max(v(P), v(Q)), \qquad v(P \wedge Q) = \min(v(P), v(Q)).$$

It is clear that $\vee$ and $\wedge$ are commutative, associative, idempotent, distributive over one another, and do not satisfy the excluded-middle laws in the sense that $v(P \vee \neg P) \neq 1$ and $v(P \wedge \neg P) \neq 0$; moreover, we have

$$v(P \vee (P \wedge Q)) = v(P); \qquad v(P \wedge (P \vee Q)) = v(P) \qquad \text{(absorption)};$$
$$v(\neg(P \wedge Q)) = v(\neg P \vee \neg Q);$$
$$v(\neg(P \vee Q)) = v(\neg P \wedge \neg Q) \qquad \text{(De Morgan)};$$
$$v[(\neg P \vee Q) \wedge (P \vee \neg Q)] = v[(P \wedge Q) \vee (\neg P \wedge \neg Q)] \qquad \text{(equivalence)};$$
$$v[(\neg P \wedge Q) \vee (P \wedge \neg Q)] = v[(P \vee Q) \wedge (\neg P \vee \neg Q)] \qquad \text{(exclusive disjunction)}.$$

Figure 6.1.2 gives the valuation of the 16 connectives that have been introduced with $v(P) = p$ and $v(Q) = q$.

Valuations for quantifiers are straightforwardly defined (coherently with $\wedge$ and $\vee$) as

$$v(\forall x P(x)) = \inf_x (v(P(x))), \qquad v(\exists x P(x)) = \sup_x (v(P(x))),$$

where x denotes an element of the universe of discourse.

This multivalent logic is usually called K-standard sequence logic (K-SEQ), first developed by Dienes. This logic is compatible with Piaget's group of transformations in the sense of Figure 6.1.1. Moreover, we have the following properties: For implication, $v[P \rightarrow (Q \rightarrow R)] = v[(P \wedge Q) \rightarrow R]$; For tautology and contradiction,

$$v(P \rightarrow P) = v(\dot{P}); \qquad v(\dot{P} \rightarrow P) = v(P);$$
$$v(P \rightarrow \dot{P}) = v(\dot{P}); \qquad v(P \leftrightarrow P) = v(\dot{P});$$
$$v(\mathring{P} \rightarrow P) = v(\dot{P}); \qquad v(P \rightarrow \mathring{P}) = v(\neg P);$$
$$v(P \leftrightarrow \neg P) = v(\mathring{P});$$

P	Q	$\dot{P}Q$	$P \vee Q$	$Q \rightarrow P$	P
p	q	$\max(p, 1-p, q, 1-q)$	$\max(p, q)$	$\max(p, 1-q)$	p

P	Q	$P \rightarrow Q$	Q	$P \leftrightarrow Q$	$P \wedge Q$
p	q	$\max(1-p, q)$	q	$\min(\max(1-p, q), \max(p, 1-q))$	$\min(p, q)$

P	Q	$P \mid Q$	$P \operatorname{ex} Q$	$\neg Q$	$Q \rightsquigarrow P$
p	q	$\max(1-p, 1-q)$	$\max(\min(1-p, q), \min(p, 1-q))$	$1-q$	$\min(p, 1-q)$

P	Q	$\neg P$	$P \rightsquigarrow Q$	$P \downarrow Q$	$\mathring{P}Q$
p	q	$1-p$	$\min(1-p, q)$	$\min(1-p, 1-q)$	$\min(p, 1-p, q, 1-q)$

Figure 6.1.2.

For Sheffer's and Peirce's connectives,

$$v(\neg P) = v(P \mid P); \quad v(P \rightarrow Q) = v(P \mid (Q \mid Q));$$
$$v(\dot{P}) = v(P \mid (P \mid P)).$$

Sheffer's connective alone (or Peirce's) is sufficient to build every binary and unary connective in standard binary logic. This result remains valid for the "extended" connectives of K-SEQ.

(ii) Logic Associated with $(\tilde{P}(X), \dot{\cup}, \dot{\cap}, ^{-})$

The disjunction and the conjunction underlying $\dot{\cup}$ and $\dot{\cap}$ are, respectively,

$$v(P \mathbin{\dot{\vee}} Q) = \min(1, v(P) + v(Q)),$$
$$v(P \mathbin{\dot{\wedge}} Q) = \max(0, v(P) + v(Q) - 1).$$

It is clear that $\dot{\vee}$ and $\dot{\wedge}$ are commutative and associative, but are not idempotent and not distributive over one another; they satisfy

$$v(\neg(P \mathbin{\dot{\wedge}} Q)) = v(\neg P \mathbin{\dot{\vee}} \neg Q);$$
$$v(\neg(P \mathbin{\dot{\vee}} Q)) = v(\neg P \mathbin{\dot{\wedge}} \neg Q) \qquad \text{(De Morgan)}$$
$$v(P \mathbin{\dot{\vee}} \neg P) = 1; \qquad v(P \mathbin{\dot{\wedge}} \neg P) = 0 \qquad \text{(excluded-middle laws)}.$$

Figure 6.1.3 gives the valuation of the 16 connectives that have been introduced ($v(P) = p$; $v(Q) = q$). (To avoid confusion, $\vee$, $\rightarrow$, $\leftrightarrow$, $\wedge$, $\mid$, ex, $\rightsquigarrow$, and $\downarrow$ are denoted in this logic $\dot{\vee}$, $\Rightarrow$, $\Leftrightarrow$, $\dot{\wedge}$, $\|$, ⓔⓧ, $\approx\!>$, $\downarrow\downarrow$.)

P	Q	$\dot{P}Q$	$P \dot{\vee} Q$	$Q \Rightarrow P$	P
p	q	1	$\min(1, p + q)$	$\min(1, p + 1 - q)$	p

P	Q	$P \Rightarrow Q$	Q	$P \Leftrightarrow Q$	$P \dot{\wedge} Q$
p	q	$\min(1, 1 - p + q)$	q	$1 - \lvert p - q \rvert$	$\max(0, p + q - 1)$

P	Q	$P \Vert Q$	$P \textcircled{ex} Q$	$\neg Q$	$Q \approx\rangle P$
p	q	$\min(1, 1 - p + 1 - q)$	$\lvert p - q \rvert$	$1 - q$	$\max(0, p - q)$

P	Q	$\neg P$	$P \approx\rangle Q$	$P \downarrow\downarrow Q$	$\mathring{P}Q$
p	q	$1 - p$	$\max(0, q - p)$	$\max(0, 1 - p - q)$	0

Figure 6.1.3.

This logic is compatible with Piaget's group of transformations in the sense of Figure 6.1.1.

Moreover, we have the following properties for tautology and contradiction:

$$v(P \Rightarrow P) = v(\dot{P}); \qquad v(\dot{P} \Rightarrow P) = v(P);$$
$$v(P \Rightarrow \dot{P}) = v(\dot{P}); \qquad v(P \Leftrightarrow P) = v(\dot{P});$$
$$v(\mathring{P} \Rightarrow \dot{P}) = v(\dot{P}); \qquad v(P \Rightarrow \mathring{P}) = v(\neg P).$$

Also,

$$v(P \rightarrow Q) = v(\neg P \dot{\vee} (P \dot{\wedge} Q))$$

and

$$v(P \Rightarrow Q) = v(\neg P \dot{\vee} (P \wedge Q)).$$

(iii) Logic Associated with $(\tilde{P}(X), \hat{+}, \cdot, ^{-})$

The disjunction and the conjunction underlying $\hat{+}$ and $\cdot$ are, respectively,

$$v(P \gamma Q) = v(P) + v(Q) - v(P) \cdot v(Q);$$
$$v(P \,\&\, Q) = v(P) \cdot v(Q).$$

It is clear that γ and & are commutative and associative, but are not idempotent and not distributive over one another; they satisfy

$$v(\neg(P \,\&\, Q)) = v(\neg P \gamma \neg Q);$$
$$v(\neg(P \gamma Q)) = v(\neg P \,\&\, \neg Q) \qquad \text{(De Morgan)}.$$

Also, with $v(P) = p$, $v(Q) = q$ we get:

$$\begin{array}{ll} \text{Implication:} & v(P \mathrel{\circ\!\!\to} Q) = 1 - p + pq; \\ \text{Tautology:} & v(\dot{P}) = 1 - p(1 - p); \\ \text{Contradiction:} & v(\mathring{P}) = p(1 - p). \end{array}$$

Note also that we have the hybrid formulas

$$v(P \mathrel{\circ\!\!\to} Q) = v(\neg P \veebar (P \& Q))$$

and

$$\begin{aligned} v[(P \mathrel{\circ\!\!\to} Q) \wedge (Q \mathrel{\circ\!\!\to} P)] &= v[(P \& Q) \veebar (\neg P \& \neg Q)] \\ &= v[(P \& Q) \Leftrightarrow (P \gamma Q)]. \end{aligned}$$

This logic is compatible with Piaget's group of transformations in the sense of Figure 6.1.1. This logic is often called *stochastic logic.*

Let us examine in what situations & coincides with $\wedge$ or $\barwedge$ (or γ with $\vee$ or $\veebar$). First, note that

$$\begin{aligned} &0 \leq \max(0, p + q - 1) \leq pq \leq \min(p, q), \\ &1 \geq \min(1, p + q) \geq p + q - pq \geq \max(p, q). \end{aligned}$$

Clearly,

$$v(P \& Q) = v(P \wedge Q);$$

that is, $pq = \min(p, q)$ iff the truth value of P or of Q is equal to 0 or to 1. Note that $v(P\gamma Q) = v(P \vee Q)$ holds under the same conditions, and it is the same for $v(P \& Q) = v(P \barwedge Q)$ and $v(P\gamma Q) = v(P \veebar Q)$. Furthermore, v is a lattice-valuation in all three cases; namely,

$$\begin{aligned} v(P) + v(Q) &= v(P \wedge Q) + v(P \vee Q) = v(P \barwedge Q) + v(P \veebar Q) \\ &= v(P \& Q) + v(P\gamma Q). \end{aligned}$$

The *modus ponens* rule allows Q to be inferred from P and $P \mapsto Q$ in propositional calculus. In multivalent logics the problem is to compute $v(Q)$ given $v(P)$ and $v(P \mapsto Q)$ where $\mapsto$ is any given multivalent implication. Several researchers have looked for a detachment operation $*$ such that

$$v(P) * v(P \mapsto Q) \leq v(Q),$$

to have $v(Q)$ as large as possible. Note that the situation is similar to probabilistic inference, where if $P(A) \geq \alpha$ and $P(B \mid A) \geq \beta$, then $P(B) \geq \alpha\beta$, since $P(B) = P(B \mid A)P(A) + P(B \mid \neg A)P(\neg A)$.

For $v(P \mapsto Q) = v(P \to Q) = \max(1 - v(P), v(Q))$, $*$ can be the min operation since we have, if $\min(v(P), v(P \to Q)) > 0.5$,

$$\min(v(P), v(P \to Q)) \leq v(Q) \leq \max(v(P), v(P \to Q)).$$

More precisely, if $v(P) \geq \alpha$ and $v(P \to Q) \geq \beta$ with $\alpha + \beta > 1$, then $v(Q) \geq \beta$. In particular, if $v(P) > \frac{1}{2}$ and $v(P \to Q) \geq \frac{1}{2}$, then $v(Q) \geq \frac{1}{2}$; but if $v(P) \geq \frac{1}{2}$ instead of $v(P) > \frac{1}{2}$, then $v(Q)$ is indeterminate. The validity of a chain of implications $\rightarrow$ when $* = \min$ is not less than the validity of the least valid element in the chain.

6.2 FUZZY MODAL LOGIC

Many researchers have used the expression "fuzzy logic" to denominate some multivalent logics, especially L_{x_1}, which underlies Zadeh fuzzy-set theory. Zadeh employs "fuzzy logic" to designate a logic on which a theory of approximate reasoning is based. However, multivalent logics may be viewed as fuzzy logics in the sense that there are no longer only crisp truth values like 0 or 1, but also intermediate ones. Lakeoff [1973a], in a brilliant exposition, generalizes this point of view when he proposes assigning to each proposition a 3-tuple (α, β, γ) such that $\alpha + \beta + \gamma = 1$, where α, β, γ are interpreted as degrees of truth, falsity, and nonsense, respectively.

He has obtained a fuzzy modal logic by adding to a set of semantic truth functions for connectives and quantifiers the following valuations for the modal operators $\Box$ and $\Diamond$.

$$v(\Box P, w) = \inf_{wRw'} v(P, w'), \qquad v(\Diamond P, w) = \sup_{wRw'} v(P, w'),$$

where w, $w' \in W$, and where $v(P, w)$ is the truth value of P in the world w, and R is an alternativeness (or accessibility) reflexive relation between the "possible worlds." We indicate that W is the set of "possible worlds." Note that the valuations are coherent with the identity $\Box P = \neg\Diamond\neg P$, $(v(\neg P, w) = 1 - v(P, w))$. We interpret $v(\Box P, w)$ as the degree of necessary truth of P in w; $v(\Box P, w) = \alpha$ means that the truth value of P never falls below α in any world alternative to w. Lakeoff gives the following example of a statement that is necessarily true to a degree: "Approximately half of the prime numbers are of the form $4N + 1$."

In a very interesting paper Schotch [1975] has applied fuzzy set theory to modal logic in the following way: Let us consider the relational model consisting of a binary relation R on W (intuitively the set of possible worlds) and a valuation V that assigns to each elementary proposition P the set of worlds in which P is true. By definition, we have

$$V(\neg P) = W - V(P); \qquad V(P \wedge Q) = V(P) \cap V(Q);$$
$$V(\Diamond P) = \{w \in W, wRw' \text{ and } w' \in V(P)\};$$
$$V(\Box P) = \{w \in W, wRw' \text{ implies } w' \in V(P)\},$$

which is coherent with $\Box P = \neg\Diamond\neg P$.

This model can be fuzzified in two ways: using fuzzy valuations and/or fuzzy relations.

A fuzzy valuation $\tilde{V}$ assigns to each elementary proposition P the fuzzy set $\tilde{V}(P)$ of worlds in which P is more or less true. Then $\chi_{\tilde{V}(P)}(w)$ is the degree to which P is true in the world $w \in W$. Now $\tilde{V}$ is extended by

$$\tilde{V}(\neg P) = \overline{\tilde{V}(P)} \qquad (\chi_{\tilde{V}(\neg P)}(w) = 1 - \chi_{\tilde{V}(P)}(w)),$$
$$\tilde{V}(P \wedge Q) = \tilde{V}(P) \cap \tilde{V}(Q) \qquad (\chi_{\tilde{V}(P \wedge Q)}(w)) = \min(\chi_{\tilde{V}(P)}(w), \chi_{\tilde{V}(Q)}(w)).$$

Also, $\chi_{\tilde{V}(\Diamond P)}$ is the two-valued characteristic function of the set

$$\{w \in W, wRw' \quad \text{and} \quad \chi_{\tilde{V}(P)}(w') \neq 0\}$$

and $\chi_{\tilde{V}(\Box P)}$ is the two-valued characteristic function of the set

$$\{w \in W, wRw' \quad \text{implies} \quad \chi_{\tilde{V}(P)}(w') = 1\}.$$

If R is a fuzzy relation, $V(\Diamond P)$ is defined as

$$V(\Diamond P) = \{w \in W, \chi_R(w, w') = 1 \qquad \text{and} \qquad w' \in V(P)\}.$$

Another modal operation, M, may be defined as

$$V(MP) = \{w \in W, \chi_R(w, w') \neq 0 \qquad \text{and} \qquad w' \in V(P)\},$$

where MP means "It might be possible that P." Then we have

$$\Diamond P \mapsto MP; \qquad \neg M \neg P \mapsto \Box P.$$

Note that $MP \neq \Diamond\Diamond P$.

6.3 FUZZY-VALUED LOGICS

A fuzzy-valued logic is a many-valued logic where the truth space is the set of the fuzzy numbers (i.e., convex normalized piecewise-continuous fuzzy sets) on the real interval $[0, 1]$; that is, the *truth value of a proposition* is a fuzzy number whose support is included in $[0, 1]$. Such fuzzy numbers may model linguistic truth values whose names are "true," "very true," "borderline," "false," etc.

In order to develop the concept of a fuzzy-valued logic we will again use the extension principle introduced by L. A. Zadeh [1975a] which provides a general method for extending nonfuzzy mathematical concepts in order to deal with fuzzy and imprecise structures.

Let X be a Cartesian product of universes, $X = X_1 \times \cdots \times X_r$, and $A_1, \ldots, A_r$ be r fuzzy sets in $X_1, \ldots, X_r$, respectively. The *Cartesian product* of $A_1, \ldots, A_r$ is defined as

$$A_1 \times \cdots \times A_r = \int_{X_1 \times \cdots \times X_r} \min(\chi_{A_1}(x_1), \ldots, \chi_{A_r}(x_r))/(x_1, \ldots, x_r).$$

Let f be a mapping from $X_1 \times \cdots \times X_r$ to a universe Y such that

$$y = f(x_1, \ldots, x_r).$$

The extension principle allows us to induce from r fuzzy sets A_i a fuzzy set B

on Y through f such that

$$\chi_B(y) = \sup_{\substack{x_1, \ldots, x_r \\ y = f(x_1, \ldots, x_r)}} \min\left(\chi_{A_1}(x_2), \ldots, \chi_{A_r}(x_r)\right),$$

and

$$\chi_B(y) = 0 \quad \text{if } f^{-1}(y) = \emptyset,$$

where $f^{-1}(y)$ is the inverse image of y, and $\chi_B(y)$ is the greatest among the membership values $\chi_{A_1 \times \cdots \times A_r}(x_1, \ldots, x_r)$ of the realizations of y using r-tuples $(x_1, \ldots, x_r)$.

The extended operations on fuzzy sets on $\mathbb{R}$ are defined as follows: Let $*$ be a binary operation in $\mathbb{R}$. We say that $*$ is *increasing* iff:

$$\text{If } x_1 > y_1 \quad \text{and} \quad x_2 > y_2, \qquad \text{then } x_1 * x_2 > y_1 * y_2.$$

In the same way, $*$ is said to be *decreasing* iff $x_1 > y_1$ and $x_2 > y_2$ imply $x_1 * x_2 < y_1 * y_2$.

Using the extension principle, $*$ can be extended into $\circledast$ to combine two fuzzy numbers (i.e., convex and normalized fuzzy sets in $\mathbb{R}$) M and N. Moreover, χ_M and χ_N are assumed to be continuous functions of $\mathbb{R}$;

$$\chi_{M \circledast N}(z) = \sup_{z = x * y} \left\{\min\left[\chi_M(x), \chi_N(y)\right]\right\}.$$

Clearly, if $*$ is commutative, so is $\circledast$, and if $*$ is associative, so is $\circledast$.

Distributivity of $\circledast$ *over* $\cup$ is defined by the following:

$$\forall (M, N, P) \in [\tilde{P}(\mathbb{R})]^3, \qquad M \circledast (N \cup P) = \left(M \circledast N\right) \cup \left(M \circledast P\right).$$

On the contrary, $\cup$ is not distributive over $\circledast$, and $\circledast$ is not distributive over $\cap$.

Unary Operations

Let ϕ be a unary operation; the extension principle implies that

$$\forall M \in \tilde{P}(\mathbb{R}), \qquad \chi_{\phi(M)}(z) = \sup_{\substack{x \\ z = \phi(x)}} \chi_M(x).$$

Opposite of a Fuzzy Number

$\phi(x) = -x$. Also $\phi(M)$ is denoted by $-M$ and is such that

$$\forall x \in \mathbb{R}, \qquad \chi_{-M}(x) = \chi_M(-x).$$

Here M and $-M$ are symmetrical with respect to the axis $x = 0$.

Inverse of a Fuzzy Number

$\phi(x) = 1/x$. Also $\phi(M)$ is denoted by M^{-1} and is such that

$$\forall x \in \mathbb{R} - \{0\}, \qquad \chi_{M^{-1}}(x) = \chi_M\left(\frac{1}{x}\right).$$

Let us call a fuzzy number M *positive* (*negative*) if its membership function is such that $\chi_M(x) = 0$, $\forall x < 0$ ($\forall x > 0$). This is denoted $M > 0$ ($M < 0$).

Scalar Multiplication

$$\chi_{\lambda \cdot M}(x) = \chi_M(x/\lambda), \qquad \forall \lambda \in \mathbb{R} - \{0\}.$$

Exponential of a Fuzzy Number

$\phi(x) = e^x$. $\phi(M)$ is denoted e^M and is given by

$$\chi_{e^{M(x)}} = \begin{cases} \chi_M(\ln x), & x > 0, \\ 0, & \text{otherwise}, \end{cases}$$

where e^M is a positive fuzzy number. Moreover, $e^{-M} = (e^M)^{-1}$.

Absolute Value of a Fuzzy Number

The absolute value of M is denoted abs(M);

$$\text{abs}(M) = \begin{cases} M \cup (-M) & \text{on } \mathbb{R}^+, \\ 0 & \text{on } \mathbb{R}^-. \end{cases}$$

Clearly, abs(M) is positive.

Extended Operations—Addition

$\oplus$ denotes the extended sum. To illustrate the noncommutativity of support discretization and extended operations, let us use the following example:

$$\left(\frac{0.5}{4} + \frac{1}{5} + \frac{0.5}{6}\right) \oplus \left(\frac{0.5}{1} + \frac{1}{2} + \frac{0.5}{3}\right)$$
$$= \left(\frac{0.5}{5} + \frac{1}{7} + \frac{0.5}{9}\right)$$
$$\neq \left(\frac{0.5}{5} + \frac{0.5}{6} + \frac{1}{7} + \frac{0.5}{8} + \frac{0.5}{9}\right),$$

where the latter was obtained by a direct application of sup-min composition, and $\oplus$ denotes the extended addition, since addition is an increasing operation. Hence, the extended addition ($\oplus$) of fuzzy numbers gives a fuzzy number. Note that $-(M \oplus N) = (-M) \oplus (-N)$. Note that $\oplus$ is commutative and associative but has no group structure. The identity of $\oplus$ is the nonfuzzy number 0. But M has no symmetrical element in the sense of a group structure. In particular, $M \oplus (-M) \neq 0, \forall M \in \tilde{P}(\mathbb{R}) - \mathbb{R}$.

Multiplication

Multiplication is an increasing operation on $\mathbb{R}^+$ and a decreasing operation on $\mathbb{R}^-$. Hence, the product of fuzzy numbers ($\odot$) that are all either positive or negative gives a positive fuzzy number. Note that $(-M) \odot N = -(M \odot N)$, so that the factors can have different signs. $\odot$ is commutative and associative. The set of positive fuzzy numbers is not a group for $\odot$: although $\forall M$, $M \odot 1 = M$, the product $M \odot M^{-1} \neq 1$ as soon as M is not a real number. Moreover, M has no inverse in the sense of group structure.

Clearly,

$$\forall (M, N) \in [\tilde{P}(\mathbb{R})]^2, \qquad (M \odot N)^{-1} = (M^{-1}) \odot (N^{-1}).$$

Also, if M is either a positive or a negative fuzzy number, and N and P are *together* either positive or negative fuzzy numbers, then

$$M \odot (N \oplus P) = (M \odot N) \oplus (M \odot P).$$

Also,

$$e^M \odot e^N = e^{M \oplus N}, \qquad \forall (M, N) \in [\tilde{P}(\mathbb{R})]^2.$$

This is obvious since e^{x+y} is an increasing binary operation.

Subtraction ($\ominus$)

Subtraction is neither increasing nor decreasing. However, it is clear that $M \ominus N = M \oplus (-N), \forall (M, N) \in [\tilde{P}(\mathbb{R})]^2$, so that $M \ominus N$ is a fuzzy number whenever M and N are.

Division ($\oslash$)

Division is neither increasing nor decreasing. But since

$$M \oslash N = M \odot (N^{-1}), \forall (M, N) \in [\tilde{P}(\mathbb{R}^+) \cup \tilde{P}(\mathbb{R}^-)]^2,$$

$M \oslash N$ is a fuzzy number when M and N are positive or negative fuzzy numbers. The division of ordinary fuzzy numbers can be performed similarly to multiplication, by decomposition.

Max and Min

Max and min are increasing operations in $\mathbb{R}$. The maximum (minimum) of k fuzzy numbers $M_1, \ldots, M_k$, denoted $\widetilde{\max}(M_1, \ldots, M_k)(\widetilde{\min}(M_1, \ldots, M_k))$, is a fuzzy number.

Let M, N, P be three fuzzy numbers, then the following properties hold: $\widetilde{\max}$ and $\widetilde{\min}$ are commutative and associative operations. They are mutually distributive,

$$\widetilde{\min}\left(M, \widetilde{\max}(N, P)\right) = \widetilde{\max}\left[\widetilde{\min}(M, N), \widetilde{\min}(M, P)\right],$$

$$\widetilde{\max}\left(M, \widetilde{\min}(N, P)\right) = \widetilde{\min}\left[\widetilde{\max}(M, N), \widetilde{\max}(M, P)\right].$$

They obey the absorption laws,

$$\widetilde{\max}\left(M, \widetilde{\min}(M, N)\right) = M, \qquad \widetilde{\min}\left(M, \widetilde{\max}(M, N)\right) = M;$$

and the De Morgan laws,

$$1 \ominus \widetilde{\min}(M, N) = \widetilde{\max}(1 \ominus M, 1 \ominus N),$$

$$1 \ominus \widetilde{\max}(M, N) = \widetilde{\min}(1 \ominus M, 1 \ominus N).$$

Note that $1 \ominus M$ is the "dual" of M: indeed, $1 \ominus (1 \ominus M) = M$. With regard to idempotence,

$$\widetilde{\max}(M, M) = M = \widetilde{\min}(M, M);$$

$$M \oplus \widetilde{\max}(N, P) = \widetilde{\max}(M \oplus N, M \oplus P); \text{ the same with } \widetilde{\min};$$

and

$$\widetilde{\max}(M, N) \oplus \widetilde{\min}(M, N) = M \oplus N.$$

Thus, $(\tilde{\pi}(\mathbb{R}), \widetilde{\max}, \widetilde{\min})$ is a noncomplemented distributive lattice, where $\tilde{\pi}(\mathbb{R})$ denotes the set of real fuzzy numbers. Going back to the truth value of propositions, we get the semantic truth functions for the connectives of negation, conjunction, and disjunction:

$$\tilde{v}(\neg P) = 1 \ominus \tilde{v}(P), \qquad \tilde{v}(P \wedge Q) = \widetilde{\min}(\tilde{v}(P), \tilde{v}(Q)),$$

$$\tilde{v}(P \vee Q) = \widetilde{\max}(\tilde{v}(P), \tilde{v}(Q)),$$

where $\tilde{v}(P)$ is a fuzzy number on $[0,1]$, and $\tilde{v}(\neg P)$ is generally called the *antonym of* $\tilde{v}(P)$. Thus, "false" will be defined as the antonym of "true."

Other semantic truth functions for different connectives can be extended, by means of the extension principle, in a similar way. For instance,

$$\tilde{v}(P \rightarrow Q) = \widetilde{\max}(1 \ominus \tilde{v}(P), \tilde{v}(Q));$$

$$\tilde{v}(P \Rightarrow Q) = \widetilde{\min}(1, 1 \oplus \tilde{v}(P) \ominus \tilde{v}(Q));$$

$$\tilde{v}(P \veebar Q) = \widetilde{\min}(1, \tilde{v}(P) \oplus \tilde{v}(Q));$$

$$\tilde{v}(P \Leftrightarrow Q) = \operatorname{abs}(\tilde{v}(P) \ominus \tilde{v}(Q)).$$

It is clear that with these extended valuations, $\vee$ and $\wedge$ are still commutative, associative, idempotent, and mutually distributive, and that they satisfy the absorption and De Morgan laws; negation is still involutive; and

$$\tilde{v}(P \wedge Q) \oplus \tilde{v}(P \vee Q) = \tilde{v}(P) \oplus \tilde{v}(Q),$$

provided that $\forall P$, Q, $\tilde{v}(P)$, $\tilde{v}(Q)$ are fuzzy numbers, i.e., convex and normalized. However,

$$\tilde{v}[(P \wedge Q) \vee (\neg P \wedge \neg Q)] \neq \tilde{v}[(\neg P \vee Q) \wedge (P \vee \neg Q)],$$

$$\tilde{v}[(P \vee Q) \wedge (\neg P \vee \neg Q)] \neq \tilde{v}[(\neg P \wedge Q) \vee (P \wedge \neg Q)].$$

Clearly, $\widetilde{\max}$, $\widetilde{\min}$, and extended negation $1 \ominus (\cdot)$ are used to evaluate composite propositions whose elementary propositions are fuzzily valued, whereas the fuzzy-set operations max, min, and negation ($\vee$, $\wedge$, $^{-}$) do create linguistic truth values from the values of given membership functions.

6.4 THE THEORY OF APPROXIMATE REASONING AND LINGUISTIC APPROXIMATION

We will now discuss the basic concepts of the theory of approximate reasoning following Zadeh [1983a, 1983b, 1983c, 1983d].

Let P and Q be two propositions and let Π_P and Π_Q be the possibility distributions induced by P and Q. P and Q are said to be *semantically equivalent* if $\Pi_P = \Pi_Q$, which is denoted by $P \Leftrightarrow Q$. This definition could be weakened by means of approximate equalities or similarities, relations discussed in Chapter 5.

While the concept of semantic equivalence relates to the equality of possibility distributions, that of *semantic entailment* relates to inclusion. More specifically, denoting that P *semantically entails* Q by $P \Rightarrow Q$, we have

$$P \Rightarrow Q \qquad \text{iff } \Pi_P \subseteq \Pi_Q.$$

We will use the concept of a *translation rule* to express a set of rules that yield the translation of a modified composite proposition from the translations of its constituents. We will now illustrate four types of such rules:

Type 1: Modifier Rules for Simple Propositions

Given the proposition $P \equiv$ "X is A" such that $\Pi_X = \chi_A$, find $\Pi'_X = \chi_{A^*}$ related to "X is ηA" where η is a modifier such as "not," "very," . . . ; A^* is given by $A^* = \Pi' A$. Each modifier is related to a function f such that $\chi_{A^*} = f \circ \chi_A$.

The generic term *fuzzy quantifier* is employed in order to denote the collection of quantifiers in natural languages whose representative elements are: *several*, *most*, *much*, *not many*, *very many*, *not very many*, *few*, *quite a few*, *large number*, *small number*, *close to five*, *approximately ten*, *frequently*, etc. Such quantifiers may be treated as fuzzy numbers which may be manipulated through the use of fuzzy arithmetic AND, and more generally, fuzzy logic.

For example, "NOT" will be expressed by

$$f(x) = 1 - x,$$

while "VERY" is given by $f(x) = x^2$.

Type 2: Composition Rules

Composition rules pertain to the translation of a proposition P that is a composition of propositions Q and R, such as conjunction, disjunction, implication. For instance:

$$\text{"If } X \text{ is } A\text{, then } Y \text{ is } B\text{"} \rightarrow \Pi_{(Y|X)} = \chi_{c(\bar{A}) \cup c(B)}$$

if we employ for the implication

$$v(Q \Rightarrow R) = v(\neg Q \vee R).$$

Here $\chi_{c(\bar{A}) \cup c(B)}(t, t') = \min(1, 1 - \chi_A(t) + \chi_B(t'))$ where c denotes the cylindrical extension and $t \in T$, $t' \in T'$; T and T' are the respective universes of A and B.

Type 3: Quantification Rules

These rules work on propositions of the form $P \equiv$"ηX are A" where η is a fuzzy quantifier. A fuzzy quantifier may be viewed as a fuzzy characterization of absolute or relative cardinality. Thus, in the proposition $p \triangleq QA$'s *are* B's, where Q is a fuzzy quantifier and A and B are labels of fuzzy or nonfuzzy sets, Q may be interpreted as a fuzzy characterization of the relative cardinality of B in A. The fuzzy set A will be referred to as the *base set*.

When both A and B are nonfuzzy sets, the relative cardinality of B in A is a real number and Q is its possibility distribution. The same is true if A and/or B are fuzzy sets and the sigma-count is employed to define the relative cardinality. The situation becomes more complicated, however, if an FCount is employed for this purpose, since Q, then, is the possibility distribution of a conjunctive fuzzy number.

To encompass these cases, we shall assume that the following propositions are semantically equivalent:

$$\textit{There are } QA\textit{'s} \leftrightarrow \text{Count}(A) \textit{ is } Q,$$

$$QA\textit{'s are } B\textit{'s} \leftrightarrow \text{Prop}(B/A) \textit{ is } Q,$$

where the more specific term *Proportion* (or Prop, for short) is used in place of Count to underscore that Prop(B/A) is the relative cardinality of B in A, with the understanding that both Count and Prop may be fuzzy or nonfuzzy counts. In the sequel, we shall assume for simplicity that, except where stated to the contrary, both absolute and relative cardinalities are defined via the sigma-count.

The equation above may be translated into possibility assignment equations:

$$\text{Count}(A) \textit{ is } Q \to \Pi_{\text{Count}(A)} = Q$$

and

$$\text{Prop}(B/A) \textit{ is } Q \to \Pi_{\text{Prop}(B/A)} = Q;$$

in which $\Pi_{\text{Count}(A)}$ and $\Pi_{\text{Prop}(B/A)}$ represent the possibility distributions of Count(A) and Prop(A/B), respectively. Furthermore, we have

$$\textit{There are } QA\textit{'s} \to \Pi_{\text{Count}(A)} = Q,$$

$$QA\textit{'s are } B\textit{'s} \to \Pi_{\text{Prop}(B/A)} = Q.$$

These translation rules provide a basis for deriving a variety of syllogisms for propositions containing fuzzy quantifiers, an instance of which is the *intersection/product* syllogism given by

$$\frac{\begin{array}{c} Q_1 A\textit{'s are } B\textit{'s} \\ Q_2(A \textit{ and } B)\textit{'s are } C\textit{'s}, \end{array}}{Q_1 \otimes Q_2\ A\textit{'s are } (B \textit{ and } C)\textit{'s},}$$

in which Q_1, Q_2, A, B, and C are assumed to be fuzzy, as in

$$\frac{\begin{array}{c} \textit{Most tall men are heavy,} \\ \textit{Many tall and heavy men are bald,} \end{array}}{\textit{Most} \otimes \textit{many tall men are heavy and bald.}}$$

To establish the validity of syllogisms of this form, we shall rely, in the main, on the semantic entailment principle and on the special case of this principle which will be referred to as the *quantifier extension principle.*

Stated in brief, the semantic entailment principle asserts that a proposition p entails proposition q, which we shali express as $p \to q$ or

$$\frac{p}{q},$$

iff the possibility distribution that is induced by p,

$$\Pi^p_{(X_1,\dots,X_n)},$$

is contained in the possibility distribution induced by q,

$$\Pi^q_{(X_1,\dots,X_n)}.$$

Thus stated in terms of possibility distribution functions of Π^p and Π^q, we have

$$\frac{p}{q} \text{ iff } \Pi^p_{(X_1,\dots,X_n)} \le \Pi^q_{(X_1,\dots,X_n)}$$

for all points in the domain of Π^p and Π^q.

Informally, this means that p entails q only if q is less specific than p. For example, the proposition

$$p \triangleq \textit{Don is } 32 \textit{ years old}$$

entails the proposition

$$q \triangleq \textit{Don is in his early thirties},$$

because q is less specific than p, which in turn is a consequence of the containment of the nonfuzzy set "32" in the fuzzy set "*early thirties.*"

It should be noted that the inequality of possibilities may be expressed as a corresponding inequality of overall test scores. Thus τ^p and τ^q are some test scores associated with p and q, respectively, then

$$\frac{p}{q} \text{ iff } \tau^p \leq \tau^q.$$

We shall be concerned, for the most part, with an entailment relation between a collection of propositions $p_1, \ldots, p_n$ and a proposition q that is entailed by the collection. Under the assumption that the propositions that constitute the premises are noninteractive, the statement of the entailment principle becomes:

$$\begin{array}{c} p_1 \\ \vdots \\ p_n \\ \hline q \end{array} \quad \text{iff } \pi^{p_1} \wedge \cdots \wedge \pi^{p_n} \leq \pi^q$$

where $\pi^{p_1}, \ldots, \pi^{p_n}$, π^q, are the possibility distribution functions induced by $p_1, \ldots, p_n$, q, respectively.

We are now in a position to formulate an important special case of the entailment principle which will be referred to as the *quantifier extension principle.*

Specifically, assume that each of the propositions $p_1, \ldots, p_n$ is a fuzzy characterization of an absolute or relative cardinality which may be expressed as $p_i \triangleq C_i$ is Q_i, $i = 1, \ldots, n$, in which C_i is a count and Q_i is a fuzzy quantifier; that is,

$$p_i \triangleq \Sigma\text{Count}(B/A) \textit{ is } Q_i$$

or, more concretely,

$$p_i \triangleq \textit{Most A's are B's}.$$

Now, in general, a syllogism involving fuzzy quantifiers has the form of a collection of premises of the form $p_i \triangleq C_i$ is Q_i, $i = 1, \ldots, n$, followed by a conclusion of the same form, that is, $q \triangleq C$ is Q, where C is a count that is related to $C_i, \ldots, C_n$, and Q is the fuzzy quantifier that is related to $Q_i, \ldots, Q_n$. The quantifier extension principle makes these relations explicit, as represented in the following inference scheme:

$$\begin{array}{c} C_1 \textit{ is } Q_1, \\ \vdots \\ C_n \textit{ is } Q_n, \\ \hline C \textit{ is } Q, \end{array}$$

where Q is given by

$$\textit{If } C = g(C_1, \ldots, C_n), \textit{ then } Q = g(Q_1, \ldots, Q_n),$$

in which g is a function that expresses the relation between C and the C_i, and the meaning of $Q = g(Q_1, \ldots, Q_n)$ is defined by the extension principle.

A somewhat more general version of the quantifier extension principle which can also be readily deduced from the extension principle is the following:

$$\begin{array}{c} C_1 \textit{ is } Q_1, \\ \vdots \\ C_n \textit{ is } Q_n, \\ \hline C \textit{ is } Q, \end{array}$$

where Q is given by

$$\textit{If } f(C_1, \ldots, C_n) \leq C \leq g(C_1, \ldots, C_n),$$
$$\textit{then } f(Q_1, \ldots, Q_n) \leq Q \leq g(Q_1, \ldots, Q_n).$$

The meaning of the inequalities that bound Q is defined by the extension principle. In more concrete terms, these inequalities imply that Q is a fuzzy interval which may be expressed as

$$Q = (\geq f(Q_1, \ldots, Q_n)) \cap (\leq g(Q_1, \ldots, Q_n)),$$

where the fuzzy s-number $\geq f(Q_1, \ldots, Q_n)$ and the fuzzy z-number $\leq g(Q_1, \ldots, Q_n)$ should be read as "at least $f(Q_1, \ldots, Q_n)$" and "at most $g(Q_1, \ldots, Q_n)$," respectively, and are the composition of the binary relations $\geq$ and $\leq$ with $f(Q_1, \ldots, Q_n)$ and $g(Q_1, \ldots, Q_n)$.

The relation between C and Q may be expressed as:

$$\textit{If } f(C_1, \ldots, C_n) \leq C \leq g(C_1, \ldots, C_n),$$
$$\textit{then } Q = (\geq f(Q_1, \ldots, Q_n)) \cap (\leq g(Q_1, \ldots, Q_n)).$$

We are now in a position to apply the quantifier extension principle. Let

$$Q_1 A\textit{'s are } B\textit{'s} \leftrightarrow \text{Prop}(B/A) \textit{ is } Q_1,$$

$$Q_2(A \textit{ and } B)\textit{'s are } C\textit{'s} \leftrightarrow \text{Prop}(C/A \cap B) \textit{ is } Q_2,$$

and

$$QA\textit{'s are } (B \textit{ and } C)\textit{'s} \rightarrow \text{Prop}(B \cap C/A) \textit{ is } Q,$$

where

$$\text{Prop}(B/A) = \frac{\Sigma\,\text{Count}(B \cap A)}{\Sigma\,\text{Count}(A)},$$

$$\text{Prop}(C/A \cap B) = \frac{\Sigma\,\text{Count}(A \cap B \cap C)}{\Sigma\,\text{Count}(A \cap B)},$$

$$\text{Prop}(B \cap C/A) = \frac{\Sigma\,\text{Count}(A \cap B \cap C)}{\Sigma\,\text{Count}(A)}.$$

Thus the relative counts $K_1 \triangleq \text{Prop}(B/A)$, $K_2 \triangleq \text{Prop}(C/A \cap B)$, and $K \triangleq \text{Prop}(B \cap C/A)$ satisfy the identity

$$\text{Prop}(B \cap C/A) = \text{Prop}(B/A)\text{Prop}(C/A \cap B),$$

and hence

$$K = K_1 K_2.$$

We can also see that Q_1, Q_2, and Q are the respective possibility distributions of K_1, K_2, and K. Consequently, from the quantifier extension principle applied to arithmetic expressions, it follows that the fuzzy quantifier Q is the fuzzy product of the fuzzy quantifiers Q_1 and Q_2 given by

$$Q = Q_1 \otimes Q_2.$$

We can also deduce the following syllogism:

$$\begin{array}{c} Q_1 A\text{'s are } B\text{'s}, \\ \underline{Q_2(A \text{ and } B)\text{'s are } C\text{'s},} \\ (\geq (Q_1 \otimes Q_2))\ A\text{'s are } C\text{'s}, \end{array}$$

where the quantifier $(\geq (Q_1 \otimes Q_2))$, which represents the composition of the unary relation $\geq$ with the binary relation $Q_1 \otimes Q_2$, should be read as *at least* $(Q_1 \otimes Q_2)$. This is true due to the inequality

$$\Sigma\,\text{Count}(B \cap C) \leq \Sigma\,\text{Count}(C),$$

which holds for all fuzzy or nonfuzzy B and C. Thus, the syllogism may be represented compactly in the form:

$$\begin{array}{c} Q_1 A\text{'s are } B\text{'s}, \\ \underline{Q_2(A \cap B)\text{'s are } C\text{'s},} \\ \underline{(Q_1 \otimes Q_2) A\text{'s are } (B \text{ and } C)\text{'s}} \\ (\geq (Q_1 \otimes Q_2)) A\text{'s are } C\text{'s}. \end{array}$$

Type 4: Qualification Rules

(4i) Linguistic truth qualification,
(4ii) Linguistic probability qualification,
(4iii) Linguistic possibility qualification.

(4i) Truth Qualification

A truth-qualified version of a proposition such as "X is A" is a proposition expressed as "X is A is τ," where τ is a linguistic truth value. We must not confuse τ with the linguistic truth value of a proposition in a fuzzy-valued logic. Here τ is a *local* linguistic truth-value rather than an absolute one: τ is defined as the degree of compatibility of the proposition "X is A" with a reference proposition "X is R":

$$\chi_\tau(z) = \sup_{z=\chi_A(t)} \chi_R(t).$$

We want to find R from the knowledge of A and τ; the greatest R is

$$\chi_R = \chi_\tau \circ \chi_A.$$

The translation of the proposition "X is A is τ" is thus the possibility distribution induced by R.

(4ii) Probability Qualification

A probability-qualified version of a proposition such as "X is A" is a proposition expressed as "X is A is λ" where λ is a linguistic probability value such as "likely," "very likely,".... This may be interpreted as "$P(A)$ is λ" where A is viewed as a fuzzy event whose probability is $P(A)$. Thus,

$$P(A) = \int_T \chi_A(t) p(t)\, dt,$$

where p is a probability distribution. Since $P(A)$ is fuzzily restricted by λ, the probability distribution is fuzzily restricted by the possibility distribution

$$\pi(P) = \chi_\lambda\left(\int_T \chi_A(t) p(t)\, dt\right).$$

(4iii) Possibility Qualifications

A possibility-qualified version of a proposition such as "X is A" is a proposition expressed "X is A is β" where β is a linguistic possibility value such as "possible," "very possible",.... Here β is viewed as a fuzzy restriction on the nonfuzzy possibility values $\Pi(A)$ of the fuzzy event A. Recalling that

$$\Pi(A) = \sup_{t\in T} \min(\chi_A(t), \text{pi}(t))$$

where pi$(\cdot)$ is the possibility distribution associated with the possibility measure $\Pi(\cdot)$, we get

$$\pi(\text{pi}) = \chi_\beta\left\{\sup_{t\in T} \min[\chi_A(t), \text{pi}(t)]\right\}.$$

Under these four types the general modification rule is that, if η is a modifier, given proposition P, then ηP is semantically equivalent to the proposition that results from applying η to the possibility distribution induced by P.

(i) *Simple propositions.* η("X is A") $\Leftrightarrow$ "X is ηA" which is exactly a Type-1 translation rule. Examples:

$$\eta = \text{“not”}, \qquad \chi_{\eta A} = 1 - \chi_A,$$

$$\eta = \text{“very”}, \qquad \chi_{\eta A} = \chi_A^2.$$

(ii) *Composed propositions.* η("X is A and Y is B") $\Leftrightarrow$ (X, Y) is $\eta(A \times B)$. Examples:

not("X is A and Y is B") $\Leftrightarrow$ "X is not A or Y is not B",

very("X is A and Y is B") $\Leftrightarrow$ "X is very A and Y is very B".

(iii) *Quantified propositions.* η("βX are A") $\Leftrightarrow$ "$(\eta\beta)X$ are A." Example: η = "not". This formula can be employed here to generalize the standard negation rule in predicate calculus:

$$\neg(\forall x)P(x) \Leftrightarrow (\exists x)\neg P(x).$$

To see this connection, we first assert the semantic equivalence

"βX are A" $\Leftrightarrow$ "(ant β) X are not A"

where ant β is the antonym $1 \ominus \beta$ of β, since

$$\pi(\rho) = \chi_F\left(\int_T \chi_A(s)\rho(s)\,ds\right) = \chi_F\left(1 - \int_T (1 - \chi_A(s))\rho(s)\,ds\right)$$

as

$$\int_T \rho(s)\,ds = 1.$$

Thus we have

"(not β) X are A" $\Leftrightarrow$ "ant(not β) X are not A"

for which β = "all" gives

"not all X are A" $\Leftrightarrow$ "some X are not A"

with "some" defined as ant(not "all"), meaning "at least some".

(iv) *Truth-qualified propositions*

η("X is A is τ") $\Leftrightarrow$ "X is A is $\eta\tau$".

Example: η = "not", τ = "true",

not("X is A is true") $\Leftrightarrow$ "X is A is not true".

On the other hand, we have

"X is not A is τ" $\Leftrightarrow$ "X is A is ant τ";

where if τ = "true", ant τ = "false".

(v) *Possibility-qualified propositions*

not(X is A is fully-possible) ⇔ "X is A is impossible";

very(X is A is fully-possible) ⇔ "X is very A is fully-possible".

6.5 GENERAL CLASS OF FUZZY CONNECTIVES

A significant body of literature has been published regarding appropriate definitions of union and intersection of fuzzy sets. In what follows we adopt Yager's [1980c] generalization of the class of fuzzy-set connectives.

Let A and B be fuzzy subsets of X with membership grades in the unit interval [0, 1]. In a very interesting and original paper, Yager [1980b] defines a general class of fuzzy intersections to be:

$$A \cap_p B = C,$$

in which

$$C_p(x) = 1 - \min\left\{1, \left[(1 - A(x))^p + (1 - B(x))^p\right]^{1/p}\right\} \quad \text{for } p \geqq 1.$$

It is obvious from its definition that this operation is defined pointwise. In order to justify that this is a generalization of the fuzzy intersection, one can show that Zadeh's original definition of fuzzy intersection, as minimization, is a special case of the above.

Theorem 6.5.1

If $p = \infty$, then

$$C_p(x) = \min[A(x), B(x)].$$

We also note that, when $p = 1$, we get

$$C_p(x) = 1 - \min[1, 2 - (A(x) + B(x))] = \max[0, A(x) + B(x) - 1].$$

Our definition collapses, for all $p \geq 1$, to the ordinary definition of intersection when the grades of membership lie in the set $\{0, 1\}$.

Theorem 6.5.2

If A and B are crisp sets and $\cap$ is the usual set intersection, then if $A(x)$ and $B(x) \in \{0,1\}$ for all x,

$$A \cap B = A \cap_p B \qquad \text{for all } p \geq 1.$$

Theorem 6.5.3

For all $A(x)$, $B(x) \in [0,1]$,

$$C_p(x) \in [0,1].$$

Thus $C_p = A \cap_p B$ is always a fuzzy subset of X.

It is obvious from the definition of $C_p(x)$ that this form of intersection is commutative.

Theorem 6.5.4

$$A \cap_p B = B \cap_p A.$$

We can show that this definition leads to a form of intersection that is associative for all p.

Theorem 6.5.5

$$A \cap_p (B \cap_p C) = (A \cap_p B) \cap_p C = A \cap_p B \cap_p C.$$

Theorem 6.5.6
If $((1 - A(x))^p + (1 - B(x))^p)^{1/p} \geq 1$, *then*

$$A(x) \cap_p B(x) = 0.$$

If $((1 - A(x))^p + (1 - B(x))^p)^{1/p} < 1$, *then*

$$A(x) \cap_p B(x) = 1 - \left[(1 - A(x))^p + (1 - B(x))^p\right]^{1/p}.$$

This theorem indicates that the smaller $A(x)$ and $B(x)$, the smaller $C_p(x)$.
An immediate corollary of this theorem is:

Corollary 6.5.1
If at least one of the terms $A(x)$ *or* $B(x)$ *is zero, then* $A(x) \cap_p B(x) = 0$.

Theorem 6.5.7
$C_p(x)$ *is a monotonically nondecreasing function of* $A(x)$ *and* $B(x)$.

It is obvious that $C_p(x)$ is a continuous function in both arguments since all the operations involved are continuous.

From the associativity and commutative property, if $A, B, C, \ldots,$ are a finite collection of fuzzy subsets, their intersection is the same no matter in what order we connect them.

Theorem 6.5.8
Assume that A *is an arbitrary fuzzy subset of* X *and* $\emptyset$ *is the null subset of* X; *that is,* $\emptyset(x) = 0$ *for* $x \in X$; *then*

$$A \cap_p \emptyset = \emptyset \quad \text{for all} \quad p \geq 1.$$

Thus, the intersection of any set with the null set is the null set.

Theorem 6.5.9
If $B(x) = 1$, *then*

$$A(x) \cap_p B(x) = A(x) \cap_p 1 = 1 \cap_p A(x) = A(x).$$

An immediate result of this theorem is that the intersection of any fuzzy set with X is the original fuzzy set.

Theorem 6.5.10
If A is an arbitrary fuzzy subset of X, then

$$A \cap_p X = A.$$

Theorem 6.5.11

$$A(x) \cap_p A(x) = 1 - \min\left[1, 2^{1/p}(1 - A(x))\right].$$

Corollary 6.5.2
For any arbitrary $A(x)$,

$$A(x) \cap_p A(x) \leq A(x).$$

This result implies that, as we keep on intersecting or "anding" equal fuzzy subsets, the membership grades are nonincreasing.

Corollary 6.5.3
If $p' < p''$, then

$$A(x) \cap_{p'} A(x) \leq A(x) \cap_{p''} A(x).$$

Corollary 6.5.4:

$$A(x) \cap_\infty A(x) = A(x).$$

Thus, idempotency is a property of only $p = \infty$. Therefore $\cap_p$ can only be an operation in a lattice for $p = \infty$.

More generally, one can prove the following results.

Theorem 6.5.12

$$A(x) \cap_p B(x) \leq \min(A(x), B(x)).$$

Theorem 6.5.13

$$A(x) \cap_p B(x) \text{ is an increasing function of } p.$$

Thus $C_p(x)$ assumes its smallest value for $p = 1$ and its largest value for $p = \infty$.

Theorem 6.5.14
If $A(x) \cap_p (1 - A(x)) = C_p(x)$, then

1. $p = 1 \Rightarrow C_p(x) = 0$ *for all $A(x)$; or*
2. $A(x) = 0$ *or* $A(x) = 1$ *implies* $C_p(x) = 0$, *for $p \geq 1$; or*
3. $p \neq 1$ *and* $A(x) \neq 1$ *or* 0, $C_p(x) > 0$.

This theorem then implies that, if we define negation as $\bar{A} = 1 - A$, then the law of contradiction, that is,

$$\bar{A} \cap A = \emptyset,$$

holds only if $p = 1$ or if A is crisp.

Theorem 6.5.15

The general fuzzy intersection operation satisfies the law of contradiction for crisp sets or, if $p = 1$, for all sets.

In particular, if A and B are two fuzzy subsets of X, then De Morgan's law requires that

$$\overline{A \cup_p B} = \bar{A} \cap_p \bar{B},$$

where bar indicates negation. We shall use Zadeh's definition of negation, $\bar{A} = 1 - A$.

Using the pointwise definition of intersection we get

$$\bar{A}(x) \cap_p \bar{B}(x) = 1 - \min\left[1, (A^p(x) + B^p(x))^{1/p}\right].$$

Furthermore, since

$$\overline{A \cup_p B} = 1 - (A \cup_p B),$$

we get

$$1 - (A(x) \cup_p B(x)) = 1 - \min\left[1, (A^p(x) + B^p(x))^{1/p}\right];$$

and finally, the general definition of union

$$D_p(x) = A(x) \cup_p B(x) = \min\left[1, [(A(x))^p + (B(x))^p]^{1/p}\right],$$

where $p \geq 1$.

We see that for $p = \infty$, this reduces to Zadeh's definition of union.

Theorem 6.5.16

If $p = \infty$, $D_p(x) = \max[A(x), B(x)]$.

Note here that, if $p = 1$,

$$D_p(x) = \min[1, A(x) + B(x)].$$

Theorem 6.5.17

If A and B are crisp sets, then the general fuzzy union collapses to the ordinary union for all p.

The general fuzzy union always generates fuzzy sets.

Theorem 6.5.18

If $A(x)$ and $B(x) \in [0,1]$, then

$$D_p(x) \in [0,1].$$

Commutativity and associativity can also be easily proved for the union.

Theorem 6.5.19

$$A \cup_p B = B \cup_p A.$$

Theorem 6.5.20

$$A \cup_p (B \cup_p C) = (A \cup_p B) \cup_p C.$$

The following theorems can easily be proved.

Theorem 6.5.21
If $(A^p(x) + B^p(x)) \geq 1$, *then* $A(x) \cup_p B(x) = 1$ *and if* $(A^p(x) + B^p(x)) < 1$, *then*

$$A(x) \cup_p B(x) = (A^p(x) + B^p(x))^{1/p}.$$

Corollary 6.5.5
If at least one term $A(x)$ *or* $B(x)$ *is one, then* $A(x) \cup_p B(x) = 1$.

Theorem 6.5.22
$D_p(x)$ *is a monotonically nondecreasing function of* $A(x)$ *and* $B(x)$.

Theorem 6.5.23
If A *is an arbitrary fuzzy subset of* X, *then, for any* p:

$$A \cup_p X = X.$$

Theorem 6.5.24
If $A(x) = 0$, *then* $A(x) \cup_p B(x) = B(x)$.

Theorem 6.5.25
If A *is an arbitrary fuzzy subset of* X *and* $\emptyset$ *the null set, then*

$$A \cup_p \emptyset = A.$$

Theorem 6.5.26

$$A(x) \cup_p A(x) = \min[1, 2^{1/p} A(x)],$$

and

$$A(x) \cup_p A(x) \geq A(x).$$

Corollary 6.5.6
For an arbitrary $A(x)$,

$$A(x) \cup_p A(x) \geq A(x).$$

Corollary 6.5.7
If $p' < p''$,

$$A(x) \cup_{p'} A(x) \geq A(x) \cup_{p''} A(x).$$

Corollary 6.5.8

$$A(x) \cup_{\infty} A(x) = A(x).$$

Thus, idempotency again is a property of $p = \infty$.

Theorem 6.5.27

$$A(x) \cup_p B(x) \geq \max[A(x), B(x)].$$

Theorem 6.5.28

$A(x) \cup_p B(x)$ is a monotonically decreasing function of p.

Thus, $A(x) \cup_p B(x)$ assumes its largest value:

$$\min[1, A(x) + B(x)] \quad \text{at } p = 1,$$

and its smallest value

$$\max[A(x), B(x)] \quad \text{at } p = \infty.$$

Theorem 6.5.29

If $A(x) \cup_p (1 - A(x)) = D_p(x)$, then

1. if $p = 1$, $D_p(x) = 1$;
2. *if* $A(x) = 1$ *or* 0, $D_p(x) = 1$;
3. *if* $p \neq 1$ *and* $A(x) \neq 1$ *or* 0, $D_p(x) < 1$.

It should be noted that one nice property that this set of connectives is lacking is distributivity between union and intersection for all p. It can be shown that distributivity holds only for $p = \infty$ or crisp sets. However, it should be noted that the multiplication-value definition of intersection,

$$A \cdot B = C,$$

where

$$C(x) = a \cdot b,$$

and of union,

$$A \cup B = C,$$
$$C(x) = a + b - ab,$$

is also not distributive. Furthermore, idempotency, which is also not satisfied in this multiplication, is available only when $p = \infty$.

The question naturally arises as to the significance of the selection of the parameter p. Yager [1980b] suggests one possible interpretation of its significance.

Note first that, by the logical statement "S_1 and S_2", we are requiring or demanding the simultaneous satisfaction of both conditions S_1 and S_2. It is a common phenomenon in spoken language to strongly emphasize the "and" when we are demanding strong satisfaction to these two conditions. Thus, by the "strength" of an "and," we shall mean how strongly we are demanding this simultaneous satisfaction. We suggest that in the "and" operator we can consider the parameter p as inversely related to the strength of the "and." That is, p is a measure of how strongly we demand this simultaneous satisfaction; p then is an inverse measure of how strong we mean "and" in the statement S_1 "and" S_2. Note that it was shown that $a \cap_p b$ is a monotonically nondecreasing function of p. Thus, for a fixed a and b as p decreases, in our terms the strength of the "and" increases; then the measure of simultaneous satisfaction C_p decreases.

The "or" connective has a similar type of interpretation. Noting that by the statement "S_1 or S_2" we are allowing for an interchangeable or collective satisfaction to S_1 and S_2. We can consider that p is a measure of the degree of interchangeability allowed. The higher the p, the less interchangeability allowed. That is, when $p = \infty$, we are suggesting that this corresponds to the exclusive, "either $\cdots$ or $\cdots$" type "or". Whereas, when $p = 1$, we allow complete interchange. Thus, when $p = 1$, we have a very soft "hard or". This interpretation is supported by the fact that, for a fixed a and b,

$$D_p = a \cup_p b$$

is a nonincreasing function of p.

We further note the connection between the "and" and "or" is such that the stronger the "and" the more interchangeable the associated "or." That is, when $p = 1$, we have the strongest most demanding "and" and the least interchangeable "or." This interpretation would be supportive of Zadeh's suggestion that min and max, $p = \infty$, be the *default selection* in the face of no further information.

CHAPTER SEVEN

Some Applications

7.1 MANIPULATION OF KNOWLEDGE IN FUZZY EXPERT SYSTEMS VIA FUZZY DECISION TABLES

One of the underlying principles of software engineering is that the methodology for solving a problem is based on techniques that are application-independent. Among other things, this methodology should be well defined and should facilitate straightforward documentation and traceability. One tool that lends itself well to these objectives of software engineering and is application-independent is a *decision table*.

Decision tables (DTs) can be used in all phases of software engineering from system planning, through the software design process, down to software maintenance. In the system-planning stage, the amount of communication that is necessary is extremely high. Without a manageable and usable way to keep track of the information as it develops, misunderstandings, omissions, inconsistencies as well as redundancies, and out-and-out errors will soon overtake any sizeable project. If the information can be represented in a DT, then omissions, inconsistencies, and redundancies can be checked for automatically. Also, just storing the information formally in a DT will prevent misunderstandings. When maintaining software, any DTs that were used to design the software serve also as documentation. Since these DTs can be easily changed, they can also be easily maintained in parallel with the software, keeping the documentation up to date.

The fundamental problems that DTs are concerned with are those of understanding, analyzing, designing, and describing complex procedures. Although these are the same problems encountered in software design, the use of

Condition stub	Condition entry
Action stub	Action entry

Figure 7.1.1. Parts of a decision table.

DTs in this area has not been very widespread. One reason for this is that there is no natural way to incorporate DTs in bottom-up programming. So, although the concept of using DTs for programming appeared in the literature in the early 1960s, the programming practices of the day could not use them to best advantage. Today, however, the standard method of software design is top-down, and in top-down design, DTs have a natural place.

Very simply, a DT is a special form of a table that determines a set of decision rules based on a clearly identified set of conditions and resulting actions.

DTs are made up of four major parts: the condition stub and entry, and the action stub and entry. Figure 7.1.1 shows the relative positions of these sections.

The condition stub contains a row for each condition to be evaluated. Similarly, the action stub has a row for each action. The condition and action entry sections are divided into columns called rules. Each column specifies values for certain conditions and the resulting action to be taken when those conditions meet the specified values. A simple DT to determine which sort to use is given in Figure 7.1.2.

	1	2	3	4	5	6
No. of records $\leq$ 10	T	T	F	F	F	F
Records have $\leq$ 3 fields	T	–	–	–	–	–
Records have > 3 fields	–	T	–	–	–	–
Records have > 100 fields	–	F	T	–	–	–
Alphabetizing	–	–	–	T	F	F
Recursion available	–	–	–	–	F	T
Call Insertion Sort	X					
Call Selection Sort		X				
Call Heapsort					X	
Call Quicksort						X
Call Bucket Sort				X		
Sort array of pointers to records			X			
GO AGAIN			X			
EXIT	X	X		X	X	X

Figure 7.1.2. Decision table to sort N records.

In this DT, according to Rule 5, if the number of records is greater than 10, you do not want to alphabetize, and recursion is not available, then call the Heapsort procedure to sort the records. If, however, all the above is true except that recursion is available, then Rule 6 tells us to call the Quicksort procedure. In either case, the size of the records is not a consideration.

In a top-down design, DTs can be set up for each level of the design. The top level table would be very general and would call other tables for more specific processing. Besides the traditional type of condition and action entries of a DT, there should also be specific exit actions for each rule of a table. We can define four permanent exit actions and one temporary exit action as follows:

Permanent: 1) RETURN
2) GO TO (table name)
3) GO AGAIN
4) STOP
Temporary: 1) CALL (table name)

Unlike the other permanent exit actions, the GO AGAIN exit does not really cause processing to leave a table. Rather it is used to facilitate the loop concept. Therefore in this section, we will consider the GO AGAIN to be a temporary exit action like the CALL (table name) exit action. And since these exits are only temporary, they will both be treated as nonexit actions in the table.

In keeping with the general principles of structured programming, the following restrictions will be put on the use of DTs.

1. Any table that is entered by a CALL from another table must have a RETURN as its only exit.
2. GO TO exits should be avoided at all costs.
3. The top-level DT must have a STOP exit.

Another reason why DTs have seen such limited use in software engineering is their completely deterministic nature. In each rule, the values of any condition must be precisely one of a given set of possible values. The most used set of possible values is {True, False}, although it is perfectly legal for the sets to have more than two members; but even with multivalued sets, the values must still be precisely defined. In the system-planning phase of software engineering, almost nothing is precise in the beginning. Conditions do not really get to be strictly deterministic until the coding of the software. If DTs are to be of any use before this stage of a project, they must be able to handle imprecise conditions.

Some previous work has been done along these lines, using the concepts of fuzzy set theory, by Francioni and Kandel [1983]. DTs were modified so that all conditions were defined as fuzzy variables. The rules then became combinations of these variables represented as fuzzy switching functions. Although this

	1	2	3	4	5	6
No. of records	Low	Low	High	High	High	High
Size of records	Small	Med to large	Very large	–	–	–
Alphabetizing	–	–	–	T	F	F
Recursion available	–	–	–	–	F	T
Call Insertion Sort	X					
Call Selection Sort		X				
Call Heapsort					X	
Call Quicksort						X
Call Bucket Sort				X		
Sort array of pointers to records			X			
GO AGAIN			X			
EXIT	X	X		X	X	X

Figure 7.1.3. Nondeterministic decision table to sort N records.

technique can handle imprecise conditions, the method of processing the table is somewhat restrictive. In this section, we will deal with a more general, imprecise DT.

Let us define then a nondeterministic decision table (NDT) as a decision table whose condition entries may be nondeterministic. In Figure 7.1.3 we have modified the DT of Figure 7.1.2 as an example of an NDT. In this NDT, we have generalized the first condition of Figure 7.1.2 and have combined the second, third, and fourth conditions. The table is now more representative of how people think, rather than being based on how the computer works.

In order to get a feel for how sensitive an NDT is to the subjective evaluation of the user, we define a Sensitivity Index (SI) as follows:

$$SI = \frac{NC * NR}{\Sigma M * TR}$$

where NC is the number of nondeterministic conditions; NR is the number of rules dealing with nondeterministic conditions; M is the moduli* (number of possible values) of a condition, and TR is the total number of rules in the NDT.

This gives $SI \in [0, 1]$. An $SI = 0$ implies that the NDT is completely deterministic. An $SI = 1$ implies that the NDT is completely dependent on the user's evaluation of each condition. For the NDT in Figure 7.1.3,

$$SI = \frac{2 * 6}{(1 + 1 + 2 + 2) * 6} = \frac{12}{36} = 0.33.$$

*For nondeterministic conditions, $M = 1$.

Note that, even though half of the conditions are nondeterministic, the sensitivity of the NDT is only equal to a third. Since the possible values for a nondeterministic condition are infinite, we assign a moduli equal to 1 to represent one range of values, as opposed to the distinct values of a deterministic condition. For example, the set of possible values for the first condition is $[0, \infty]$, whereas the set of possible values for the third condition is $\{True, False\}$. The first set has only one member but the second set has two, and therefore carries more weight in judging the sensitivity of the NDT.

In order to further demonstrate the use of NDTs in the top-down design of software, we present a simplified stocks and bonds problem.

A narrative description of the problem is given below*:

1. We decide to spend the market day trading in either stocks or bonds or nothing, depending on market conditions and our finances. If the stocks are low (Dow Jones Industrial Average [DJIA] less than 800), we will pick stocks for the day; if bonds are low (Dow Jones Bond Average [DJBA] less than 80), we will pick bonds; and if they are both low, we will compare ten times the DJBA to the DJIA to see which is the better buy of the day. We buy nothing if both markets are high.
2. Our selections will be governed by information in either *Standard & Poor's Bond Guide* or *Standard & Poor's Stock Reports*, plus current market prices. Whichever we choose for the day, we will progress sequentially through the *S & P* listings.
3. Our purchases will each be in $2000 or $1000 amounts per security. For the larger purchases, a bond must be rated A, have a current yield of at least 10%, and a yield to maturity of at least 15%. A stock must have a yield greater than 7% and a growth rate of at least 7%. For the smaller purchases, a grade A bond must have a current yield of at least 10%, and a grade B bond must have a current yield of at least 10% plus a yield to maturity of at least 15%. Grade C bonds will not be purchased, and no more securities will be considered once our cash reserves drop below $2000 or the market closes.
4. Smaller stock purchases will be made either if the growth rate is at least 7% or if the growth rate is at least 5%, but less than 7%, and the yield is greater than 7%. Otherwise, no purchase is made. Also, no stock is purchased if the corporation's debt exceeds its annual net income or if its current price/earnings (P/E) ratio is greater than 10.
5. Sales of all holdings of any bond will be made if the current yield is less than 10% or if the bond is rated lower than B. Sales of all holdings of any stock will be made if the yield is 7% or less, or if the debt, P/E ratio, or growth rate requirements are failed.

Although deterministic DTs can be used to model this problem, there are some

*R. B. Harley, *Decision Tables in Software Engineering*. Van Nostrand Reinhold Co., Inc., N.Y. 1983.

intrinsic nondeterministic conditions involved which are not addressed and which lend themselves quite naturally to NDTs. These nondeterministic conditions are outlined below.

1. If both stocks and bonds are low, we are told to choose the better buy. But if they are both very low, then we would want to buy both.
2. With respect to current yield and yield-to-maturity of bonds, we are given very definite cutoff points of consideration. Since these are considered together, it is possible that one would be high enough to warrant accepting the other even if it is below the stated minimum. In this case our decision to buy a bond should be a function of both these variables together rather than separately.
3. There is some confusion as to whether we should make a large or a small stock purchase when the growth rate is $\geq$ 7%. The choice should be more closely related to yield as well as growth rate.
4. When the growth rate of a stock is $\geq$ 7% and the yield is $\leq$ 7%, then there is a contradiction and confusion as to whether to make a small purchase or to sell the stock.
5. There are several instances of very similar conditions resulting in quite different actions. For example

   a) For a bond rated A, with a yield to maturity $\geq$ 15%, if the current yield is 10%, then make a large purchase; if the current yield is 9.9%, then sell all holdings.
   b) For stock with a growth rate $\geq$ 5%, if the yield is 7.1%, then make a small purchase; if the yield is 7.0%, then sell all holdings.

   What we really want to do in these cases is buy or sell in proportion to the yield, and not be restricted to an arbitrary all-or-nothing situation.
6. The purchase amounts are very restrictive; they could be made more flexible if we could consider them functions of the variables involved.

Before presenting the NDTs for this problem, we will define some conventions to be used in the tables.

1. In order to be more formal, we will consider each nondeterministic condition to be a variable $x_i \in [0, 1]$. The condition entry will be some function of the x_i's as defined below. The condition entry will then specify an interval of acceptance for the action entry. Prior to determining these intervals, the specific ranges of values for the variables must be scaled to the interval [0, 1].
2. The allowable functions of the x_i's for the condition entry are:

   x_i,
   $1 - x_i$: represented as $\bar{x}_i$;
   $\max(x_i, x_j)$: represented as $x_i + x_j$;

TRADE	R1	R2	R3	R4	M
Trading possible	Y	Y	N		2
Stock average (x1)	x1 + x2	x1x2		x1/x2	1
Bond average (x2)					1
Call (STOCKS)	≤ 0.2			> 0	
Call (BONDS)	≤ 0.2			0	
Call (ACCOUNT)	≤ 0.2			X	
Process R4		≤ 0.4			
GO AGAIN	≤ 0.2			X	
STOP		> 0.4	X		

$$SI = \frac{2*3}{4*4} = \frac{6}{16} = 0.37$$

Figure 7.1.4. TRADE NDT (Top-level).

$\min(x_i, x_j)$: represented as $x_i x_j$;
$\max(x_i, x_j) - x_j$: represented as x_i/x_j;
[This function $= 0$ if $x_j = \max(x_i, x_j)$, and
> 0 if $x_i = \max(x_i, x_j)$.],

and any combination of the above.

3. Some of the rules in the tables are specified as conditional rules. This means they should be only considered as directed by a table action. These rules are set off from the others by a dashed vertical line (see Fig. 7.1.4).
4. The rules should be processed in order from left to right. Within each rule, the deterministic conditions should be evaluated first. On a GO AGAIN action, processing of the table should be started over from the first rule.
5. If none of the conditions in a table is satisfied, then a RETURN action becomes automatic.
6. An *M* column is included showing the moduli of each condition to be used in computing the *SI* of the table.

The NDTs in Figures 7.1.4, 7.1.5, 7.1.6, and 7.1.7 represent the solution to the stocks and bonds problem with the TRADE table as the top-level table.

Rule 1 states that, if the maximum of the stock average and the bond average is ≤ 0.2 (both are very low), then we should call both STOCKS and BONDS, do the accounting, and then process the table again.

Because of the GO AGAIN action in Rule 1, Rule 2 is processed only if the conditions of Rule 1 are not met. In this case either one or both of the stock and bond averages are > 0.2. If at least one of them is ≤ 0.4 (low), then trading will be done in whichever is lower, according to Rule 4. If neither is low, then no trading should be done for the day.

ACCOUNT	R1	R2	R3	M
Market closed	Y	N	N	2
Transaction awaiting processing	–	Y	N	2
Cash balance (x1)	–	–	x1	1
Update balance		X		
Set to trading not possible	X		≤ 0.2	
GO AGAIN		X		
RETURN	X		X	

$SI = 1/15 = 0.07$

$\frac{1}{5} * \frac{1}{3} = \frac{1}{15} = 0.07$

Figure 7.1.5. ACCOUNT NDT.

According to Rule 2 of the BOND NDT, if a bond is rated A and the current yield is ≥ 10%, then we should buy stock based on the yield to maturity.

For A-rated bonds where the current yield is < 10%, rather than automatically selling them, Rule 3 is processed. In this rule, based on some proportion threshold of current yield and yield-to-maturity (0.025 in this example), the decision of how much to buy or sell is made as a function of both. The functions may evaluate to 0 in which case no transaction would take place. This is realistic, since it is sometimes better just to wait rather than to buy or sell. In any case, the rule also says to report the transaction (event) if a 0 amount is involved, turn to the next listing, and then return.

Rules 4 and 5 are similar to Rules 2 and 3, except for the different functions in the action entries. These would correspond to bonds that are rated B.

BOND	R1	R2	R3	R4	R5	M
Rating	C	A	A	B	B	3
Current yield (x1)		x1	x1x2	x1	x1x2	1
Yield to maturity (x2)						1
Holding now	Y	–	–	–	–	2
Buy: How much =		$f_1(x2)$	≥ 0.025: $f_2(x1, x2)$	≥ 0.10: $f_3(x2)$	≥ 0.025: $f_4(x1, x2)$	
Sell: How much =	X: all		< 0.025: $f_2(x1, x2)$		< 0.025: $f_2(x1, x2)$	
Report transaction		≥ 0.10	X	≥ 0.10	X	
Turn to next listing		≥ 0.10	X	≥ 0.10	X	
RETURN	X	≥ 0.10	X	≥ 0.10	X	

$$SI = \frac{2 * 4}{7 * 5} = \frac{8}{35} = 0.23$$

Figure 7.1.6. BOND NDT.

STOCKS	R1	R2	R3	R4	R5	M
Debt > Annual net	Y	N	N	N	N	2
P/E Ratio (x1)		x1				1
Growth rate (x2)			x2		x2	1
Yield (x3)				x3		1
Holding now	Y	Y	Y	Y		2
Buy: How much =					≥ 0.05: $f_4(x2, x3)$	
Sell: How much =	X: all	> 0.10: $f_1(x1)$	< 0.05: $f_2(x2)$	≤ 0.07: $f_3(x3)$		
Report transaction	X	> 0.10	< 0.05	≤ 0.07	≥ 0.05	
Turn to next listing	X	> 0.10	< 0.05	≤ 0.07	≥ 0.05	
RETURN	X	> 0.10	< 0.05	≤ 0.07	≥ 0.05	

$$SI = \frac{3*4}{7*5} = \frac{12}{35} = 0.33.$$

Figure 7.1.7. STOCKS NDT.

In Figure 7.1.7, Rules 1, 2, 3, and 4 all deal with conditions for selling stock. Except for the first rule, the amount of stock that should be sold depends on how bad the situation is, and is therefore a function of the particular variable involved.

In Rule 5, if the growth rate is high enough, then stock should be purchased on the basis of yield as well as the growth rate: higher yield implies a larger purchase.

The aim of this discussion is to demonstrate also that analysis of various kinds of fuzziness encountered in knowledge-based engineering indicates strongly that fuzzy-set theory and fuzzy decision tables are appropriate formalisms both for knowledge representation and for knowledge manipulation.

Information systems can be implemented as either hardware or software. Since an expert system is a software information system, it should be designed according to the principles of software engineering. Specifically, the information of an expert system represents a subset of human knowledge and inferences made on this knowledge base. The human factor in the information implies that the premises used for the inferences will not always be precise. This also means that the conclusions inferred from these premises will be uncertain to some degree. In order for an expert system to be useful, it must associate an actual measure of this uncertainty with each conclusion. How the system computes this measure, commonly called the *certainty factor*, defines how well it will model human knowledge.

The NDTs presented earlier can be used in the design of expert systems. They not only relate premises to conclusions necessary for the inferences, but also specify a sensitivity index which can be interpreted as a certainty factor. Their main advantage, however, is that they adhere to software engineering principles.

As shown by Zadeh [1983a], the uncertainty of an expert system is exemplified in one of the following forms:

1. The fuzziness of premises and/or conclusions;
2. A partial match between the premise of a rule and a fact supplied by the user; and
3. The presence of fuzzy quantifiers in the premise and/or conclusion of a rule.

The following discussion shows how NDTs are able to deal with these uncertainties where the premise of a rule corresponds to a condition and the conclusion corresponds to a related action.

Clearly, the knowledge base of an expert system consists of a collection of propositions which represent the facts, and a collection of conditional propositions which constitute the rules. For example, a typical rule in MYCIN is exemplified by

If: (1) The route of the administration of the penicillin is oral, and
(2) There is a gastro-intestinal factor that may interfere with the absorption of the penicillin,
then: There is suggestive evidence (0.6) that the route of administration of the penicillin is not adequate.

Typical rules in PROSPECTOR are exemplified by

(a) If: Abundant quartz sulfide veinlets with no apparent alteration halos,
(b) then: (LS, LN) alteration favorable for the potassic stage.

(a) If: Volcanic rocks in the region are contemporaneous with the intrusive system (coeval volcanic rocks),
(b) then: (LS, LN) the level of erosion is favorable for a porphyry copper deposit.

In these rules, LS and LN are real numbers representing likelihood ratios. Informally, if the ratio LS that is associated with a hypothesis H and evidence E is greater than unity, then the odds on H are increased by the presence of E. On the other hand, if LN is greater than unity, then the odds on H are increased by the absence of E. In consequence of the definitions of LS and LN, they cannot be simultaneously greater than unity.

It is our assertion that most of the facts and rules in expert systems contain fuzzy predicates and thus are fuzzy propositions. This is particularly true of the heuristic rules that are encoded as production rules in what are, in effect, fuzzy algorithms.

These fuzzy elements of the expert system are inserted into the fuzzy decision table, which now becomes the integral part of a rule-based system, for making decisions in uncertain environments.

To quote Zadeh [1983a]:

> In the existing expert system, the fuzziness of the knowledge base is ignored, because neither predicate logic nor probability-based methods provide a systematic basis for dealing with it. As a consequence, fuzzy facts and rules are generally manipulated as if they were nonfuzzy, leading to conclusions whose validity is open to question.

Fuzzy decision tables provide a natural framework for the management of uncertainty in expert systems because — in contrast to traditional systems — their main purpose is to provide a systematic basis for representing and inferring from imprecise rather than precise knowledge.

The greater expressive power of fuzzy decision tables derives from the fact that they contain as special cases the traditional two-valued as well as multivalued decision tables. The main features of fuzzy decision tables that are of relevance to the management of uncertainty in expert systems are the following:

(i) In two-valued systems a proposition, p, is either true (T) or false (F). In multivalued systems, a proposition may be true (T), or have an intermediate truth-value which may be an element of a finite or infinite truth-value set. In fuzzy decision tables, the truth-values are allowed to range over the fuzzy subsets of the infinite truth-value set.

(ii) Fuzzy decision tables provide a mechanism for representing data in both fuzzy and nonfuzzy formats and modifiers, exemplified by VERY, NOT, MUCH, MORE OR LESS, etc.

(iii) In fuzzy decision tables the predicates may be crisp (as in two-valued DT) or fuzzy. A crisp predicate, for example, is represented by terms such as EVEN, ODD, etc.; fuzzy ones may be, for example, SMALL, KIND, etc.

(iv) Two-valued as well as multivalued DTs allow only two quantifiers: all and some. By contrast, fuzzy DTs allow, in addition, the use of fuzzy quantifiers exemplified by MOST, MANY, SEVERAL, FEW, MUCH OF, FREQUENTLY, OCCASIONALLY, etc. Such quantifiers may be interpreted as fuzzy numbers, which provide an imprecise characterization of the cardinality of one or more fuzzy or nonfuzzy sets. In this perspective, a fuzzy quantifier may be viewed as a second-order fuzzy predicate. Based on this view, fuzzy quantifiers may be used to represent the meaning of propositions containing fuzzy probabilities, and thereby make it possible to manipulate probabilities within fuzzy decision tables.

(v) In two-valued systems, a proposition, p, may be qualified, principally by associating with p a truth-value, TRUE or FALSE; a modal operator such as POSSIBLE or NECESSARY: and a knowledge operator such as KNOW, BELIEVE, etc. In fuzzy decision tables, the three principal modes of qualification are: (a) truth qualification, expressed as p is τ, in which τ is a fuzzy truth-value; (b) probability qualification, expressed as p is λ, in which λ is a fuzzy probability; and (c) possibility qualification, expressed as p is π, in which π is a fuzzy possibility, e.g., QUITE POSSIBLE, ALMOST IMPOSSIBLE, etc. Furthermore, KNOWING and BELIEVING are assumed to be binary fuzzy predicates.

Now if the knowledge-based system consists of a collection of propositions, some or all of which may be fuzzy in nature, there may be unconditional–unqualified propositions, unconditional–qualified propositions, conditional–qualified propositions, or conditional–unqualified propositions. Many of the rules of the third kind are dispositions; namely, propositions with implicit fuzzy quantifiers. These dispositions play a very important role in the representation and manipulation of commonsense knowledge as well as in the inference from that kind of knowledge.

The group of propositions in the expert system may be expressed as a fuzzy relation in which the entries are fuzzy sets and the entire structure is treated as a single fuzzy decision table.

This formulation suggests that fuzzy-set theory provides a natural, conceptual framework for knowledge representation and manipulation in expert systems that have imprecise components, incomplete data, or unreliable sources of information. The system is represented by the model discussed before — that of a fuzzy decision table. Generally, the use of fuzzy decision tables gives the basic issues of knowledge representation and manipulation a more realistic approach and leads to conclusions that represent a natural way of dealing with both expert systems and decisions in uncertain environments. This results, therefore, in a more solid mathematical approach to the problem of managing uncertainty and its systematic manipulation in knowledge-based systems.

7.2 FUZZY OPTIMIZATION

The purpose of this example is to present an overall picture of the field of fuzzy optimization following the exposition of Negoita [1981a].

All published work on fuzzy optimization has appeared within the last twelve years. No more than twenty people have ever worked on it. No more than fifty papers dealing explicitly with it have ever appeared. Optimization in a fuzzy environment seems to be therefore a very small corner of operational research.

Interest in this subject stems from the fact that vague concepts like "great" or "many," widely used in decision-making and nonprobabilistic in nature, can be properly modelled using the mathematical language of fuzzy sets.

Fuzzy programming originated in an attempt to extend the applicability of classical programming models. If $X \subseteq R^n$, $X \neq \emptyset$, h: $R^n \to R$, h_i: $R^n \to R$,

$$P = \{x \in X \mid h_i(x) \leq 0, \quad i = 1, 2, \ldots, m\}$$

$$= \bigcap_{i=1}^{m} \{x \in X \mid h_i(x) \leq 0\}, \quad P \neq \emptyset;$$

then the classical problem can be formulated as: find $p^0 \in P$ such that

$$h(p^0) = \sup\{h(p) | p \in P\}.$$

•

The common example is making best use of limited resources in relation to some strict criterion of value. In this way, optimality is viewed as equivalent to efficiency, and linear programming as a technique for efficient choice.

For many decision problems, such a model is deceptive. In order to make this statement clear, we shall consider the set X as the set of alternatives and the numbers $h(x)$, $h_i(x)$ as quantifying the effects of choosing an alternative $x \in X$. In fact, in many real situations, an alternative x must be selected such that $h(x)$, $h_i(x)$ are located in some intervals b, b_i. These intervals are desirable targets. Typical examples can be found in planning procedures. Such procedures can be described as finding a point $x \in X$ subject to

$$w_b \leq h(x) \leq W_b,$$
$$w_{b_i} \leq h_i(x) \leq W_{b_i}, \quad i = 1, 2, \dots, m,$$

where w, W are bounding the intervals b.

In this problem, optimality means effectiveness (in contrast to efficiency). Fuzzy programming is obviously a technique for effective choice.

In classical terms, the first problem is not always a compatible system and, therefore, the initial decision problem has no solution. This is the reason for a reformulation, for defining a new problem having a solution with properties "as good as possible." For instance, the quality of an alternative x can be defined by the position of the numbers in the intervals b, b_i. These intervals define on the real line a preference relation; and to express this relation, one can use fuzzy sets. The procedure is as follows: One defines the fuzzy sets B, B_i: $R \to L$ such that the preorder determined by them models the preference from the real process. A necessary condition is

$$B(x) = 1 \Leftrightarrow x \in b, \qquad B_i(x) = 1 \Leftrightarrow x \in b_i.$$

There is no unique methodology for establishing these fuzzy sets, but there is no doubt that good approximations can be found in this way.

Now, consider the following fuzzy subsets of X:

$$f = B * h, \qquad f_i = B_i * h_i\colon \quad X \to L,$$

where $*$ means composition. We may state our problem as: Find x such that

$$f(x) = B(h(x)), \qquad f_i(x) = B_i(h_i(x)), \quad i = 1, 2, \dots, m,$$

are as large as possible. In this way, constraints and objectives are viewed as fuzzy subsets of the set of alternatives.

A nonempty set X and a finite number of fuzzy subsets ($f_1, f_2, \dots, f_m$: $X \to L$) is a system of symmetrical fuzzy constraints denoted ($X, f_1, f_2, \dots, f_m$). Thus, a system of inequalities in R^n is generalized by identifying every inequality with a characteristic function of the set of points $x \in R^n$ that verify it.

The first question to be answered is that of the meaning of a solution for a system of fuzzy constraints. Such a notion was introduced for the first time by Bellman and Zadeh [1970].

Definition 7.2.1

A solution for the system $(X, f_1, \ldots, f_m)$ *is any point which maximizes* $D(x) = \min(f_1(x), \ldots, f_m(x))$.

This notion, based on the intersection of fuzzy subsets, corresponds to the classical one. The problem of conjunctive aggregation of fuzzy sets has been attacked also by empirical methods and psychological theories.

Another notion can be derived as a direct consequence of the nature of the system of fuzzy constraints. Let us consider the function $f\colon X \to L^m$, $f(x) = (f_1(x), \ldots, f_m(x))$, and the order on L^m defined as

$$x = (x_1, \ldots, x_m) \le (y_1, \ldots, y_m) = y \Leftrightarrow x_i \le y_i, \quad i = 1, \ldots, m.$$

Definition 7.2.2

A solution of the system $(X, f_1, \ldots, f_m)$ *is a point* $x^0 \in X$ *which, for* $f(x^0)$, *is a maximal element of* $(f(x), \le) \subseteq (L^m, \le)$.

According to this definition, the solutions are nondominated points of the function f. If the order on L^m is

$$x \prec y \Leftrightarrow x_i = y_i \quad \text{or} \quad x_i < y_i, \qquad i = 1, 2, \ldots, m,$$

the solutions are the weak nondominated points of f.

If we accept such a definition, the set of solutions is very rich and, therefore, one must indicate a procedure for selection. Generally speaking, a selection procedure associates to $(f_1, \ldots, f_m)$ a class of functions $\psi(f_1, \ldots, f_m)$ whose maximal or minimal points are solutions for the initial problem in the sense of our definition.

The obvious conclusion is that the system of fuzzy constraints represents a kind of multi-criteria decision problem. The fuzzy approach, however, is more intuitive. This seems to be the reason for the increasing interest in the identification of fuzzy constraints.

The notion of a system of fuzzy constraints was a first step towards a generalization of the classical optimization problem.

Definition 7.2.3

An optimization problem in a fuzzy environment is a system of fuzzy constraints together with an objective function $P\colon X \to R$.

Bearing in mind the first definition, a solution for our problem would correspond to those points that maximize

$$D(x) = \min(P(x), \quad f_1(x), f_2(x), \ldots, f_m(x)).$$

However, when f_i are usual characteristic functions of subsets, the usual notion of maximal point for the function $P\colon X \to R$ is not recovered. The generalization of the classical notion can be done if P is bounded.

The natural conclusion is that the notion of a nondominated point for the vector-valued function $f\colon X \to R \times L^m$,

$$f(x) = (P(x), \quad f_1(x), \ldots, f_m(x)),$$

is very convenient for an optimization problem with fuzzy constraints. Its solution can be expressed also in the language of fuzzy sets. A system of fuzzy constraints $(X, f_1, \ldots, f_m)$ can be expressed as the pair (X, D), and the optimization problem on fuzzy sets can be defined as (X, P, D), where

$$P\colon X \to R, \quad D\colon X \to L,$$

its solution being defined by the following fuzzy subsets:

$$v\colon X \to L,$$

$$v(\bar{x}) = \begin{cases} \sup\{\alpha \neq 0 \mid \bar{x} \in M_\alpha\}, & \text{if } \bar{x} \in \bigcup M_\alpha, \\ 0, & \text{otherwise,} \end{cases}$$

where

$$M_\alpha = \left\{\bar{x} \in L_\alpha(D) \mid P(\bar{x}) = \sup_{x \in L_\alpha(D)} P(x)\right\},$$

$$L_\alpha(D) = \{x \in X \mid D(x) \geq \alpha\}$$

are compact subsets of $X \subseteq R^n$.

This notion of a solution is not different from that of a nondominated point of a vector-valued function $(P, D)\colon X \to R \times L$. Clearly, if

$$X^* = \{x \in X \mid D(x) > 0\},$$

then $x \in X^*$ is the nondominated point of $f = (P, D)$, if $v(\bar{x}) > 0$.

Motivated by the fact that, in many of the multiple-criteria decision problems, the underlying domination structures are fuzzy, one can attempt to fuzzify the concept of domination structures to allow it to be applied to a larger class of situations.

So far, the multiple-objective programming problem has been discussed as such. However, every such problem is intimately related to another single-objective programming problem. This statement would be no more than an interesting mathematical curiosity, if it were not for the fact that the reduction of a multi-criteria problem to a single-criterion one proves to be of substantial significance for the decision-maker. Illustrations of this discussion can be found in Negoita [1981a] and Kickert [1979].

Much decision-making in the real world concerns problems whose structures are not known exactly. For instance, the structure of a linear programming problem is defined by the matrix A and the vector b. There are many cases when A and b cannot be precisely given. To deal quantitatively with imprecision due to the observer, one can use the concepts and techniques of fuzzy set theory. In what follows, a possible approach to this problem will be presented and commented upon.

So far, fuzziness has been modelled by membership functions that might be described as extensions of the usual characteristic functions in the setting of mathematical sets. Bearing in mind this fact, a connection between fuzzy subsets of a set X, $F(X) = L^X$, and classical subsets of X must be found. This connection can be put into evidence by introducing the concept of level-set.

Let $F_1(X)$ be the set of all families of level-sets. It is well known that between $F_1(X)$ and $F(X)$ there is a complete lattice isomorphism. This representation suggests that the mathematics of fuzzy sets and that of sets are closely intertwined and that much is to be gained by exploring this connection.

In the following, we shall refer to the general case of the optimization problem which includes the linear one. Let

$$f\colon R^n \times R^P \to R^m$$

be a mapping and $a \in R^P$ a given vector. The set of constraints is usually defined as a system of inequalities

$$f(x, a) \leq b.$$

We shall further assume that a and b are fuzzy subsets of R^P and R^m, respectively. Any point $x \in R^n$ verifies the inequality, if

$$f(x, a) \in K_b = \{ y \in R^m \mid y \leq b \}.$$

Let A: $R^P \to [0,1]$ be a fuzzy subset of R^P and K_b: $R^m \to [0,1]$ a fuzzy subset of R^m, which describe the imprecision due to the observer, i.e., the imperfect knowledge of a and b. A natural extension of the system $f(x, a) \leq b$ can be accomplished by the inclusion of fuzzy subsets. For $x \in R^n$, let f_x: $R^P \to R^m$ be defined as $f_x(z) = f(x, z)$. The mapping

$$M\colon R^n \to F(R^m), \qquad M(x) = f_x(A), \qquad f_x(A)\colon R^m \to L \in F(R^m)$$

is the image of A by f_x. Now, (R^n, f, A, K_b) will be a system of fuzzy constraints.

Definition 7.2.4

A point $x \in R^n$ is a solution of the system (R^n, f, A, K_b), if $M(x) \subseteq K_b$.

This notion is perfectly compatible with that of a solution for a system of inequalities in R^n, that is, if A: $R^P \to \{0,1\}$ is the characteristic function of a point $a \in R^P$ and K_b: $R^m \to \{0,1\}$ is the characteristic function of a subset $\{ y \in R^m \mid y < b\}$, then a point $x \in R^n$ is a solution for the system (R^n, f, A, K_b), if and only if x verifies the inequality $f(x, a) \leq b$.

We now have a framework to guide the generalization of many classical optimization models.

Definition 7.2.5

Let P: $R^n \to R$ be an objective function and (R^n, f, A, K_b) a system of fuzzy constraints. The optimization problem is: Find $x^0 \in R^n$ such that $P(x^0) = \sup\{ P(x) \mid x \in Q\}$, where Q is the set of all solutions for the given system.

This generalization is different from the one presented before, since in this one constraints now have a special structure.

One can define a solution of a system (R^n, f, A, K_b) using the strong levels of A and K_b. Let

$$A_s = \{ y \in R^P \mid A(y) > s \} \quad \text{and} \quad K_b^s = \{ y \in R^m \mid K_b(y) > s \}.$$

Then $x \in R^n$ is a solution of the given system, if and only if, for all $s \in [0,1]$,

$$f(x, A_s) \subseteq K_b^s.$$

It is not so easy to find algorithms in order to solve this problem. Some additional hypothesis on K_b — for instance, K_b a step function — may change the problem into one having inclusion constraints.

Similar extensions can be imagined also for other optimization problems, where fuzzy parameters are supposed to be present. The presence of such parameters involves a mathematical structure such that our approach, the previous one, seems to be the forerunner of a general trend in the foundation of algorithms which will necessarily be forced to take much greater technical cognizance of theoretical results dealing with the representation.

Most of the actually classical procedures do not tackle the problem of dynamic structure. They implicitly assume that the structure is fixed. In fact, the decision-maker needs flexible or robust models able to face evolutionary situations. Fuzzy constraints of fuzzy parameters can reflect the capacity to absorb changes encountered in real life. Thus, fuzziness may be interpreted as being a measure of the complexity. In other words, fuzziness is a measure of the variety and surprise with which a manager must deal. If fuzziness increases, then the variety of a system is increasing, and hence the system describes a greater number of situations.

7.3 ANALYSIS OF TRANSIENT BEHAVIOR IN DIGITAL SYSTEMS

The problems of determining whether or not a digital system operates correctly, and of ensuring its perfect operation even when some of its elements are fuzzy, are of both practical concern and theoretical interest. We now discuss the fuzzy treatment of the transient behavior of switching systems, and then demonstrate a diagnostic procedure for detecting static hazards.

In the combinational system which demonstrates a hazardous behavior, it is possible for the output signals to behave in an unpredicted manner under certain input transitions. Namely, the system belongs to a set of ill-defined systems, some of which cannot admit any precise analysis.

Perhaps the major reason for the ineffectiveness of classical techniques in dealing with static hazard and obtaining a logical explanation of the existence of static hazard lies in their failure to come to grips with the issue of fuzziness. This is due to the fact that the hazardous variable implies imprecision in the binary system, which stems not from randomness but from a lack of sharp transition between members in the class of input states. It is the same type of imprecision that arises when one is dealing, for example, with the class of systems that are approximately linear and other systems and classes that admit the possibility of partial membership.

A transition between a pair of adjacent input combinations to the switching system, producing an identical output, contains a static hazard if a spurious pulse may be present on the output line of the system.

A technique for generating the ternary function from the binary switching system has been quite useful for detecting hazards, since the third value, designated as one-half, is used to represent a signal that may be either one or zero during the transition. However, it is our feeling that this third fixed value which represents transient states does not represent the actual signal very well. Even though it may appear incongruous to mention logic design and fuzziness in the same breath, this incongruity becomes less paradoxical if we examine the nature of hazards in combinational systems. Indeed, it is quite possible that during the transient state the input signal has a fuzzy-structure representation. Intuitively, fuzziness is a type of imprecision which stems from a grouping of elements into classes that do not have sharply defined boundaries — that is, in which there is no *sharp transition* from membership to nonmembership. Thus the transition of a state has a fuzzy behavior during the transition time, since this is a member in an ordered set of operations, some of which are fuzzy in nature; e.g., "Switch x is closed slightly at each unit of time until y approximately is open." Treating such operations in a precise mathematical way provides a means of designing combinational systems when the conventional nonfuzzy techniques become infeasible. The success of the fuzzy procedure to be described lies in its philosophical and mathematical description of the hazard in combinational systems and in extending the ternary method of detecting it.

Intuitively, a fuzzy n-variable function is a mapping on Z^n to Z, $Z = [0, 1]$, and Z^n is the set of all fuzzy vectors

$$(x_1, x_2, \ldots, x_n), \quad x_i \in Z, \quad i = 1, 2, \ldots, n.$$

Definition 7.3.1

Let $x = (x_1, x_2, \ldots, x_n)$ and $y = (y_1, y_2, \ldots, y_n)$ be two binary n-dimensional vectors. Two binary n-dimensional vectors are adjacent iff they differ in exactly one component [that is, $y = (x_1, x_2, \ldots, x_{j-1}, \bar{x}_j, x_{j+1}, \ldots, x_n)$].

Definition 7.3.2

Variable x_j is called a perfect fuzzy variable iff its grade membership $\chi_A(x_j)$ is a number in the open interval $(0, 1)$.

Definition 7.3.3

The fuzzy transmission vector $T_{x_j}^{y} \in Z^n$ is defined as the transition vector from vector x to its adjacent vector y (that is, $T_{x_j}^{y} = x$ such that x_j is a perfect fuzzy variable).

Definition 7.3.4

A B-fuzzy n-variable function $f(x)$, $x = (x_1, x_2, \ldots, x_n)$, is any fuzzy function over n variables that is either constantly one or zero, or is obtained from its arguments $x_1, x_2, \ldots, x_n$ by successive application of the fuzzy operations of max, min, *and* complement (comp).

We now present the main theorem regarding hazard detection.

Theorem 7.3.1

Let $f(x)$, $x = (x_1, x_2, \ldots, x_n)$ be a B-fuzzy n-variable function and ξ and ρ any adjacent binary n-dimensional vectors. Then

$$f(T^{\rho}_{\xi_j}) \neq \overline{f(\xi)}\,.$$

In a similar way we define now a V-fuzzy function as a fuzzy function $f(x)$ such that $f(\xi)$ is a binary function for every binary n-dimensional vector ξ. It is clear that a V-fuzzy function f induces a binary function F such that

$$F\colon \{0,1\}^n \to \{0,1\},$$

determined by $F(\xi) = f(\xi)$ for every binary n-dimensional vector ξ.

If the B-fuzzy function f describes the complete behavior of a binary combinational system, its steady-state behavior is represented by F, the binary function induced by f. Let $f(x)$ be an n-dimensional V-fuzzy function, and let ξ and ρ be adjacent binary n-dimensional vectors. The vector $T^{\rho}_{\xi_j}$ is a *static hazard* of f iff $f(\xi) = f(\rho) \neq f(T^{\rho}_{\xi_i})$.

If $f(\xi) = f(\rho) = 1$, $T^{\rho}_{\xi_i}$ is a 1-hazard.

If $f(\xi) = f(\rho) = 0$, $T^{\rho}_{\xi_j}$ is a 0-hazard.

If f is B-fuzzy and $T^{\rho}_{\xi_i}$ is a static hazard, then $f(T^{\rho}_{\xi_j})$ has a perfect fuzzy value, that is, $f(T^{\rho}_{\xi_i}) \in (0,1)$.

Definition 7.3.5

A combinational system is a static-hazard-free (SHF) system if and only if its B-fuzzy function f is SHF.

Consider the static hazard as a malfunction represented by an actual or potential deviation from the intended behavior of the system. We can detect all static hazards of the V-fuzzy function $f(x)$ by considering the following extension of Shannon normal form.

Let $f(x)$, $x = (x_1, x_2, \ldots, x_n)$, be a fuzzy function and denote the vector

$$(x_1, x_2, \ldots, x_{j-1}, x_{j+1}, \ldots, x_n) \text{ by } x^j.$$

By successive applications of the rules of fuzzy algebra, the function $f(x)$ may be expanded about, say, x_j as follows:

$$f(x) = x_j f_1(x^j) + \bar{x}_j f_2(x^j) + x_j \bar{x}_j f_3(x^j) + f_4(x^j),$$

where f_1, f_2, f_3, and f_4 are also fuzzy functions.

It is clear that the same expansion holds when the fuzzy functions are replaced by B-fuzzy functions of the same dimension.

Let ξ and ρ be two adjacent n-dimensional binary vectors that differ only in their jth component. Treating ξ_j as a perfect fuzzy variable during transition time implies that $T^{\rho}_{\xi_j}$ is a 1-hazard of f iff $f(\xi) = f(\rho) = 1$ and $f(T^{\rho}_{\xi_j}) \in [0, 1)$.

We will show now that the above conditions for the vector $T^{\rho}_{\xi_j}$ to be 1-hazard yield the following result.

Theorem 7.3.2

The vector $T^{\rho}_{\xi_j}$ is a 1-hazard of the B-fuzzy function $f(x)$ given above iff the binary vector ξ^j is a solution of the following set of Boolean equations:

$$f_1(x^j) = 1, \qquad f_2(x^j) = 1, \qquad f_4(x^j) = 0.$$

Proof

State 1: $\xi_j = 1$ *and* $\bar{\xi}_j = 0$ *imply* $f_1(\xi^j) + f_4(\xi^j) = 1$.
State 2: $\xi_j = 0$ *and* $\bar{\xi}_j = 1$ *imply* $f_2(\xi^j) + f_4(\xi^j) = 1$.
Transition state: $\xi_j \in (0,1)$ [*which implies* $\bar{\xi}_j \in (0,1)$], *and thus*

$$0 \le \max\left\{\min\left[\xi_j, f_1(\xi^j)\right], \min\left[\bar{\xi}_j, f_2(\xi^j)\right], \min\left[\xi_j, \bar{\xi}_j, f_3(\xi^j)\right], f_4(\xi^j)\right\} < 1.$$

It is clear from the transition state that $f_4(\xi^j)$ cannot be equal to one, and thus

$$f_4(\xi^j) = 0, \qquad f_1(\xi^j) = f_2(\xi^j) = 1.$$

Similarly, the vector $T^{\rho}_{\xi_j}$ is 0-hazard iff $f(\xi) = f(\rho) = 0$ and $f(T^{\rho}_{\xi_j}) \in (0, 1]$. These conditions imply the following:
State 1: $\xi_j = 1$ *and* $\bar{\xi}_j = 0$ *imply* $f_1(\xi^j) + f_4(\xi^j) = 0$.
State 2: $\xi_j = 0$ *and* $\bar{\xi}_j = 1$ *imply* $f_2(\xi^j) + f_4(\xi^j) = 0$.
Transition state: $\xi_j \in (0,1)$ *implies*

$$0 < \max\left\{\min\left[\xi_j, f_1(\xi^j)\right], \min\left[\bar{\xi}_j, f_2(\xi^j)\right], \min\left[\xi_j, \bar{\xi}_j, f_3(\xi^j)\right], f_4(\xi^j)\right\} \le 1.$$

These simultaneous conditions are equivalent to

$$f_1(\xi^j) = f_2(\xi^j) = f_4(\xi^j) = 0, \qquad f_3(\xi^j) = 1.$$

Hence the following can be stated.

Theorem 7.3.3

The vector $T^{\rho}_{\xi_j}$ is a 0-hazard of the B-fuzzy function $f(x)$ iff the binary vector ξ^j is a solution of the following set of Boolean equations:

$$f_1(x^j) = f_2(x^j) = f_4(x^j) = 0, \qquad f_3(x^j) = 1.$$

Dual procedures can be obtained if one expands the B-fuzzy function $f(x)$ by the dual Shannon form, about, say, x_i, as

$$f(x) = \left[x_i + g_1(x^i)\right]\left[\bar{x}_i + g_2(x^i)\right]\left[x_j + \bar{x}_i + g_3(x^i)\right]\left[g_4(x^i)\right].$$

In order to make practical use of the above results, one has to use some algebraic means to solve a set of simultaneous Boolean equations, such as the resolution principle.

If f is represented in disjunctive normal form (dnf), it is sufficient to convert f_1, f_2, and f_4 to conjunctive normal form (cnf) in order to detect a 1-hazard (f_1, f_2, f_3, and f_4 if detection of 0-hazard is desired). Each of these is a function of at most $(n-1)$ variables and therefore we allocate a matrix $M(i,j)$ with $(n-1)$ columns (one column for each variable) and each row of the matrix corresponds to a clause in cnf. Within a row an entry of 1, -1, or 0 represents occurrence, negated occurrence, or nonocurrence of the corresponding variable. Two rows i_1 and i_2 clash in column j if

$$M(i_1, j) \times M(i_2, j) = -1.$$

Two rows i_1 and i_2 are resolvable if there is *exactly one* column j such that i_1 and i_2 clash in column j. The generation of the resolvent row i_3, given two resolvable rows i_1 and i_2 which clash in column j, can be done as follows:

$$M(i_3, k) = \begin{cases} 0, & \text{if } k = j, \\ M(i_2, k), & \text{if } k \neq j \text{ and } M(i_2, k) = M(i_1, k), \\ M(i_2, k) + M(i_1, k), & \text{if } k \neq j \text{ and } M(i_2, k) \neq M(i_1, k). \end{cases}$$

With the generation of a row consisting entirely of zeros, the contradiction is established.

The functions $f_1, \ldots, f_4$ can be transformed under De Morgan laws. That is, if a function has a true (1) value, it is sufficient to convert it to cnf; in case it has a false (0) value, represent it by its negation.

■ ***Example 7.3.1***

Let $f(x_1, x_2, x_3, x_4) = x_1x_2x_3 + x_1\bar{x}_4 + \bar{x}_1\bar{x}_2\bar{x}_3\bar{x}_4 + \overline{\bar{x}_2\bar{x}_3x_4}$. Checking for 1-hazard generated by x_1, we get

$$f_1(x_2, x_3, x_4) = x_2x_3 + \bar{x}_4, \qquad f_2(x_2, x_3, x_4) = \bar{x}_2\bar{x}_3\bar{x}_4,$$

$$f_4(x_2, x_3, x_4) = x_2 + x_3 + \bar{x}_4$$

Under 1-hazard assumption, $f_1 = f_2 = 1$, $f_4 = 0$.

		x_2	x_3	x_4
f_1	(1)	1	0	-1
	(2)	0	1	-1
f_2	(3)	-1	0	0
	(4)	0	-1	0
	(5)	0	0	-1
f_4	(6)	-1	0	0
	(7)	0	-1	0
	(8)	0	0	1

Rows (5) and (8) are resolvable since they clash only in column x_4 and thus the resolvent row consists entirely of zeros. Therefore we have a contradiction to our assumption and the function has no 1-hazard generated by x_1. ■

7.4 MODELLING OF SYSTEMS IN UNCERTAIN ENVIRONMENTS

Many problems arising in scientific investigations generate data incorporating nonstatistical uncertainty. In such instances a fuzzy axiomatic structure for dealing with such problems usually increases both their mathematical tractability and physical realism. We now deal with the extension of analytical concepts based on real analysis to the theory of fuzzy sets. Specifically, we discuss the application of *fuzzy expectation* to the investigation of dynamical systems that are basically represented by differential equations with fuzzy coefficients. General principles can be applied to a variety of fields dealing with the problem of decision-making in an imprecise environment in both science and engineering. The problem of representing and solving such a system described by the death process due to cigarette smoking is discussed in detail. The technique illustrates that the concepts introduced may be very useful in analyzing economic, urban, social, biological, and other human-oriented systems. In all these areas there are physically interesting processes for which an average probability per unit time cannot be found even *a posteriori* without excessive amounts of effort. What is needed in such a case is an estimate of the average. Consider, for example, a system with a linear operator L and $\langle \lambda \rangle$ being the average of some parameter λ. Then, even if we have an estimate of the average, it is not generally true that the solution i to an equation of the form

$$Li + \langle \lambda \rangle i = 0$$

is equal to $\langle i \rangle$, the solution

$$Li + \lambda i = 0$$

averaged over the distribution of λ's.

As we saw, an average is a *typical* value representing a data set accumulated by an observer. Since such typical values tend to lie centrally within an arranged set of data, the average is nothing but a *measure of central tendency*. Also, it is well known that for large populations the sampling distribution of means is approximately a Gaussian distribution irrespective of the population, which is an approximated special case of the central-limit theorem in classical probability, a refinement of the Chebyshev inequality. Logically, since the FEV is a form of central tendency, the examination of data must be a crucial point in the evaluation of the FEV. However, the data set cannot be examined at a single point but over the range of t on the time axis. We have to remember that a nondeterministic statement (probability statement) is taken as making an assertion about the data set. It may be right or wrong—and it is generally held that we never really know with full certainty which it is—but it is a statement for which the evidence is chiefly observational. However, when an observer makes certain statements regarding a collection of data, his statements represent in essence his degree of belief regarding the data (distribution, etc.), and he is using this degree of belief to infer statistical results.

Philosophical arguments apart, the solution obtained by applying fuzzy statistics to the above problem has two main advantages:

1. Instead of constructing a distribution structure from the data and then proceeding to find an average value or a solution to a stochastic differential equation, the observer has to use the collection of data as it stands and infer from it the needed results via the techniques of fuzzy statistics.
2. As shown here, the process described is not only logical but very easy to apply, whereas the stochastic approaches are tedious, if at all possible, to solve.

We will now illustrate the above discussion by investigating the following problem, titled "the smoker's problem."

Let us explore the death process as a realistic model for decrease in the size of populations due to natural causes and smoking. The notion of a "smoker" is a good example of a fuzzy structure. For a better understanding of this phenomenon, assume that A is a population whose cardinality is n_s at time t. The question is: How many of them are there at time $t + \Delta t$, assuming a pure death process caused by natural death and by smoking? The Boltzman equation for the assumed stochastic process is

$$n_s(t + \Delta t) = n_s(t)[1 - p(\Delta t)],$$

where $p(\Delta t)$ is the probability that a member of A will die in time Δt. Now, all members of A will be examined with regard to their smoking habits. Clearly, all of them are also members of the set of people with a grade of membership equal to unity in that set. But "smokers" cannot be defined uniquely in terms of probabilities. Questions such as "How long has the person smoked" or "How many cigarettes has he smoked per unit time" are involved in any decisions about classification of members in the set of smokers.

Let us define a time-dependent grade of membership $\chi_s(\alpha, t)$ in the set of smokers, where $0 \leq \chi_s(\alpha, t) \leq 1$ [$\chi_s(\alpha, t) = 1$ denotes a full member and $\chi_s(\alpha, t) = 0$ denotes a nonsmoker]. Then

$$p(\Delta t) = \Delta t(\lambda_1 + \chi_s(\alpha, t)\lambda_s)$$

is the probability that a smoker will die in a time Δt, first because he is a person and people die at a rate of λ_1 per unit time, and secondly because he has a membership grade $\chi_s(\alpha, t)$ in the set of smokers who die at a rate of $\lambda_1 + \lambda_s$ per unit time. The parameter α is proportional to the number of cigarettes per unit time used by the smoker. Clearly, $\lambda_1 + \lambda_s$, which is the death rate of a "full" smoking population, is larger than λ_1, the death rate of a nonsmoking population. We shall assume also that $\chi_s(\alpha, t)$ is defined only for a finite t since after a certain t_0 (for example, $t_0 = 120$ years), the entire population is dead.

As $\Delta t \to 0$, we get

$$\frac{dn_s}{dt} = -[\lambda_1 + \chi_s(\alpha, t)\lambda_s]n_s$$

or

$$\frac{d\ln(n_s)}{dt} = -[\lambda_1 + \chi_s(\alpha, t)\lambda_s]$$

with the solution

$$n_s(t) = n_s(0)e^{-\lambda_1 t}e^{-\lambda_s \int_0^t \chi_s(\alpha, \eta)\,d\eta}.$$

Define $\bar{\chi}_s(\alpha)$ as

$$\bar{\chi}_s(\alpha) \triangleq \frac{1}{t}\int_0^t \chi_s(\alpha, \eta)\,d\eta$$

so that

$$n_s(t) = n_s(0)e^{-\lambda_1 t}e^{-\lambda_s \bar{\chi}_s(\alpha)t}.$$

The quantity $\bar{\chi}_s(\alpha)$ is called a membership grade in the set of smokers who have been damaged. Thus,

$$n_s(t) = n_s(0)e^{-t/\tau},$$

where $\tau = 1/(\lambda_1 + \lambda_s\bar{\chi}_s(\alpha))$ is the time it takes a population of smokers to decay to e^{-1} of their number at $t = 0$. Now $\tau_1 = 1/\lambda_1$ is the e-folding time of nonsmokers, so that

$$\tau_1 - \tau = \frac{1}{\lambda_1} - \frac{1}{\lambda_1 + \lambda_s\bar{\chi}_s(\alpha)} = \frac{\lambda_s\bar{\chi}_s(\alpha)}{\lambda_1[\lambda_1 + \bar{\chi}_s(\alpha)\lambda_s]}$$

is the decrease in the life expectancy of a smoker. We assume $\tau_1 - \tau \ll \tau_1$ (that is, decrease in life expectancy of a smoker is less than the life expectancy of a nonsmoker),

$$\tau_1 - \tau = \frac{\tau_1^2\lambda_s\bar{\chi}_s(\alpha)}{\dfrac{\lambda_s\bar{\chi}_s(\alpha)}{\lambda_1} + 1}$$

becomes, to an approximation,

$$\tau_1 - \tau = \tau_1^2\lambda_s\bar{\chi}_s(\alpha)[1 - \tau_1\lambda_s\bar{\chi}_s(\alpha) + \cdots].$$

We now turn to the question of choices for $\chi_s(\alpha, t)$ and the determination of $\bar{\chi}_s(\alpha)$. It is reasonable to require that

$$\chi_s(\alpha, t) \to 0 \quad \text{as } \alpha \to 0,$$

since by definition, α is proportional to the number of cigarettes smoked per unit time. Obviously $\chi_s(\alpha, t) \equiv 0$ for nonsmokers ($\alpha = 0$). If one smokes, for example, five cigarettes per day for ten years, it is clear that one is a smoker. One might classify such a person as a light smoker, but there is no doubt that he smokes. However, $\chi_s(\alpha, t) \in [0,1]$ and must reach unity sooner in time for someone who smokes forty cigarettes a day than for someone who smokes five cigarettes a day. Requiring $\chi_s(\alpha, t)$ to be a function of the dimensionless variable αt is a choice which assumes that the damage from smoking is cumulative in time. We now show that there are certain choices of $\chi_s(\alpha, t)$ which lead to plausible results.

■ ***Example 7.4.1***

Let $\chi_s(\alpha, t) = \alpha t/(1 + \alpha t)$ in the interval $[0,1]$. Then

$$\bar{\chi}_s(\alpha) = \frac{1}{t}\int_0^t \frac{\alpha\eta}{1 + \alpha\eta}\,d\eta = 1 - \frac{1}{\alpha t}\ln(1 + \alpha t),$$

and thus

$$n_s(t) = n_s(0)\,e^{-\lambda_1 t}e^{-\bar{\chi}_s(\alpha)\lambda_s t}$$

becomes

$$n_s(t) = n_s(0)\,e^{-\lambda_1 t}e^{-\lambda_s t[1-(1/\alpha t)\ln(1+\alpha t)]}.$$

Now as $\alpha \to 0$ we have to find the limit of $n_s(t)$; since the term $(1/\alpha t)\ln(1 + \alpha t)$ is undefined for $\alpha = 0$, we shall use l'Hôpital's rule, namely,

$$\lim_{\alpha \to 0}\left[\frac{\ln(1 + \alpha t)}{\alpha t}\right] = \lim_{\alpha \to 0}\left[\frac{\dfrac{1}{1 + \alpha t}\cdot t}{t}\right] = 1,$$

and thus

$$\lim_{\alpha \to 0} n_s(t) = n_s(0)\,e^{-\lambda_1 t}.$$

Clearly, as $\alpha \to \infty$ we get

$$n_s(t) = n_s(0)\,e^{-(\lambda_1 + \lambda_s)t}.$$

■

■ ***Example 7.4.2***

Let $\chi_s(\alpha, t) = \text{erf}(\alpha t)$, so that, again, $\chi_s \to 0$ as $\alpha \to 0$, and for finite t, as $\alpha \to \infty$ we get $\chi_s(\alpha, t) \to 1$. We find that

$$\bar{\chi}_s(\alpha) = \frac{1}{t}\int_0^t \text{erf}(\alpha\eta)\,d\eta = \text{erf}(\alpha t) - \frac{1}{2\alpha t}\left(1 - e^{-\alpha^2 t^2}\right)$$

whereupon

$$n_s(t) = n_s(0)\,e^{-\lambda_1 t}e^{-\lambda_s t\,\text{erf}(\alpha t)}e^{(\lambda_s/2\alpha)(1-e^{-\alpha^2 t^2})}.$$

As $\alpha \to 0$ we get that $n_s(t) \to n_s(0)e^{-\lambda_1 t}$, which is the rate of decline of a population consisting of nonsmokers. If α is very large for a finite t, we get

$$n_s(t) = n_s(0)\,e^{-(\lambda_1 + \lambda_s)t}.$$

The expression

$$n_s(t) = n_s(0)\,e^{-\lambda_1 t}e^{-\alpha\lambda_s t^2} \approx n_s(0)\,e^{-\lambda_1 t}\left(1 - \alpha\lambda_s t^2\right),$$

which is a valid approximation for small αt, shows that the effect of smoking has a quadratic dependence on time for early times. ■

Next, we consider the problem of the average value of $n_s(t)$ over the parameter α. If $\bar{\chi}_s(\alpha)$, a pure number in the interval [0, 1] is called the grade of membership in the set of damaged smokers (i.e., as $\bar{\chi}_s(\alpha) \to 1$, irreversible damage has certainly been done and life expectancy is thereby shortened) one has a solution $n_s(t, \alpha)$ depending on membership in a fuzzy set of "damage." How are we to determine the expected value of $n_s(t)$ over the distribution of α's?

If we use a probabilistic distribution of α's and apply the integration, we get

$$\langle n_s(t) \rangle = \int n_s(0) e^{-\lambda_1 t} e^{-\bar{\chi}_s(\alpha)\lambda_s t}\, dp_\alpha,$$

which is not easy to compute analytically, so some approximation is needed, assuming some exact distribution is known.

The methods of stationary phase and the steepest descent can be used to obtain an approximate formula for $\langle n_s(t) \rangle$. In general, the expression

$$f(k) = \int_a^b g(\tau) e^{kh(\tau)}\, d\tau$$

with $k \gg 0$, can be approximated by

$$f(k) \approx g(\tau_0) e^{kh(\tau_0)} \left[\frac{-\Pi}{2kh''(\tau)|_{\tau=\tau_0}} \right]^{1/2},$$

where τ_0 is the value of τ at which $h'(\tau)_{\tau=\tau_0} = 0$.

We shall illustrate this approximation in the following example.

■ ***Example 7.4.3***

Let

$$\chi_s(\alpha, t) = \alpha t_0 \text{ for } 0 \le t \le \frac{1}{\alpha},$$
$$= 1 \quad \text{for } t \ge \frac{1}{\alpha},$$

as shown in Figure 7.4.1.

The slope αt_0 can represent a least-square fit to some experimental data. Thus for $t \ge t_0$, we get

$$\bar{\chi}_s(\alpha) = \frac{1}{t}\left[\int_0^{t_0} \alpha\tau\, d\tau + \int_{t_0}^{t} d\tau \right] = \frac{1}{t}\left[t - t_0 + \frac{\alpha t_0^2}{2} \right]$$

and for $t_0 = 1/\alpha$, the equation becomes

$$\bar{\chi}_s(\alpha) = \frac{1}{t}\left[t - \frac{1}{\alpha} + \frac{1}{2\alpha} \right] = 1 - \frac{1}{2\alpha t},$$

and therefore

$$n_s(t) = n_s(0) e^{-\lambda_1 t} e^{-\lambda_s t(1-(1/2\alpha t))}.$$

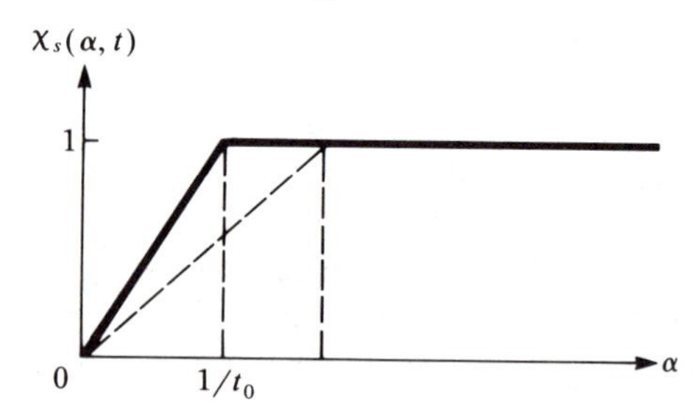

Figure 7.4.1. Membership function, Ex. 7.4.3.

If α is very large ($\alpha \to \infty$), then $n_s(t) \to n_s(0)e^{-(\lambda_1+\lambda_s)t}$; as a matter of fact, for finite α, if $t \to \infty$, then $\bar{\chi}_s(\alpha) \to 1$.

For $\alpha = 0$ we see that $t_0 = \infty$ and thus $t\bar{\chi}_s(\alpha)$ is an integration of a zero function that is zero.

Assume normal distribution of α's; we get that

$$\langle n_s(t)\rangle = \int n_s(0)\, e^{-\lambda_1 t} e^{-\lambda_s t(1-(1/2\alpha t))} \frac{1}{\sqrt{2\Pi}\,\sigma} e^{-(\alpha-\alpha_0)^2/2\sigma^2}\, d\alpha$$

$$= \frac{1}{\sqrt{2\Pi}\,\sigma} n_s(0)\, e^{-(\lambda_1+\lambda_s)t} \int e^{\lambda_s/2\alpha-(\alpha-\alpha_0)^2/2\sigma^2}\, d\alpha.$$

Let

$$h(\alpha) = \lambda_s(2\alpha)^{-1} - (\alpha-\alpha_0)^2(2\sigma^2)^{-1},$$

and

$$h'(\alpha) = -\frac{\lambda_s}{2\alpha^2} - \frac{2(\alpha-\alpha_0)}{2\sigma^2} = \frac{-\lambda_s\sigma^2 - 2(\alpha-\alpha_0)\alpha^2}{2\alpha^2\sigma^2}.$$

Here

$$h'(\alpha) = 0 \text{ implies } 2\alpha^2(\alpha-\alpha_0) = -\lambda_s\sigma^2$$

or

$$\alpha^3 - \alpha_0\alpha^2 + \frac{\lambda_s\sigma^2}{2} = 0.$$

The real root is thus given by $\alpha = \beta\alpha_0$ where β is a function of λ_s and σ and is always smaller than or equal to 1. As a first-order approximation, assume that σ^2 is small enough so that we get $\beta \approx 1$ and $\alpha \approx \alpha_0$; thus

$$h(\alpha_0) = \frac{\lambda_s}{2\alpha_0}.$$

Also,

$$h''(\alpha) = \frac{\lambda_s}{\alpha^3} - \frac{1}{\sigma^2},$$

and

$$h''(\alpha)|_{\alpha=\alpha_0} = \frac{\lambda_s}{\alpha_0^3} - \frac{1}{\sigma^2}.$$

Hence,

$$\langle n_s(t)\rangle \approx \frac{1}{\sqrt{2\Pi}\,\sigma} n_s(0)\, e^{-(\lambda_1+\lambda_s)t} e^{\lambda_s/2\alpha_0}\left[\frac{-\Pi}{\dfrac{2\lambda_s}{\alpha_0^3}-\dfrac{2}{\sigma^2}}\right]^{1/2}. \qquad \blacksquare$$

There is no question that this approximation is not easy to perform; it is based on the assumption that the exact distribution of α's is known, as well as many assumptions in the approximation method itself. However, we can easily compute the average "decrease in the life expectancy of a smoker," since

$$\mathrm{FEV}\{\tau_1 - \tau\} \cong \{\tau_1^2\lambda_s\}\mathrm{FEV}\{\bar{\chi}_s(\alpha)\}.$$

Clearly, the average over α's of $\bar{\chi}_s(\alpha)$ is an indicator of the expected life span of a smoker, and it is a fuzzy function by itself. Hence, for a given t,

$$\mathrm{FEV}\{\bar{\chi}_s(\alpha)\} = \operatorname*{Sup}_{T\in[0,1]} \{\min[T, \mu(\xi_T)]\},$$

where $\xi_T = \{\alpha \mid \bar{\chi}_s(\alpha) \geq T\}$. Since

$$\left|\int \bar{\chi}_s(\alpha)\, dp_\alpha - \mathrm{FEV}\{\bar{\chi}_s(\alpha)\}\right| \leq \tfrac{1}{4},$$

our result is a reasonable approximation and very easy to obtain.

Proposition 7.4.1

Let x be a fuzzy variable and let β be a positive real number in the interval $[0, \infty]$. Then

$$e^{-\beta x} \in [0,1].$$

In our case it is clear that both t and λ_s are positive real numbers, and thus

$$e^{-\bar{\chi}_s(\alpha)\lambda_s t} \in [0,1].$$

Clearly,

$$e^{-\bar{\chi}_s(\alpha)\lambda_s t} = \begin{cases} 1 & \text{when } \bar{\chi}_s(\alpha) = 0 \ (\text{or } t = 0), \\ 0 & \text{when } \bar{\chi}_s(\alpha) \neq 0 \text{ and } t \to \infty. \end{cases}$$

Thus, for a given t,

$$\mathrm{FEV}\{n_s(t)\} = n_s(0)\, e^{-\lambda_1 t}\,\mathrm{FEV}\{e^{-\bar{\chi}_s(\alpha)\lambda_s t}\},$$

where

$$\mathrm{FEV}\{e^{-\bar{\chi}_s(\alpha)\lambda_s t}\} = \operatorname*{Sup}_{T\in[0,1]} \{\min[T, \mu(\xi_T)]\}$$

and

$$\xi_T = \{\alpha \mid e^{-\bar{\chi}_s(\alpha)\lambda_s t} \geq T\}.$$

■ ***Example 7.4.4***

In Example 7.4.3, let $\sigma = 0.1$, $\lambda_s = 9$, and $\alpha_0 = 0.5$. Then

$$\langle n_s(t)\rangle \approx \frac{1}{\sqrt{2\Pi} \times 0.1} n_s(0)\, e^{-(\lambda_1+\lambda_s)t} e^{\lambda_s/2\alpha_0} \left[\frac{-\Pi}{2\left(\dfrac{9}{(0.5)^3} - 100\right)} \right]^{1/2}$$

$$\approx \frac{5}{\sqrt{28}} e^{-\lambda_s(t-(1/2\alpha_0))} n_s(0)\, e^{-\lambda_1 t}.$$

For the same example, consider the data collected with the following populations, from which the normal distribution has been assumed:

$$x_1 = 100@\alpha_0 = 0.5;\ \left[e^{-\bar{X}_s(\alpha)\lambda_s t} = e^{-9(t-1)}\right];$$

$$x_2 = 5@\alpha_1 = 1.5;\ \left[e^{-\bar{X}_s(\alpha)\lambda_s t} = e^{-9(t-(1/3))}\right];$$

$$x_3 = 2@\alpha_2 = 2.5;\ \left[e^{-\bar{X}_s(\alpha)\lambda_s t} = e^{-9(t-(1/5))}\right].$$

Now

$$A = x_1 + x_2 + x_3 = 107; \quad \frac{x_1}{A} = 0.934; \quad \frac{x_2 + x_1}{A} = 0.981.$$

Then

$$\text{FEV}\{e^{-\bar{X}_s(\alpha)\lambda_s t}\}$$
$$= \text{Median}_{\text{I.O.}}\{0.934, 0.981, e^{-9(t-1)}, e^{-9(t-(1/3))}, e^{-9(t-(1/5))}\},$$

(where the subscript "I.O." indicates "increasing order"). The result obviously depends upon t, so, for example, let $t = 1$. Then

$$\text{FEV}\{e^{-\bar{X}_s(\alpha)\lambda_s t}\} = \text{Median}_{\text{I.O.}}\{0.934, 0.981, 1, e^{-6}, e^{-36/5}\} = 0.934.$$

Hence,

$$\text{FEV}\{n_s(t)\}|_{t=1} = 0.934 n_s(0)\, e^{-\lambda_1},$$

whereas

$$\langle n_s(t)\rangle|_{t=1} \cong 0.943 n_s(0)\, e^{-\lambda_1}.$$

In other words,

$$[\langle n_s(t)\rangle - \text{FEV}\{n_s(t)\}]|_{t=1} = 0.009 n_s(0)\, e^{-\lambda_1}.$$ ■

We do not claim this small difference is always true, or even that it is true for this particular example where exact computations are performed, rather than approximations. It is quite difficult, however, if at all possible, to compute the exact value of $\langle n_s(t)\rangle$, even if one assumes that the data-set has an exact distribution of α's.

The FEV, however, is obtained without approximating any data-set or any technique. No assumptions are incorporated into the method, it is easy to find, and it is a reasonable approximation of an "average," since we always have, for $\Delta A(t)$ (differences in averages),

$$\Delta A(t) \triangleq |\langle n_s(t)\rangle - \text{FEV}\{n_s(t)\}|$$

$$= \left[n_s(0)e^{-\lambda_1 t} \left| \int e^{-\bar{\chi}_s(\alpha)\lambda_s t}\, dp_\alpha - \text{FEV}\{e^{-\bar{\chi}_s(\alpha)\lambda_s t}\} \right| \right] \le \frac{n_s(0)e^{-\lambda_1 t}}{4}.$$

Clearly, as t increases, $\Delta A(t)$ decreases, and

$$\lim_{t\to\infty} \Delta A(t) = 0.$$

In general it is clear that, if $\Delta A(t)$ represents a monotonic decreasing function in time, then

$$\langle n_s(t)\rangle = \lim_{t\to\infty} [\text{FEV}\{n_s(t)\}].$$

It is clear in Example 7.4.4 that after some t_0, we will get

$$\text{FEV}\{e^{-\bar{\chi}_s(\alpha)\lambda_s t}\} = e^{-9(t-1)}, \quad t > t_0,$$

and for these t's the difference between

$$\frac{5}{\sqrt{28}} e^{-\lambda_s(t-(1/2\alpha_0))} n_s(0)e^{-\lambda_1 t}$$

and

$$e^{-9(t-1)} n_s(0)e^{-\lambda_1 t}$$

is negligible.

It should be noted that extensions of the above analysis to birth and death processes involving several fuzzy attributes can be obtained, but they will not be dealt with here. However, the significance and ease of treating such problems via fuzzy statistics have been clearly illustrated. Not only is this technique simple to perform but it is useful in the analysis of such problems.

The fuzzy system illustrated here is represented by a differential equation with fuzzy coefficients, and the aim of this modelling process is to show how fuzzy-set theory can be applied to an imprecise modelling scheme where some parameters of the system are fuzzy by nature.

The search for a typical solution to a fuzzy differential equation, describing a fuzzy process or a fuzzy environment, outlined above represents an attempt to generalize the concept of a nondeterministic differential equation and its solution via the FEV. Fuzzy-set theory has been applied here to represent a system that has a fuzzy parameter as its main feature, whereas fuzzy statistics has been applied to obtain a solution to this model.

The solution obtained by treating the problem with techniques developed in fuzzy-set theory shows that philosophically we can model an uncertain environment or a fuzzy behavior by a fuzzy differential equation, and find meaningful and typical solutions to this model.

In conclusion, the technique can be applied to many other nondeterministic dynamic processes, since it has the virtues of simplicity and, where comparison with physical experiments can be made, accuracy, with minimal complexity of computations.

7.5 PATTERN CLUSTERING

Once we describe the clustering problem as one of finding natural groupings in a data-set, we have to investigate measures of similarity between samples, as well as the evaluation of a partitioning of a set of samples into clusters. It is this partitioning that we are interested in when we discuss the subject of *hierarchical clustering*, which is the topic of this section.

Let us consider a sample space of k samples that we want to partition into q classes. The first stage is to partition the data-set into k clusters, each containing exactly one sample. Then we partition the sample space into $k-1, k-2, \ldots, j$ clusters where, at level j of the sequence, we have

$$q = k - j + 1.$$

Thus, level one corresponds to k clusters and level k to one. Given any two samples $\vec{x}_1$ and $\vec{x}_2$, at some level they will be grouped together in the same cluster. If the sequence has the property that, whenever two samples are in the same cluster at level k, they remain together at all higher levels, then the sequence is said to be a *hierarchical clustering*. Examples of hierarchical clustering appear in biological taxonomy, where individuals are grouped into species, species into genera, genera into families, and so on.

For every hierarchical clustering there is a corresponding tree, called a *dendrogram*, that shows how the samples are grouped.

Hierarchical clustering procedures are divided into two distinct classes, *agglomerative* and *divisive*. Agglomerative (bottom-up, clumping) procedures start with q singleton clusters and form the sequence by successively merging clusters. Divisive (top-down, splitting) procedures start with all of the samples in one cluster and form the sequence by successively splitting clusters. The computation needed to go from one level to another is usually simpler for the agglomerative procedures. However, when there are many samples and one is interested in only a small number of clusters, this computation will have to be repeated many times. At any level, the distance between nearest clusters can provide the dissimilarity value for that level. It should be noted that we have not said how to measure the distance between two clusters. The considerations here are much like those involved in selecting a criterion function. Basic distance measures include:

$$d_{\min}(\rho_i, \rho_j) = \min_{\alpha \in \rho_i, \beta \in \rho_j} \|\alpha - \beta\|,$$

$$d_{\max}(\rho_i, \rho_j) = \max_{\alpha \in \rho_i, \beta \in \rho_j} \|\alpha - \beta\|,$$

$$d_{\text{avg}}(\rho_i, \rho_j) = \frac{1}{k_i k_j} \sum_{\alpha \in \rho_i,} \sum_{\beta \in \rho_j} \|\alpha - \beta\|,$$

$$d_{\text{mean}}(\rho_i, \rho_j) = \|M_i - M_j\|.$$

All of these measures have a minimum-variance flavor, and they usually yield the same results if the clusters are compact and well separated. However, if the clusters are close to one another, or if their shapes are not basically hyperspherical, quite different results can be obtained.

Interestingly enough, if we define the dissimilarity between two clusters by

$$\delta_{\min}(\rho_i, \rho_j) = \min_{\alpha \in \rho_i, \beta \in \rho_j} \delta(\alpha, \beta)$$

or

$$\delta_{\max}(\rho_i, \rho_j) = \max_{\alpha \in \rho_i, \beta \in \rho_j} \delta(\alpha, \beta),$$

then the hierarchical clustering procedure will induce a distance function for the given set of n samples. Furthermore, the ranking of the distances between samples will be invariant to any monotonic transformation of the dissimilarity values.

The graph-theoretic principle involves the selection of a threshold distance. Once the threshold distance d_0 is selected, two elements are said to be in the same cluster if the distance between them is less than d_0. This procedure can easily be generalized to apply to arbitrary similarity measures. Suppose that we pick a threshold value d_0 and say that α is similar to β if $s(\alpha, \beta) > d_0$:

$$S_{ij} = \left\{ \begin{array}{ll} 1 & \text{if } s(\alpha_i, \alpha_j) > d_0, \\ 0 & \text{otherwise,} \end{array} \right\} i, j = 1, \ldots, n.$$

This matrix defines a *similarity graph* in which nodes correspond to points and an edge joins node i and node j if and only if $S_{ij} = 1$.

The clusterings produced by the single-linkage algorithm and by a modified version of the complete-linkage algorithm are readily described in terms of this graph. With the single-linkage algorithm, two samples α and β are in the same cluster if and only if there exists a chain $\alpha_1, \alpha_2, \ldots, \alpha_k$ such that α is similar to α_1, α_1 is similar to α_2, and so on for the whole chain. Thus this clustering corresponds to the connected components of the similarity graph. With the complete-linkage algorithm, all samples in a given cluster must be similar to one another, and no sample can be in more than one cluster. If we drop this second requirement, then this clustering corresponds to the maximal complete subgraphs of the similarity graph, the "largest" subgraphs with edges joining all pairs of nodes. In general, the clusters of the complete-linkage algorithm will be found among the maximal complete subgraphs, but they cannot be determined without knowing the unquantized similarity values.

It is clear that the nearest-neighbor algorithm could be viewed as an algorithm for finding a minimal spanning tree. Conversely, given a minimal spanning tree, we can find the clusterings produced by the nearest-neighbor algorithm. Removal of the longest edge produces the two-cluster grouping, removal of the next longest edge produces the three-cluster grouping, and so on. This amounts to an inverted way of obtaining a divisive hierarchical procedure, and suggests other ways of dividing the graph into subgraphs. For example, in selecting an edge to remove, we can compare its length to the length of other edges incident upon its nodes. Let us say that an edge is inconsistent if its length ξ is significantly larger than $\bar{\xi}$, the average length of all other edges incident on its vertices.

When the data points are strung out into long chains, a minimal spanning tree forms a natural skeleton for the chain. If we define the diameter path as the longest path through the tree, then a chain will be characterized by the shallow depth of branching off the diameter path. In contrast, for a large, uniform cloud of data points, the tree will usually not have an obvious diameter path, but rather several distinct, near-diameter paths. For any of these, an appreciable number of nodes will be off the path. While slight changes in the locations of the data points can cause major rerouting of a minimal spanning tree, they typically have little effect on the recognition process.

To estimate subjectively the resemblance between pairs of data points, we adopt the convention of arranging data for numerical classification in the form of a matrix. Each entry $[ij]$ in such a matrix is the score of the proximity relation (subjective similarity) between data points i and j. It should be noted that the numerical values in the proximity matrix are in general only quantitatively descriptive numbers whose significance usually cannot be evaluated by conventional statistical techniques, and thus are determined subjectively.

Since the proximity relation is not necessarily transitive, we must utilize the theory of fuzzy matrices as described in Section 5.4, in order to formulate a transitive closure structure that will enable us to separate the data-set into mutually exclusive clusters which are, in essence, equivalent classes under a certain threshold.

Definition 7.5.1

Let R be a fuzzy relation on data-set D and let r be a subset of D. Then R is said to refine r iff xRy implies that $x \in r$ iff $y \in r$. Symbolically,

$$xRy \to (x \in r \leftrightarrow y \in r).$$

Theorem 7.5.1

Let $T_1 \geq T_2$. Then

$$R_{T_1} \text{ refines } R_{T_2}.$$

Proof

$xR_{T_1}y \to \bar{\chi}_s(x, y) \geq T_1 \geq T_2 \to xR_{T_2}y$, using Def. 5.3.2. Q.E.D.

It is clear that if R_T is a threshold relation induced by $\chi_s(x, y)$ and R'_T is a threshold relation induced by $\chi'_s(x, y)$ and $\chi_s(x, y) \leq \chi'_s(x, y)$ for all x, $y \in \Omega$, then R_T refines R'_T.

It is our assumption that if $x \neq y$, then $\bar{\chi}_s(x, y) \in [0, 1)$ and thus the function $\bar{\eta}_s(x, y) = 1 - \bar{\chi}_s(x, y)$ acts as a distance function. This is very obvious, since

(i) $\bar{\eta}_s(x, y) > 0$ for $x \neq y$ and $\bar{\eta}_s(x, x) = 0$;
(ii) $\bar{\eta}_s(x, y) = \bar{\eta}_s(y, x)$;

and

(iii) $\bar{\eta}_s(x, z) \leq \bar{\eta}_s(x, y) + \bar{\eta}_s(y, z)$

as $\bar{\chi}_s(x, z) \geq \min[\bar{\chi}_s(x, y), \bar{\chi}_s(y, z)] \geq \bar{\chi}_s(x, y) + \bar{\chi}_s(y, z) - 1$.

We will assume, for the simplicity of the analysis, that we deal with only a finite number of patterns, and hence we shall consider our threshold relation on finite sets only.

Theorem 7.5.2

Let $x_1, x_2, \ldots, x_n \in \Omega$ where n is a finite number. Then

$$R_T = \psi(R_T^*) = (R_T^*)^{n-1}$$

where

$$xR_T^*y \text{ iff } \chi_s(x, y) \geq T$$

and

$$\psi(R_T^*) = \operatorname{Sup}_i (R_T^*)^i = \operatorname{adj}(R_T^*).$$

Proof

(i) Clearly, $\psi(R_T^*)$ refines R_T since $\psi(R_T^*) = \bigcup_j (R_T^*)^j$ and, if $x[\psi(R_T^*)]y$, then

$$\exists x_1, x_2, \ldots, x_{n-1} \in \Omega \text{ such that } \chi_s(x, x_1) \geq T, \ldots, \chi_s(x_{n-1}, y) \geq T$$

and hence

$$\chi_s^n(x, y) \geq \min[\chi_s(x, x_1), \ldots, \chi_s(x_{n-1}, y)] \geq T,$$

which implies that $\bar{\chi}_s(x, y) \geq \chi_s^n(x, y) \geq T$ and thus xR_Ty.

(ii) Assume xR_Ty. Then

$$\chi_s(x, y) = \chi_s^{n-1}(x, y) = \max_{x_1, x_2, \ldots, x_{n-2} \in \Omega} \{\min[\chi_s(x, x_1), \ldots, \chi_s(x_{n-2}, y)]\} \geq T$$

Therefore

$$\exists x_1, x_2, \ldots, x_{n-2} \in \Omega$$

such that

$$\begin{aligned} \chi_s\ (x, x_1) &\geq T, \\ \chi_s\ (x_1, x_2) &\geq T, \\ &\vdots \\ \chi_s\ (x_{n-2}, y) &\geq T, \end{aligned}$$

and thus we have

$$xR_T^*x_1, x_1R_T^*x_2, \ldots, x_{n-2}R_T^*y \rightarrow x(R_T^*)^{n-1}y \text{ exists } \rightarrow x[\psi(R_T^*)]y \text{ exists.}$$

(iii) Let $R_T^* = [\chi_{ij}]$. The ij entry of $(R_T^*)^2$ is

$$\sum_{k=1}^{n} \chi_{ik}\chi_{kj},$$

and this term has the grade membership of

$$\max_k \left[\min(\chi_{ik}, \chi_{kj})\right]$$

iff there is a direct path between vertices i and j or there is a path from i to j through one intermediate vertex. Extending this argument to $(R_T^*)^i$, it is clear that no path requires more than $(n-2)$ intermediate vertices, since there are only n vertices and internal loops are excluded. Hence, the ij entry of $(R_T^*)^{n-1}$ has the grade membership of $\max_{\text{subterms}}\{ij \text{ terms of } (R_T^*)^{n-1}\}$ iff i and j are connected; that is, $(R_T^*)^{n-1} = \psi(R_T^*)$. Q.E.D.

Corollary 7.5.1

$R_T^* = \psi(R_T^*) \leftrightarrow (R_T^*)^2 = R_T^*$.

Proof of Corollary 7.5.1

Assume $(R_T^*)^2 = R_T^*$. By induction,

$$R_T^* = (R_T^*)^{n-1} \quad \text{and} \quad (R_T^*)^{n-1} = \psi(R_T^*).$$

Conversely, if

$$R_T^* = \psi(R_T^*),$$

then

$$(R_T^*)^2 = R_T^*(R_T^*)^{n-1},$$

$$(R_T^*)^n = \psi(R_T^*) = R_T^*.$$

Q.E.D.

Based on the above result, Algorithm A is presented to compute $\psi(R_T^*)$.

Algorithm A

Given the matrix R_T^* constructed from the inexact patterns $x_1, x_2, \ldots, x_n$. Generate the matrix $(R_T^*)^l = (R_T^*)^{l+1}$ for some l, under the operation of fuzzy matrix multiplication.

The repeated matrix multiplication makes Algorithm A unattractive from an efficiency viewpoint. Algorithm B achieves the same result, and requires only a single scan over the matrix. In fact, Algorithm B works correctly on a wider range of input, since it is not required that the diagonal elements of the input matrix = 1, as with algorithm A.

Algorithm B

1. *Label vertices of R_T^* by the integers* $1, \ldots, n$.
2. *Generate the matrix R_T^*.*
3. DO $K = 1$ TO n
4. DO $I = 1$ TO n
5. IF $R_T^*(I, K) \neq 0$ THEN
6. DO $J = 1$ TO n
7. $R_T^*(I, J) = \max(R_T^*(I, J), \min(R_T^*(I, K), R_T^*(K, J)))$
8. END
9. END
10. END

The basic idea is to scan down *column* K, and for each nonzero element encountered (say, in row I), each element in *row* I (say, element $R_T^*(I, J)$) is possibly improved by comparing $R_T^*(I, J)$ to $\min(R_T^*(I, K), R_T^*(K, J))$. A rigorous proof of correctness is achieved by attaching the following inductive assertion W between statements (7) and (8):

$$(W) \quad R_T^*(I, J) = G(I, J, K),$$

where $G(I, J, K) \overset{\text{def}}{=} \max\{\min(\text{all chains from } I \text{ to } J \text{ such that each intermediate element has a label } \leq K)\}$.

Before proving that assertion W is true whenever control leaves step (7), it is noted that the relation

$$R_T^*(I, J) = G(I, J, n)$$

is the desired relation at the termination of the algorithm, since $G(I, J, n) = \max\{\min(\text{all chains from } I \text{ to } J)\}$.

Assertion W is proved by induction on K.

(i) $K = 1$.

The first time W is reached, K has the value 1, and analysis shows that

$$R_T^*(I, J) = \max\left(R_{T_0}^*(I, J), \min\left(R_{T_0}^*(I, 1), R_{T_0}^*(1, J)\right)\right)$$

where $R_{T_0}^*$ represents the original matrix, and the righthand side of the equation $= G(I, J, 1)$.

(ii) Assume $R_T^*(I, J) = G(I, J, K)$, $1 \leq K < n$. Then we have to show that

$$R_T^*(I, J) = G(I, J, K + 1).$$

There are two subcases to consider. If $G(I, J, K + 1)$ does not involve element $K + 1$, then no change is made to the matrix and the desired result is true. If $G(I, J, K + 1)$ does involve element $(K + 1)$, then we can guarantee that element $(K + 1)$ appears only once, since loops do not increase the max of any chain. Thus we can break the optimal chain into two subchains, $R_T^*(I, K + 1)$ and $R_T^*(K + 1, J)$. Since both subchains involve intermediate elements numbered $\leq K$, the inductive hypothesis applies to each subchain, and the desired result follows.

It is interesting to note that during the process of computing the characteristic fuzzy matrix, minimization of the fuzzy structures is possible. In general, one cannot apply binary minimization techniques, and thus more specific methods, directed toward the minimization of fuzzy functions, should be developed.

Generally, an indexed collection

$$\lambda = \{D_\alpha; \alpha \in J\}$$

$$R_T^* = \begin{array}{c} \\ x_1 \\ x_2 \\ x_3 \\ x_4 \end{array} \begin{array}{c} \begin{array}{cccc} x_1 & x_2 & x_3 & x_4 \end{array} \\ \begin{bmatrix} 1 & 0 & T_2 & T_1 \\ 0 & 1 & T_4 & T_3 \\ T_2 & T_4 & 1 & T_1 \\ T_1 & T_3 & T_1 & 1 \end{bmatrix} \end{array}.$$

Exhibit 7.5.1

of subsets of a set S, satisfying

$$S = \bigcup_{\alpha \in J} D_\alpha$$

$$D_\beta \cap D_\gamma = \emptyset \quad \text{for all } \beta \neq \gamma \in J,$$

is said to be a *partition* of S.

Clearly, we can apply the notions discussed above for the partition of our pattern set into disjoint classes which depends on the *a priori* assigned threshold.

■ ***Example 7.5.1***

Let $X = \{x_1, x_2, x_3, x_4\}$ and $\chi_s(x_i, x_j)$, $i, j = 1,2,3,4$ be as shown in Exhibit 7.5.1.

The minimized form of $\psi(R_T^*)$ is represented by Exhibit 7.5.2.

Let $T_1 = 0.3$, $T_2 = 0.5$, $T_3 = 0.6$, and $T_4 = 0.9$. Then

$$\psi(R_T^*) = \begin{array}{c} \\ x_1 \\ x_2 \\ x_3 \\ x_4 \end{array} \begin{array}{c} \begin{array}{cccc} x_1 & x_2 & x_3 & x_4 \end{array} \\ \begin{bmatrix} 1 & 0.5 & 0.5 & 0.5 \\ 0.5 & 1 & 0.9 & 0.7 \\ 0.5 & 0.9 & 1 & 0.7 \\ 0.5 & 0.7 & 0.7 & 1 \end{bmatrix} \end{array},$$

and we have the partitions:

$$R_{T=1} = \{[x_1],[x_2],[x_3],[x_4]\},$$

$$R_{0.7<T\leq 0.9} = \{[x_1],[x_2, x_3],[x_4]\},$$

$$R_{0.5<T\leq 0.7} = \{[x_1],[x_2, x_3, x_4]\},$$

$$R_{0\leq T\leq 0.5} = \{[x_1, x_2, x_3, x_4]\}.$$

■

■ ***Example 7.5.2***

Let

$$\chi_s(x, y) = \frac{1}{1 + |x - y|}, \qquad x, y \in n.$$

See Exhibits 7.5.3 and 7.5.4.

Clearly for $1 \geq T_1 \geq T_2 \geq 0$, R_{T_1} refines R_{T_2}; therefore, for every monotone, nonincreasing, finite sequence of thresholds,

$$0 \leq T_j \leq T_{j-1} \leq \cdots \leq T_2 \leq T_1 \leq 1,$$

$$\psi(R_T^*) = \begin{array}{c} \\ x_1 \\ x_2 \\ x_3 \\ x_4 \end{array} \begin{array}{c} \begin{array}{cccc} x_1 & x_2 & x_3 & x_4 \end{array} \\ \left[\begin{array}{cccc} 1 & T_2T_4 + T_1T_3 + T_1\bar{T}_1T_4 + \bar{T}_1T_2T_3 & T_2 + T_1\bar{T}_1 + T_1T_3T_4 & T_1 + \bar{T}_1T_2 + T_2T_3T_4 \\ T_2T_4 + T_1T_3 + T_1\bar{T}_1T_4 + \bar{T}_1T_2T_3 & 1 & T_4 + \bar{T}_1T_3 + T_1T_2T_3 & T_3 + \bar{T}_1T_4 + T_1T_2T_4 \\ T_2 + T_1\bar{T}_1 + T_1T_3T_4 & T_4 + \bar{T}_1T_3 + T_1T_2T_3 & 1 & \bar{T}_1 + T_1T_2 + T_3T_4 \\ T_1 + \bar{T}_1T_2 + T_2T_3T_4 & T_3 + \bar{T}_1T_4 + T_1T_2T_4 & \bar{T}_1 + T_1T_2 + T_3T_4 & 1 \end{array}\right] \end{array}$$

Exhibit 7.5.2

$R^*_T =$

	0	1	2	3	4	5	6	7	8	...
0	1	$\frac{1}{2}$	$\frac{1}{3}$	$\frac{1}{4}$	$\frac{1}{5}$	$\frac{1}{6}$	$\frac{1}{7}$	$\frac{1}{8}$	$\frac{1}{9}$	...
1	$\frac{1}{2}$	1	$\frac{1}{2}$	$\frac{1}{3}$	$\frac{1}{4}$	$\frac{1}{5}$	$\frac{1}{6}$	$\frac{1}{7}$	$\frac{1}{8}$	...
2	$\frac{1}{3}$	$\frac{1}{2}$	1	$\frac{1}{2}$	$\frac{1}{3}$	$\frac{1}{4}$	$\frac{1}{5}$	$\frac{1}{6}$	$\frac{1}{7}$	...
3	$\frac{1}{4}$	$\frac{1}{3}$	$\frac{1}{2}$	1	$\frac{1}{2}$	$\frac{1}{3}$	$\frac{1}{4}$	$\frac{1}{5}$	$\frac{1}{6}$	...
4	$\frac{1}{5}$	$\frac{1}{4}$	$\frac{1}{3}$	$\frac{1}{2}$	1	$\frac{1}{2}$	$\frac{1}{3}$	$\frac{1}{4}$	$\frac{1}{5}$	...
5	$\frac{1}{6}$	$\frac{1}{5}$	$\frac{1}{4}$	$\frac{1}{3}$	$\frac{1}{2}$	1	$\frac{1}{2}$	$\frac{1}{3}$	$\frac{1}{4}$	...
6	$\frac{1}{7}$	$\frac{1}{6}$	$\frac{1}{5}$	$\frac{1}{4}$	$\frac{1}{3}$	$\frac{1}{2}$	1	$\frac{1}{2}$	$\frac{1}{3}$	...
7	$\frac{1}{8}$	$\frac{1}{7}$	$\frac{1}{6}$	$\frac{1}{5}$	$\frac{1}{4}$	$\frac{1}{3}$	$\frac{1}{2}$	1	$\frac{1}{2}$	...
8	$\frac{1}{9}$	$\frac{1}{8}$	$\frac{1}{7}$	$\frac{1}{6}$	$\frac{1}{5}$	$\frac{1}{4}$	$\frac{1}{3}$	$\frac{1}{2}$	1	...
	⋮	⋮	⋮	⋮	⋮	⋮	⋮	⋮	⋮	⋱

Exhibit 7.5.3

we can obtain a corresponding j-level hierarchy of clusters

$$J_i = \left\{ \text{equivalence classes of } R_{T_i} \text{ in } X \mid 1 \leq i \leq j \right\}.$$

For each i, J_i is a partition of X and every class in J_{i+1} is the union of some nonempty class of subsets in J_i. It is interesting to note that, if we define recursively

$$\Gamma_1 = R^*_T,$$

$$\Gamma_n = \Gamma^2_{n-1},$$

then

$$\Gamma_j = (R^*_T)^{2^{j-1}};$$

$\psi(R^*_T) =$

	0	1	2	3	4	5	6	7	8	...
0	1	$\frac{1}{2}$	$\frac{1}{2}$	$\frac{1}{2}$	$\frac{1}{2}$	$\frac{1}{2}$	$\frac{1}{2}$	$\frac{1}{2}$	$\frac{1}{2}$	...
1	$\frac{1}{2}$	1	$\frac{1}{2}$	$\frac{1}{2}$	$\frac{1}{2}$	$\frac{1}{2}$	$\frac{1}{2}$	$\frac{1}{2}$	$\frac{1}{2}$	...
2	$\frac{1}{2}$	$\frac{1}{2}$	1	$\frac{1}{2}$	$\frac{1}{2}$	$\frac{1}{2}$	$\frac{1}{2}$	$\frac{1}{2}$	$\frac{1}{2}$	...
3	$\frac{1}{2}$	$\frac{1}{2}$	$\frac{1}{2}$	1	$\frac{1}{2}$	$\frac{1}{2}$	$\frac{1}{2}$	$\frac{1}{2}$	$\frac{1}{2}$	...
4	$\frac{1}{2}$	$\frac{1}{2}$	$\frac{1}{2}$	$\frac{1}{2}$	1	$\frac{1}{2}$	$\frac{1}{2}$	$\frac{1}{2}$	$\frac{1}{2}$	...
5	$\frac{1}{2}$	$\frac{1}{2}$	$\frac{1}{2}$	$\frac{1}{2}$	$\frac{1}{2}$	1	$\frac{1}{2}$	$\frac{1}{2}$	$\frac{1}{2}$	...
6	$\frac{1}{2}$	$\frac{1}{2}$	$\frac{1}{2}$	$\frac{1}{2}$	$\frac{1}{2}$	$\frac{1}{2}$	1	$\frac{1}{2}$	$\frac{1}{2}$	...
7	$\frac{1}{2}$	$\frac{1}{2}$	$\frac{1}{2}$	$\frac{1}{2}$	$\frac{1}{2}$	$\frac{1}{2}$	$\frac{1}{2}$	1	$\frac{1}{2}$	...
8	$\frac{1}{2}$	$\frac{1}{2}$	$\frac{1}{2}$	$\frac{1}{2}$	$\frac{1}{2}$	$\frac{1}{2}$	$\frac{1}{2}$	$\frac{1}{2}$	1	...
	⋮	⋮	⋮	⋮	⋮	⋮	⋮	⋮	⋮	⋱

Exhibit 7.5.4

and since, by Theorem 7.5.2,

$$\psi(R_T^*) = (R_T^*)^{n-1},$$

we can obtain the following relationship:

$$2^j \geq n - 1 \text{ will determine } \Gamma_j[\bar{\chi}_s(x, y)], \text{ or}$$

$$j \geq \lceil \log_2(n-1) \rceil \text{ will determine } \Gamma_j[\bar{\chi}_s(x, y)].$$

It should be noted that this is not a necessary condition, since we can find certain cases for smaller j's to imply

$$\Gamma_j[\bar{\chi}_s(x, y)].$$ ■

It is also interesting to note that we can present the same structure through a simple use of graph theory.

Definition 7.5.2

An inexact graph IG *is a pair* $[V, S]$ *where* V *is a set of data vertices, and* S *is a proximity relation on* V. *A vertex* v_i *is said to be T-accessible from another vertex* v_j *for some* $0 < T \leq 1$, *if and only if*

$$\chi_s(v_i, v_j) \geq T.$$

IG *is called strongly T-connected if and only if every pair of vertices is mutually accessible.*

A *T-cluster* in V is a maximal subset U of V such that each pair of elements in U is mutually T-accessible. Thus the construction of T-clusters of V is really nothing but constructing all maximal strongly T-connected subgraphs of IG.

This method can be related to existing nonfuzzy techniques that make use of link-node plots or spatial-dependence rearrangements of matrices. The link-node plot relates directly to the graph-theoretic approach using minimal spanning trees. The rearrangement of matrices represents a transformation on a matrix giving numbers of variables differing significantly between pairs of categories. The transformation usually brings similar rows and columns together, and so makes the organization of the matrix more obvious. In a large data-set, an algorithmic approach which systematically rearranges rows and corresponding columns in order to optimize $|j-k|A^2(j,k)$, where j and k are the row and the column, respectively, and $A(j,k)$ is the number of different variables between category j and category k, will accomplish such a rearrangement. In the following subsections, we shall investigate several closely related techniques in which the rules of the game have been altered in order to allow the reader to envision far more options in matrix clustering than have been available before.

Optimistic Transitions

Following Santos and Wee [1976], the matrix composition under the operation "max-min" is known as a *pessimistic* grade of transition. Here we investigate several different structures and their applicability to the pattern classification problem.

First we shall call the matrix composition with the grade of transition under the operation "min-max" an *optimistic* model. We shall denote it as

$$C = A\#B \leftrightarrow c_{ij} = \left[\min_k \left\{\max\left(a_{ik}, b_{kj}\right)\right\}\right].$$

We can construct the properties of this operation in a similar way to the construction in the "max-min" case. The most interesting results are:

1. $A\#(B\#C) = (A\#B)\#C$;
2. $A\#A^\circ = A^\circ\#A = A$ where $A^\circ = [a_{ij}^\circ]$ and $a_{ij}^\circ = \begin{cases} 0 & \text{if } i = j, \\ 1 & \text{if } i \neq j; \end{cases}$
3. $A^\alpha\#A^\beta = A^{\alpha+\beta}$;
4. $(A^\alpha)^\beta = A^{\alpha\beta}$;
5. If $A < C$ and $B < D$, then $A\#B < C\#D$.

It is clear that Theorem 7.5.2 does not hold under the operation #, as can be seen from the following examples:

■ ***Example** 7.5.3*

Let

$$R_T^* = \begin{bmatrix} 1 & 0 & 0.5 & 0.3 \\ 0 & 1 & 0.9 & 0.6 \\ 0.5 & 0.9 & 1 & 0.7 \\ 0.3 & 0.6 & 0.7 & 1 \end{bmatrix}.$$

Then

$$(R_T^*)_\#^2 = R_T^*\#R_T^* = \begin{bmatrix} 0 & 0.6 & 0.7 & 0.6 \\ 0.6 & 0 & 0.5 & 0.3 \\ 0.7 & 0.5 & 0.5 & 0.5 \\ 0.6 & 0.3 & 0.5 & 0.3 \end{bmatrix},$$

$$(R_T^*)_\#^3 = (R_T^*)_\#^2\#R_T^* = \begin{bmatrix} 0.6 & 0 & 0.5 & 0.3 \\ 0 & 0.6 & 0.6 & 0.6 \\ 0.5 & 0.6 & 0.7 & 0.6 \\ 0.3 & 0.6 & 0.6 & 0.6 \end{bmatrix},$$

$$(R_T^*)_\#^4 = (R_T^*)_\#^3\#R_T^* = \begin{bmatrix} 0 & 0.6 & 0.6 & 0.6 \\ 0.6 & 0 & 0.5 & 0.3 \\ 0.6 & 0.5 & 0.5 & 0.5 \\ 0.6 & 0.3 & 0.5 & 0.3 \end{bmatrix},$$

$$(R_T^*)_\#^5 = (R_T^*)_\#^4\#R_T^* = \begin{bmatrix} 0.6 & 0 & 0.5 & 0.3 \\ 0 & 0.6 & 0.6 & 0.6 \\ 0.5 & 0.6 & 0.6 & 0.6 \\ 0.3 & 0.6 & 0.6 & 0.6 \end{bmatrix};$$

$$(R_T^*)_\#^{2n} = (R_T^*)_\#^4,$$

$$(R_T^*)_\#^{2n+1} = (R_T^*)_\#^5, \qquad \text{for } n > 2.$$ ■

However, we can prove the following theorem.

Theorem 7.5.3

Let $x_1, x_2, \ldots, x_n \in \Omega$, where n is a finite number. Then

$$\hat{R}_T = \hat{\psi}(\hat{R}_T^*) = (\hat{R}_T^*)_{\#}^{n-1},$$

where $x_i \hat{R}_T^* x_j$ iff $\eta_s(x_i, x_j) \leq T$, $i, j \in \{1, 2, \ldots, n\}$ and $\hat{\psi}(\hat{R}_T^*) = \text{adj}(\hat{R}_T^*)$ under # operation.

Proof

By Theorem 7.5.2 and De Morgan laws. Q.E.D.

Clearly, $\hat{\psi}(\hat{R}_T^*) = E - \psi(R_T^*)$ where E is an $n \times n$ matrix of 1's.

■ ***Example 7.5.4***

Let

$$\hat{R}_T^* = \begin{bmatrix} 0 & 1 & 0.5 & 0.7 \\ 1 & 0 & 0.1 & 0.4 \\ 0.5 & 0.1 & 0 & 0.3 \\ 0.7 & 0.4 & 0.3 & 0 \end{bmatrix},$$

$$(\hat{R}_T^*)^2 = \hat{\psi}(\hat{R}_T^*) = \begin{bmatrix} 0 & 0.5 & 0.5 & 0.5 \\ 0.5 & 0 & 0.1 & 0.3 \\ 0.5 & 0.1 & 0 & 0.3 \\ 0.5 & 0.3 & 0.3 & 0 \end{bmatrix}.$$ ■

Hence the partitions described in this section can be performed directly on the matrix representation of the distance relation.

Max-Product Structures

Another possibility is to derive a max-product relation instead of the max-min relation; namely,

$$\chi_{\underset{\sim}{s}}(x, z) = \max_y \left[\chi_s(x, y) \cdot \chi_s(y, z)\right]$$

and thus

$$C = A \S B \leftrightarrow c_{ij} = \left[\max_k \left(a_{ik} \cdot b_{kj}\right)\right],$$

where

$$A^\circ = [a_{ij}^\circ] \qquad \text{and} \qquad a_{ij}^\circ = \begin{cases} 1 & \text{if } i = j, \\ 0 & \text{if } i \neq j. \end{cases}$$

The transitive closure in this case is defined similarly as

$$\tilde{\psi}(\tilde{R}_T^*) = \tilde{R}_T^* \cup (\tilde{R}_T^*)^2 \cup (\tilde{R}_T^*)^3 \cup \cdots$$

where

$$(\tilde{R}_T^*)^k = \underbrace{\tilde{R}_T^* \S \tilde{R}_T^* \S \cdots \S \tilde{R}_T^*}_{k \text{ times}}, \qquad k = 1, 2, 3, \ldots$$

Since $\forall \alpha, \beta \in [0,1]$, $\alpha \cdot \beta \leq \min(\alpha, \beta)$, it is clear that

$$R_T^* \circ R_T^* \subset R_T^* \Rightarrow \tilde{R}_T^* \S \tilde{R}_T^* \subset \tilde{R}_T^*,$$

where

$$(R_T^*)^2 = R_T^* \circ R_T^*$$

represents

$$\chi_{R_T^*} \cdot R_T^*(x,z) = \max_y \left\{ \min\left[\chi_{R_T^*}(x,y), \chi_{R_T^*}(y,z) \right] \right\}$$

and

$$(\tilde{R}_T^*)^2 = \tilde{R}_T^* \S \tilde{R}_T^*$$

represents

$$\chi_{\tilde{R}_T^*} \cdot \tilde{R}_T^*(x,z) = \max_y \left[\chi_{\tilde{R}_T^*}(x,y) \cdot \chi_{\tilde{R}_T^*}(y,z) \right].$$

■ ***Example 7.5.5***

Let

$$\tilde{R}_T^* = \begin{bmatrix} 1 & 0 & 0.5 & 0.3 \\ 0 & 1 & 0.9 & 0.6 \\ 0.5 & 0.9 & 1 & 0.7 \\ 0.3 & 0.6 & 0.7 & 1 \end{bmatrix},$$

$$(\tilde{R}_T^*)^2 = \begin{bmatrix} 1 & 0.45 & 0.5 & 0.35 \\ 0.45 & 1 & 0.9 & 0.63 \\ 0.5 & 0.9 & 1 & 0.7 \\ 0.35 & 0.63 & 0.7 & 1 \end{bmatrix},$$

$$(\tilde{R}_T^*)^3 = \begin{bmatrix} 1 & 0.45 & 0.5 & 0.35 \\ 0.45 & 1 & 0.9 & 0.63 \\ 0.5 & 0.9 & 1 & 0.7 \\ 0.35 & 0.63 & 0.7 & 1 \end{bmatrix} = \tilde{\psi}(\tilde{R}_T^*).$$ ■

It is trivial to verify that the rates of convergence of both the max-min and max-product operations are identical, but the results of the transitive-closure matrix are different.

Similarly, one could repeat the process with a *min-sum* operation, which will produce the following result:

$$\chi_{\text{min-sum}}(x,y) = 1 - \chi_{\text{max-product}}(x,y) \text{ for all } (x,y) \textit{ in } \tilde{\psi}(\tilde{R}_T^*).$$

It is clear that $\tilde{\psi}(\tilde{R}_T^*) \subset \psi(R_T^*)$, since

$$\max_y \left\{ \min\left[\chi_s(x,y), \chi_s(y,z) \right] \right\} \geq \max_y \left[\chi_s(x,y) \cdot \chi_s(y,z) \right]$$

and

$$\hat{\psi}(\hat{R}_T^*) \subset \psi_{\text{ms}}(R_T^*)\text{ms}$$

where $\psi_{\text{ms}}(R_T^*)\text{ms}$ is the transitive closure under the min-sum relation.

Regarding the graph-theoretic approach, we can use the inexact graph to construct the T-clusters in V. This can be illustrated by the following example.

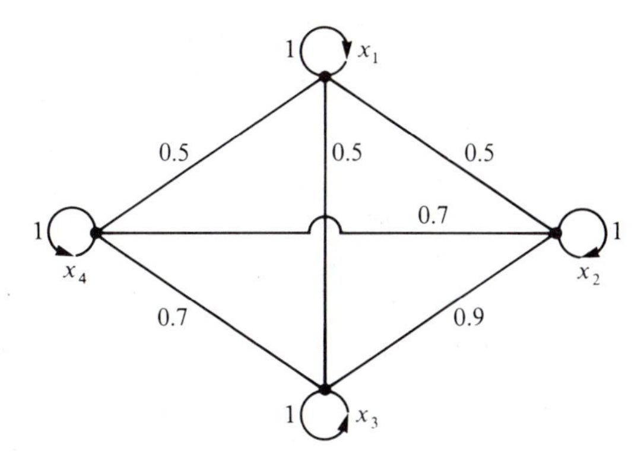

Figure 7.5.1.

■ ***Example 7.5.6***

Let R_T^* be as in Example 7.5.1, and let the graph shown in Fig. 7.5.1 represent $\psi(R_T^*)$. On every level we can construct the algebraic sum of the proximity values of the edges to a specific vertex. Thus,

$$W(x_3) = W(x_2) = 3.1,$$
$$W(x_4) = 2.9,$$
$$W(x_1) = 2.5,$$

and we have the partitions

$$R_{T \le 2.5} = \{[x_1, x_2, x_3, x_4]\}$$
$$R_{2.5 < T \le 2.9} = \{[x_1], [x_2, x_3, x_4]\}$$
$$R_{2.9 < T \le 3.1} = \{[x_1], [x_2, x_3], [x_4]\}$$
$$R_{T > 3.1} = \{[x_1], [x_2], [x_3], [x_4]\},$$

which are identical with those of Example 7.5.1. ■

We can extend these ideas to present a solution to the classification problem of time-varying patterns, since, to date, the literature on pattern recognition and classification has not considered the notion of patterns as central to the problem of classification. In fact, traditionally, the patterns have been assumed to be a set of time-invariant features.

Let pattern x be presented at intervals of Δt. By assumption, the pattern represents a dynamic system and we can introduce a time function $x(t)$ defined as

$$x(t) = x(k)$$

for

$$(k-1)\Delta t \le t < k\,\Delta t, \qquad k = 1, 2, \dots .$$

Let $X(t)$ be the space of fuzzy sets describing all possible properties η at time t, where property η is a combinational structure of dynamic and history (denoted as D/H). The subset $B(t)$ is the set of reasonable alternatives at time t. Then a characteristic function for $B(t)$ is given by the mapping

$$\chi_{B(t),t}\colon X(t) \to [0,1].$$

Moreover, each membership function $\chi_{B(t),t^*}$ represents the ranking given to all fuzzy sets at time t^* with respect to the set of criteria and constraints imposed at time t, which have produced the subset $B(t)$ of $X(t)$. Thus, for $t^* > t$ classification is considered at a later time with respect to previous criteria and constraints.

Definition 7.5.3

A dynamic fuzzy relation (DFR), $R(t,t^*)$, *is a fuzzy subset of Cartesian product*

$$X(t) \times X(t^*) = \{\langle x(t), x(t^*)\rangle | x(t) \in X(t) \wedge x(t^*) \in X(t^*)\}$$

with the characteristic function

$$\chi_{R(t,t^*)}\colon X(t) \times X(t^*) \to [0,1].$$

Clearly, $\chi_{R(t,t^*)}$ can be used as some measure of strength of relationship over a period of time. The operations described for static relations can be applied here in the time context.

■ ***Example 7.5.7***

Let $R(t,t^*)$ and $R(t^*,t^{**})$ be two time-dependent DFR's. Namely, $R(t,t^*) \subseteq X(t) \times X(t^*)$ and $R(t^*,t^{**}) \subseteq X(t^*) \times X(t^{**})$. The composition $R(t,t^*) \circ R(t^*,t^{**}) \subseteq X(t) \times X(t^{**})$ can be defined by the characteristic function

$$\chi_{R(t,t^*) \circ R(t^*,t^{**})}(\alpha,\beta) = \max_{\gamma}\left\{\min\left[\chi_{R(t,t^*)}(\alpha,\gamma), \chi_{R(t^*,t^{**})}(\gamma,\beta)\right]\right\}$$

for $\alpha \in X(t)$ and $\beta \in X(t^{**})$. ■

Definition 7.5.4

The fuzzy elements $\alpha(t)$ *and* $\gamma(t)$, *such that* $\alpha(t) \in X(t)$ *and* $\gamma(t) \in X(t^*)$, *are said to be* λ*-compatible with respect to* $R(t,t^*)$ *iff* $\chi_{R(t,t^*)}[\alpha(t), \gamma(t)] \geq \lambda$.

Obviously we can now collect all λ-*compatible* pairs $[\alpha(t), \lambda(t)]$ into a well-defined pattern.

Definition 7.5.5

Let $\alpha(t) \in X(t_1)$ *and* $\gamma(t) \in X(t_2)$ *at distinct times* t_1, t_2 *such that* $t_1 < t_2$. *Then* $\alpha(t)$ *and* $\gamma(t)$ *are chained* iff *for any* $t \in [t_1, t_2]$ *there exists* $\delta(t) \in X(t)$ *such that*

$$\min\left\{\chi_{R(t_1,t)}[\alpha(t), \delta(t)], \chi_{R(t,t_2)}[\delta(t), \gamma(t)]\right\} > 0.$$

Theorem 7.5.4

Let $\alpha(t) \in X(t_1)$ *and* $\gamma(t) \in X(t_2)$ *be* λ*-compatible with respect to* $R(t_1,t) \circ R(t,t_2)$, $\forall t \in [t_1, t_2]$, *then* $\alpha(t)$ *and* $\gamma(t)$ *are chained.*

Proof

Clearly, by the definitions of λ-compatible and fuzzy composition,

$$\max_{\delta(t)}\left\{\min\left[\chi_{R(t_1,t)}[\alpha(t),\delta(t)],\chi_{R(t,t_2)}[\delta(t),\gamma(t)]\right]\right\} \geq \lambda, \quad \forall t \in [t_1,t_2].$$

Therefore, there exists $\delta(t) \in X(t)$ such that

$$\min\left\{\chi_{R(t_1,t)}[\alpha(t),\delta(t)],\chi_{R(t,t_2)}[\delta(t),\gamma(t)]\right\} \geq \lambda > 0$$

since $0 < \lambda \leq 1$. (The case $\lambda = 0$ is not interesting since it shows only a zero measure of connectiveness between elements.) Q.E.D.

Theorem 7.5.5

Let $B(t) = [b_{ij}(t)]$ be a $p \times p$ fuzzy matrix at a given time t, and let $b_{ii}(t) \geq \max[b_{ij}(t), b_{ji}(t)]$, $\forall i, j \mid i, j \in \{1,2,\ldots,p\}$.

Then we have

$$\forall i,j\left[b_{ij}(t) = \max_k\left\{\min\left[b_{ik}(t),b_{kj}(t)\right]\right\}\right]$$

iff

$$\forall i,j,k\left[b_{ij}(t) \geq \min\left\{b_{ik}(t),b_{kj}(t)\right\}\right].$$

Proof

(i) Clearly, because the maximum is taken over all k, the left-hand side implies the right-hand side.

(ii) The statement

$$\forall i,j,k\left[b_{ij}(t) \geq \min\left\{b_{ik}(t),b_{kj}(t)\right\}\right]$$

clearly implies that

$$\forall i,j\left[b_{ij}(t) \geq \max_k\left\{\min\left[b_{ik}(t),b_{kj}(t)\right]\right\}\right].$$

Also

$$\max_k\left\{\min\left[b_{ik}(t),b_{kj}(t)\right]\right\} \geq \min\left[b_{ii}(t),b_{ij}(t)\right].$$

However,

$$b_{ii}(t) \geq \max\left[b_{ij}(t),b_{ji}(t)\right], \quad \forall i,j$$

implies

$$b_{ij}(t) = \min\left[b_{ii}(t),b_{ij}(t)\right].$$

This can be easily seen from the following:

$$\begin{aligned}\min\left[b_{ii}(t),b_{ij}(t)\right] &= b_{ii}(t)\cdot b_{ij}(t)\\ &= \left[b_{ij}(t)+b_{ji}(t)\right]\cdot b_{ij}(t)\\ &= b_{ij}(t).\end{aligned}$$

Hence

$$\max_k\left\{\min\left[b_{ik}(t),b_{kj}(t)\right]\right\} \geq b_{ij}(t),$$

and thus the equality is established. Q.E.D.

In conclusion, the basis for the similarity matrix is the assumption that pairwise-similarity coefficients between elements, or vectors in a vector space, can be subjectively found; based on these values we can partition the data-set into fuzzy partitions ranging between the *conjoint partition* (one category) and disjoint partition (singletons). The pairwise-similarity measures can be obtained with the help of some feature extraction process which we assume to be available. It is important to note that the selection of this function (or the features) is a subjective process that is quite fuzzy by itself. The reason for that is that one can always argue that the *similarity measure* is "good" or "bad" and that it is never objective and never the best possible. Therefore, the partition suggested by this criterion should compensate for such possible perturbations in the similarity values. That is, the partition proposed should detect the *internal structure* of the given data-set, and put less emphasis on the absolute values of the similarity measures.

Once the grouping has been achieved, we can evaluate the partition, especially in cases where the number of categories is unknown *a priori*. However, it should be noted again that data clustering is not a rigorously or uniformly defined concept. The division of a data-set into classes whose elements are similar in some sense is confronted with decompositions into subsets exhibiting a cohesiveness that is not entirely related to point-to-point similarity.

The value judgment of the user is the ultimate criterion for evaluating the meaning of the classification. If using these techniques produces an answer of value, no more need be asked of them.

7.6 POSSIBILISTIC SEARCH TREES

Before defining the concept of a possibilistic search tree, we need to investigate ways to arrive at possibilistic measures for record-request frequencies. In this section, based on a paper by Flowers and Kandel [1985], we will examine the plausible techniques for finding these frequencies when the request will be fuzzy by nature, and then use these frequencies to build "good" search trees, in terms of average searching times. The concepts of conjunctive and disjunctive possibilities, conditional possibilities, fuzzy inference, and necessity sets will be analyzed. The type of search to be done affects the expected record-request frequencies and thereby should dictate which technique is utilized in determining these anticipated request frequencies.

Throughout this discussion, we will assume that searches will be made in a file F containing records with information pertaining to students. Each record $r \in F$ is an ordered k-tuple

$$(r_1, r_2, \ldots, r_k)$$

of attribute values, where r_i corresponds to the attribute value of the ith attribute. Some considered attributes will be student number, age, height, weight, class, semester grade-point average, and grade-point average (GPA). We will look for possibilistic and probabilistic record-request frequencies relevant to specified searches.

Suppose we are going to search the file for students who are 21 years old. This search is straightforward and involves no fuzziness. There is no need to utilize possibilistic measures here. To accurately access the request frequencies using probability, we merely count the number of records with age attribute values equaling 21, that is, age(r) = 21, and divide 1 by this number to obtain the request frequencies for these records. All other records receive a 0 retrieval frequency.

The applicability of possibilistic techniques becomes apparent when the type of search to be initiated involves fuzziness. We will define the concept of a fuzzy search.

Definition 7.6.1

A fuzzy search is a specified search where the expected record-request frequencies are based on one or more fuzzy concepts.

Suppose we desire to search file F for "good students," in terms of their grade-point average. Now, we have a fuzzy search, i.e., the notion of a "good student" is clearly imprecise and subjective. We can translate the associated proposition "Y is a good student" to a relational assignment equation of the form

$$R(\text{GPA}(Y)) = \text{good}.$$

Here, Y denotes a particular record $r \in F$. This indicates that the proposition "Y is a good student" has the effect of assigning the fuzzy set representing the concept "good" to the fuzzy restriction on the values of GPA(Y), where GPA(Y) is the implied attribute of Y. In this example, the universe U consists of the real numbers in the interval from 0.0 to 4.0.

Thus we associate a possibility distribution, $\Pi_{\text{GPA}(Y)}$, with GPA(Y) that is equal to R(GPA(Y)). The possibility distribution function associated with GPA(Y), that is, $\pi_{\text{GPA}(Y)}$, is equal to the membership function of the fuzzy set "good" (meaning good student in terms of GPA here).

The possibilistic record request frequency for a record r will be defined as follows:

Definition 7.6.2

Given a relational assignment equation of the form

$$R(A(Y)) = G,$$

which associates a possibility distribution, $\Pi_{A(Y)}$, with $A(Y)$ that is equal to $R(A(Y))$ and a possibility distribution function associated with $A(Y)$, $\pi_{A(Y)}$, which is numerically equal to the membership function of G, then the possibilistic record request frequency (PRRF) for a record r, represented by Y, is numerically equal to $\chi_G(u)$, where $u = A(Y)$.

Thus, in this search, the possibilistic request frequency for a record r will be equal to $\chi_{\text{Good}}(\text{GPA}(r))$. It can be interpreted as meaning the possibility that the record r will be requested given that we will be looking for "good students."

The possibilistic record-request frequencies are defined in a manner similar to possibility distribution functions. The PRRFs are numerically equal to the value obtained by applying the applicable membership function to each applicable attribute value. Thus, the PRRFs are directly related to the membership function of the fuzzy concept involved in the specified search. So the PRRFs are subjectively determined when the membership function is defined. Therefore, care must be taken in defining this function. The definer should subjectively declare a function that reflects, as closely as possible, the "meaning" of the search.

As an illustration, assume that the file contains the five records with associated GPA attribute values listed in Table 7.6.1. Then, by subjectively defining the membership function of the fuzzy concept "good student" as

$$\chi_{\text{Good}}(u) = \frac{u-2}{2},$$

where $u \in U$ and u denotes $\text{GPA}(r)$, we can obtain the possibilistic record-request frequencies for these records. They will be equal to $\chi_{\text{Good}}(\text{GPA}(r^i)$, $i = 1, \ldots, 5$, and are listed in Table 7.6.1.

Relative to the same searching problem, how can the probabilistic record-request frequencies be determined? Since the search is fuzzy by nature, we will have to translate the fuzzy concept of "good student" to a hard (nonfuzzy) concept, in order to work with probabilities. Note that this translation process will be somewhat subjective, and one may see the benefit of using possibilities here. To be consistent with χ_{Good}, let's translate the concept of a "good student" to mean a student whose grade-point average is greater than 2.0.

Then, to find the probabilistic record-request frequencies (PBRRFs), we count the number of records with GPA attribute values greater than 2.0, divide 1 by this, and assign the resulting value as the PBRRF for these records. All other records receive a 0 PBRRF. The PBRRFs for the records in Table 7.6.1 are listed in Table 7.6.2.

Now, if the search is defined such that it is known that, of the students who are good, "the better students will be requested more," then the PRRFs in

Table 7.6.1

Record	$\text{GPA}(r^i)$	$\text{PRRF}(r^i)$
r^1	4.0	1
r^2	3.6	0.8
r^3	3.5	0.75
r^4	2.9	0.45
r^5	2.0	0

Table 7.6.2

Record	GPA(r^i)	PBRRF(r^i)
r^1	4.0	0.25
r^2	3.6	0.25
r^3	3.5	0.25
r^4	2.9	0.25
r^5	2.0	0

Table 7.6.1 seem to depict this view better than the PBRRFs in Table 7.6.2. In this case, of course, we could obtain similar measures for the PBRRFs by dividing each

$$\text{GPA}(r^i) > 2.0 \text{ by } \sum_i \text{GPA}(r^i) > 2.0;$$

but this is not intuitive by nature. Also, generally, the results do not reveal, at least at first glance, the relative differences between the GPAs. For example, in Table 7.6.1, a 4.0 has a 1 PRRF and a 2.9 has a 0.45 PRRF. If the PBRRFs are adjusted as described above, a 4.0 has a 0.286 PBRRF and a 2.9 has a 0.207 PBRRF. The relative difference seems to be more apparent with the PRRFs.

Possibility seems to be more intuitive in searches of this type, i.e., fuzzy searches. Many times the probability is not known or is difficult to assess. Possibilistic techniques could be useful in these cases by saving the trouble and expense of calculating probabilities. Also, generally, probabilities are more accurate when large samples are used. If large samples are not available, possibilistic techniques could again be utilized.

Consider now a composite fuzzy search involving the connectives "and" or "or." For example, suppose we want to search the file for good students and students who are useful for summer jobs. This example illustrates a conjunction of two fuzzy propositions. Conjunctive and disjunctive searches are common in many applications.

We will use a student's class to aid in defining the fuzzy concept "useful for a summer job." If we view the two propositions above as being noninteractive, then we can apply the conjunction translation rule to obtain the PRRFs for this search. The membership function of the fuzzy concept "useful for a summer job" will be defined as

$$\chi_{\text{useful}}(u) = \begin{cases} 0.4, & u = 1.0, \\ 0.8, & u = 2.0, \\ 1, & u = 3.0, \\ 0, & u = 4.0, \end{cases}$$

where $u \in U$, u denotes CLASS(r), and $u = 1.0$ signifies a freshman; $u = 2.0$ signifies a sophomore; $u = 3.0$ signifies a junior; and $u = 4.0$ signifies a senior. The PRRFs for the five sample records with indicated attribute values for CLASS(r^i) can be found by determining the PRRFs for each proposition and

Table 7.6.3

Record	GPA(r^i)	CLASS(r^i)	PRRF(r^i) good	PRRF(r^i) useful	PRRF(r^i) good and useful
r^1	4.0	4.0	1	0	0
r^2	3.6	1.0	0.8	0.4	0.4
r^3	3.5	2.0	0.75	0.8	0.75
r^4	2.9	3.0	0.45	1	0.45
r^5	2.0	2.0	0	0.8	0

then using the conjunctive translation rule, i.e., select the min. They are listed in Table 7.6.3.

Correspondingly, the PBRRFs for this search can be found by employing the multiplicative law of probability for independent events. This law states that given two events, A and B, the probability of the intersection, AB, when A and B are independent, is

$$P(AB) = P(A)P(B).$$

Therefore, to find the PBRRFs for these five records, we find the PBRRFs relative to each proposition and take their product.

Again, we will have to translate the fuzzy concept "useful for a summer job" to a hard concept in order to use probabilistic techniques. This concept will be interpreted in a manner such that all classes except the senior class will be considered useful. The calculated PBRRFs for this search are listed in Table 7.6.4.

Similar searches using disjunctive propositions could utilize these same techniques. Also, the fuzzy search could involve proposition(s) with modifiers, such as very, quite, not, etc. The modifier rule presented by Zadeh in [1983d] could then be utilized to find the PRRFs.

If G is a fuzzy restriction associated with Z, then a proposition of the form "Z is dG" associates a modification of G, G^+, defined by d with Π_Z. For example, define $G^+ = G^2$ if $d \triangleq$ very; $G^+ =$ complement of G if $d \triangleq$ not; and $G^+ = \sqrt{G}$ if $d \triangleq$ more or less.

Table 7.6.4

Record	GPA(r^i)	CLASS(r^i)	PBRRF(r^i) good	PBRRF(r^i) useful	PBRRF(r^i) good and useful
r^1	4.0	4.0	0.25	0	0
r^2	3.6	1.0	0.25	0.25	0.0625
r^3	3.5	2.0	0.25	0.25	0.0625
r^4	2.9	3.0	0.25	0.25	0.0625
r^5	2.0	2.0	0	0.25	0

Table 7.6.5

Record	GPA(r^i)	PRRF(r^i) more or less good
r^1	4.0	1
r^2	3.6	0.89
r^3	3.5	0.87
r^4	2.9	0.67
r^5	2.0	0

As an illustration, suppose we want to search the file for all the "more or less good students." Then, using the membership function

$$\chi_{\text{Good}}(u) = \frac{u-2}{2}$$

and the aforementioned modifier rule, we can obtain the PRRFs listed in Table 7.6.5

Now suppose we want to search the file using a fuzzy search involving conditional possibilities. For example, suppose we want to search for the students with "good present semester grade-point averages, given that they have a good overall GPA." This search can involve conditional possibilities, and the techniques relating to conditional possibilities can be applied to determine the PRRFs.

$\Pi(A|B)$ denotes the possibility that A will occur, given that B has occurred. The definition of $\Pi(A|B)$ requires that A and B be noninteractive, but they do not have to be independent. In this example, A denotes the elastic constraint represented by the concept "good semester grade-point average" and B denotes "good overall GPA." Here, A and B are noninteractive.

The following definition provides a means of finding the conditional possibility. The conditional possibility for a particular record will become the PRRF for that record relative to the specified search.

Definition 7.6.3

Given that u and v_i are noninteractive, but not necessarily independent:

$$\Pi(v_i|u) = \begin{cases} 1.\ \Pi(v_i) & \textit{for } \Pi(v_i) < \Pi(u), \\ 2.\ [\Pi(u), 1] & \textit{for } \Pi(v_i) \geq \Pi(u). \end{cases}$$

To illustrate, suppose the membership function of B is defined as before, that is,

$$\chi_{\text{GoodGPA}}(u) = \frac{u-2}{2}.$$

Suppose we refer to the third record of Table 7.6.1, r^3, which belongs to B with a grade of membership of 0.75. If the membership function of A is defined as

$$\chi_{\text{GoodSGPA}}(u) = \frac{u - 2.5}{1.5}$$

and r^3 has a semester GPA attribute value equaling 3.0, then r^3 belongs to A with a grade of membership of $0.\bar{3}$. Therefore, by Definition 7.6.3, the conditional possibility is equal to $0.\bar{3}$ and this would be r^3's PRRF for the search.

Suppose we want to search the file for all the short, fat students. Also, suppose we know that "if y is short, then (0.4) y is heavy," where 0.4 means, roughly, that we have 40% of confidence in the conclusion. If we also have a possibility distribution for "y is short," then we can use the fuzzy inference rule to obtain a possibility distribution for "y is heavy" and from this obtain the PRRFs.

Of course, all students who have a 0 possibility of being short will receive a 0 PRRF for their record. The fuzzy *modus ponens* rule described in Chapter 6 can be used to find the possibility distribution for "y is heavy," given the possibility distribution for "y is short" and the above implication. Once we have this distribution, we can use it to determine the PRRFs for the "short" students based on their weight.

Let us now develop the basic idea of introducing a degree of confidence in the conclusion, by altering the membership function of the fuzzy set that represents the conclusion through a suitable operator to reflect this degree of confidence in the conclusion.

The modifying function is a function h defined by

$$h(K, y_i) = y_i^K,$$

where $K \in [0, 1]$ denotes the degree of confidence in the conclusion and y_i is the degree of membership in the considered fuzzy set.

As an example, let Q represent the assertion "y is short." Let

$$\begin{aligned} Q = {} & 1/48 + 0.9/50 + 0.8/52 + 0.7/54 + 0.5/56 + 0.4/58 + 0.3/60 \\ & +0.2/62 + 0.1/64 + 0/66 \end{aligned}$$

be the corresponding fuzzy set where height is depicted in inches. The conclusion "y is heavy" can be given by:

$$\begin{aligned} R = {} & 0/100 + 0.3/115 + 0.4/120 + 0.5/125 + 0.6/130 \\ & +0.7/135 + 0.9/140 + 1/145, \end{aligned}$$

where weight is expressed in pounds. The implication "If y is short, then (0.4) y is heavy," because of the consistency between F and h, can be replaced by "If y is short, then y's weight is $R^{0.4}$" with full confidence in the implication, where

$$\begin{aligned} R^{0.4} = h(0.4, R) = {} & 0/100 + 0.62/115 + 0.69/120 + 0.76/125 \\ & +0.82/130 + 0.87/135 + 0.96/140 + 1/145. \end{aligned}$$

After the fuzzy inference rule is applied, the PRRFs can be found as described previously.

The previously presented techniques are plausible ways to attain PRRFs relevant to specified fuzzy search queries. There is a basic concept underlying these techniques. What we are actually doing, in a sense, is dividing the records into two sets — a set of records that will be requested (along with their PRRFs) and a set of records that will not be requested (those with 0 PRRFs), relevant to a specified search query. Necessity sets can be used to reflect this concept.

If Y is a set of objects, B an attribute associated with all (or some) of the elements of Y, β a universal set of values, and S an ordinary subset of β, then a necessity set, relative to S, is the set of elements in Y whose attribute value necessarily (or certainly) belongs to S. Thus, relative to a certain search, by appropriately selecting the subset S, we can define a necessity set that contains the records that will be requested in our search. The remaining records form the set of elements that will not be searched for. Note that Y can be an ordinary, multi-valued, or fuzzy mapping. Also, S could be a fuzzy subset.

To illustrate this idea, suppose we want to search the file for all the "old" students. Here, Y is the set of student records, B is the attribute "age," and let β be the set of integers from 15 up to 100. By some subjective measure, S will be defined to be the set of integers from 25 to 100, inclusive. Then, given the records in Table 7.6.6, with their corresponding age attribute values, we can procure the set of elements in Y whose attribute value necessarily belongs to S, that is,

$$B^{-1}*(S) = \{y \in Y, B(y) \neq \emptyset, B(y) \subset S\} = \{r^2, r^4\}.$$

Therefore we can assert that the elements of the necessity set, $B^{-1}*(S)$, are the records that will be requested in this search. The remaining records form the set of records that will not be requested in this search, i.e.

$$\{r^1, r^3, r^5\}.$$

Thus, we have actually "negated" these records from the searching process and will consider only those records in the necessity set.

Note that B is an ordinary mapping in this example; that is, $B(y)$ is a singleton. This necessity-set concept can be extended to the cases where B is a multi-valued or fuzzy mapping. Also, this idea can be applied when finding necessity sets for conjunctive or disjunctive propositions in a search. The

Table 7.6.6

Record	Age(r^i)
r^1	19
r^2	30
r^3	21
r^4	26
r^5	22

process involves finding a necessity set relative to each proposition. For example, suppose we want to search for all the "old" and "tall" students. We find a necessity set for the fuzzy concept "old" as before, find a necessity set for the fuzzy concept "tall," and then create a necessity set from the intersection of these two sets. The resulting necessity set contains the records that will be requested in this search.

What we have actually done by establishing the necessity set is to impose a necessity threshold on set Y. In other words, the set we "select" is a subset of Y containing $y \in Y$ such that $B(y) \geq T$, where $T = 25$. At the same time we are "negating" within set Y by placing a threshold on it when we define the set.

In dealing with the fuzzy propositions involved in various search queries, one might wonder about the degree of truth of these propositions in relation to what is known about the real situation. The truth of a proposition can be regarded as the conformity of what is stated with what is known about reality or with what reality is perceived to be. Obtaining the degree of truth of a proposition results from a comparison procedure, matching a representation of the reality against a representation of the contents of the proposition. Note that the degree of truth is not absolute — it may depend on the representation and on the comparison procedure that is used. We can now discuss this matter of the degree of truth of a proposition, i.e., matching statement against reality, by defining two independent scalar comparison indices to estimate the agreement of a possibility distribution function π_S with another one π_T, where these two possibility distribution functions are supposed to represent, respectively, the contents of a proposition S and the perception of the corresponding reality T modelled on the universe associated with S. They are

$$\mathrm{POSS}(S|T) = \operatorname*{Sup}_{u \in U} \min(\pi_S(u), \pi_T(u))$$

and

$$\mathrm{NES}(S|T) = \operatorname*{Inf}_{u \in U} \max(\pi_S(u), 1 - \pi_T(u)),$$

and denote an approximation of the possibility that the reality is in agreement with the contents of the proposition S taking into account the uncertainty on T, and the necessity measure of the same event, respectively.

The techniques examined above provide plausible means of finding possibilistic record-request frequencies relative to a particular fuzzy search. The type of search to be invoked should dictate which technique (or techniques) is applied. Other techniques could also be utilized. Necessity sets provide a means of depicting an underlying concept relative to the selection of record-request frequencies which involves "selection and negation" of records. The degree of truth of propositions involved in search queries can be questioned and investigated.

Having obtained the possibilistic record-request frequencies for a search, the goal now is to organize the records in such a manner that an efficient search can be performed. This section examines the use of tree structures for this organization, in particular, a possibilistic search tree. It will be assumed, as is generally the case, that the records are ordered on a specified attribute value designated as the key. Thus, the objective is to organize the records in a tree structure so as to utilize the possibilistic record-request frequencies to improve the average search time, while at the same time keeping the records ordered by the key so that other, more general, searches can be performed.

Records with a 0 PRRF relative to a particular search must still be considered when building a possibilistic search tree. The reason is that we want the tree to be useful for other, more general, searches and hence we do not want to omit any records that may be needed in these "other" searches. For example, we may want to search the tree for a specified record to obtain a value of a requested attribute. Therefore, we want to include records with a 0 PRRF relative to a particular search used in building the tree.

Definition 7.6.4

The possibilistic internal weight (PIW_i) *of a record* r^i *is defined as the possibilistic record request frequency of the record, where* $\mathrm{PIW}_i = \mathrm{PRRF}(r^i) \in [0,1]$ *and* $i = 1, \ldots, n$.

Note that a node considered here will consist of the key attribute value of the record associated with the node. The question of where to actually store the other attribute values of a record is not addressed, but can be resolved by established techniques.

Definition 7.6.5

Let the records r^i *be ordered in ascending order by key attribute values, where* $i = 1, \ldots, n$. *The possibilistic external weights* (PEW_i) *of these records is defined as the possibility that the search argument's key attribute value lies between those of* r^i *and* r^{i+1}, *where* PEW_0 *is the possibility that the search argument's key attribute value is less than that of* r^1 *and* PEW_n *is the possibility that it is greater than that of* r^n, *where* $\mathrm{PEW}_i \in [0,1]$.

Definition 7.6.6

A possibilistic search tree T_r *is a binary search tree with root* r *constructed using the possibilistic internal weights and the possibilistic external weights of the records such that the possibilistic average search time is minimized, relative to a specified search.*

As is evident, a possibilistic search tree is a generalization of an optimum binary search tree with possibilities in the place of probabilities. Note that, by this definition, the records will be in ascending order by key attribute values when the tree is traversed in symmetric order. Thus, the records will remain in key order, subsequent to symmetric traversal, as is desired. Other definitions pertaining to possibilistic search trees can be made.

Definition 7.6.7

The search path (s, t) is the path of the possibilistic search tree from node s to node t. The search path s is the path $(root, s)$. The search path $(s, t, u, v, \ldots, y, z)$ is the path from node s via nodes $t, \ldots, y$ to node z. The search path from node s is determined such that its left subtree is searched if the search argument key is less than the key value associated with node s and the right subtree otherwise.

Thus, the searching process is the same as in a binary search tree.

Definition 7.6.8

Two possibilistic search trees, relevant to the same set of records and PIWs *and* PEWs, *are equivalent if and only if all of the records appear on the same level in both of the trees.*

The notions of *complete*, *full*, and *balanced* trees can be applied to possibilistic search trees. The procedure for building a possibilistic search tree is similar to that for an optimum binary search tree. The input consists of the key attribute values of the records entered in ascending order, followed by the corresponding PIWs for the records (in this same order), and lastly the PEWs in order from PEW_0 to PEW_n. The output produced is the possibilistic search tree.

The definition and procedure for building a possibilistic search tree involve the possibilistic external weights. The PEWs represent the possibility that the search argument's key value is between two records' key attribute values. The PEWs are of importance in calculating average search times in that they reflect the possibilities of unsuccessful searches.

This brings up the categorization of two types of searches that we will examine. The first is a search in which we know beforehand that all retrieval requests will be successful, i.e., we know that what we are searching for is in the tree. The second case is the one where there will be successful and unsuccessful searches. Note that the average search times should be slower in the second case.

How should the PEWs be determined? Previously, techniques for determining the PIWs were presented, but no mention was made of ways to obtain the PEWs. Obviously, in a search where all search queries will be successfully answered, the PEWs should be set to zero. In the case where there will be successful and unsuccessful queries, the PEWs will be very data-dependent. They can be found using some of the previous techniques and/or subjective knowledge regarding what will be searched for.

As an illustration of a possibilistic search tree, suppose we are going to invoke a search where the students who are good and useful for a summer job will be requested more often. We assume that we will have only successful queries. The key attribute will be the students' grade-point averages. Then, using the key attribute values with their corresponding PIWs listed in Table 7.6.7, we obtain the possibilistic search tree pictured in Figure 7.6.1.

Table 7.6.7

Record	GPA(r^i)	PIW_i
r^1	0.2	0
r^2	1.4	0
r^3	2.0	0
r^4	2.7	0
r^5	2.9	0.45
r^6	3.0	0.5
r^7	3.2	0.4
r^8	3.5	0.75
r^9	3.6	0.4
r^{10}	4.0	1

This tree is in symmetric order by GPA and could be easily searched for other queries related to student GPAs. The corresponding optimum binary search tree, constructed using probabilities, for this search is given in Figure 7.6.2. The probabilities used are listed in Table 7.6.8.

In this example, PIWs and probabilities were found using conjunction techniques. The probabilities were scaled to sum to one.

Average search time measures are used to evaluate the performance of a data structure in a particular search and are used to compare different data structures. The possibilistic average search-time formula for possibilistic search trees is defined as follows.

Definition 7.6.9

The possibilistic average search time (PAST) *for a possibilistic search tree is equal to*:

$$\frac{\sum_{1 \le j \le n} \max \text{PIW}_{\text{level } j}\, x(j) + \sum_{1 \le j \le n} \max \text{PEW}_{\text{level } j}\, x(j-1)}{\sum_{1 \le j \le n} \max \text{PIW}_{\text{level } j} + \sum_{1 \le j \le n} \max \text{PEW}_{\text{level } j}},$$

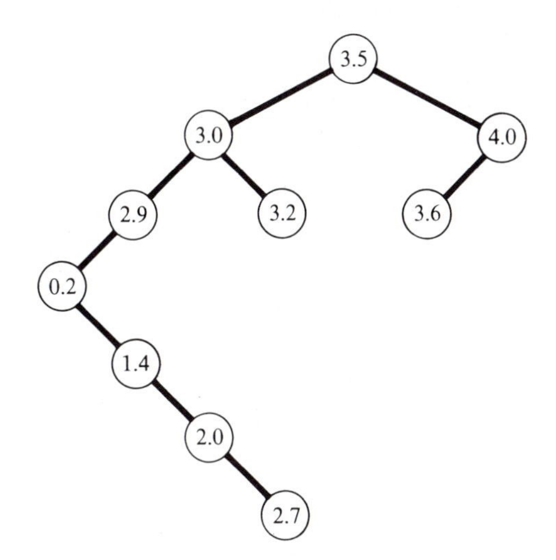

Figure 7.6.1. Possibilistic search tree.

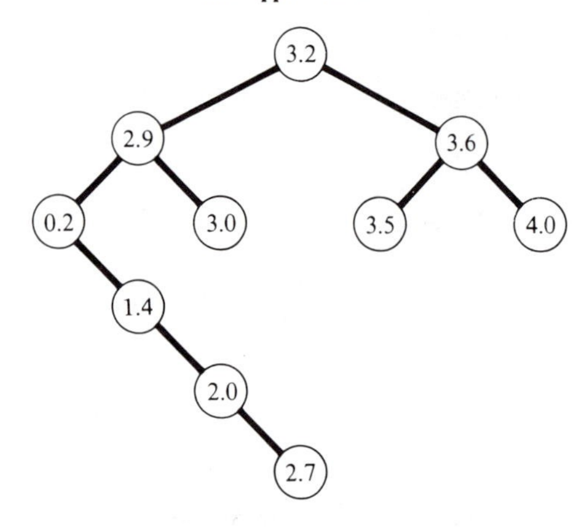

Figure 7.6.2. Optimum binary search tree.

where n denotes the number of levels in the tree and the level of a vertex w is equal to one plus the length of the path between the root and w.

Note that this formula can be used in the case of successful and unsuccessful retrievals, as well as the case of all successful retrievals. In the case of all successful retrievals, the maximum PEWs on all levels are zero and the second terms in the numerator and denominator become zero. For example, the PAST for the possibilistic search tree in Figure 7.6.1 is 1.864 (rounded to three decimal places).

In comparison, the average search time for an optimum binary search tree is defined as

$$\sum_{1 \leq i \leq n} \alpha_i(\text{level of } K_i) + \sum_{0 \leq i \leq n} \beta_i((\text{level of } i) - 1),$$

where α_i denotes the probability of using key K_i in a successful search and β_i

Table 7.6.8

Record	GPA(r^i)	$\alpha(r^i)$
r^1	0.2	0
r^2	1.4	0
r^3	2.0	0
r^4	2.7	0
r^5	2.9	0.16
r^6	3.0	0.16
r^7	3.2	0.16
r^8	3.5	0.16
r^9	3.6	0.16
r^{10}	4.0	0.16

denotes the probability of using some key K between K_i and K_{i+1} in an unsuccessful search. (Note that here β_0 is the probability that $K < K_1$ is used and β_n is the probability that $K > K_n$ is used.) Again, an "all successful" search, the second term becomes zero. The average search time for the optimum binary search tree in Figure 7.6.2 is $2.\bar{3}$.

This definition of the possibilistic average search time can be extended to answer other searching-time questions relevant to the specified search. For instance, pertaining to the search for which the possibilistic search tree in Figure 7.6.1 was built, we can ask: What is the average search time for the most possible group? Other similar questions could be asked and answered by modifying Definition 7.6.9 to interpret the question.

For example, to answer the question "What is the average search time for the most possible group in Figure 7.6.1?" we can define a possibilistic expected value (PEV) to describe the possibility distribution (that is, PIWs) in Table 7.6.7 as an entity. In defining a PEV, it seems that we should give a higher weight to the higher possibilities, since anything associated with them is much more conceivable to occur than anything associated with a small possibility. For example, if the distribution

$$\frac{0.8}{1} + \frac{0.1}{2}$$

occurred, the possibility of the distribution should be closer to 0.8, since the value associated with it is most conceivable to occur. It should be closer to 0.8 than the average of the two possibilities would be. Also, when we get a PEV for the PIWs, we see reasons to ignore 0 possibilities. We do not care about the records that will not be searched for.

Definition 7.6.10

The possibilistic expected value (PEV) *of a given distribution*

$$x_1/X_1 + \cdots + x_n/X_n$$

is defined as

$$\text{PEV} = \frac{\sum_{i=1}^{n} x_i^2}{\sum_{i=1}^{n} x_i}.$$

This definition satisfies the desired properties and PEV $\in [0, 1]$. We can use this PEV measure to categorize "the most possible group" as consisting of those records whose PIWs are greater than or equal to the PEV for the distribution. For example, the PEV for the distribution in Table 7.6.7 is 0.667 (rounded to three decimal places). The corresponding "most possible group," using the above categorization, consists of r^8 and r^{10}.

To find the average search time for this most possible group, we can apply Definition 7.6.9 to r^8 and r^{10} as positioned in the possibilistic search tree of Figure 7.6.1. The resulting average search time is equal to 1.57. Definitions 7.6.9 and 7.6.10 could also be used to find the average search time for the least possible group, i.e., those records whose PIWs are less than the PEV. In this example, the average search time for the least possible group is 2.47.

Searches were categorized into two types, "successful and unsuccessful" searches and "all successful" searches, and trees were built and average search times evaluated accordingly. PASTs for possibilistic search trees were faster than the probabilistic average search times of correspondingly built optimum binary search trees. Also, in all tests performed, the PASTs were faster than the corresponding probabilistic average search times relevant to the same possibilistic search tree. It was demonstrated how the possibilistic average search-time formula could be extended to answer more general searching-time questions, such as those involving the average search time of the most possible group of records in a search.

The primary difference between a possibilistic search tree and an optimum binary search tree is the use of possibilities in the place of probabilities. Possibility, which is based on fuzziness, will describe the record-request frequencies relevant to a fuzzy search better than probability would.

It is difficult to assess the advantages and disadvantages of using possibilities in the place of probabilities to build a search tree in an actual situation. The probabilities used in most cases are determined by subjectively translating a fuzzy concept into a hard concept. Tests could be performed with the probabilities determined empirically. The results could provide more insight into the advantages and disadvantages of using possibilities and the question of "how accurately" the PRRFs reflect request frequencies in an actual situation.

A problem inherent in building a possibilistic search tree is the time required to build it. The algorithm requires $\mathcal{O}(n^3)$ time to construct the tree, where n is the number of records. So, for a large number of records, it may not be practical to build a possibilistic search tree. Other structures could be sought and examined which employ possibilistic techniques in organizing and searching data.

The possibilistic average search-time formula presented here proved plausible for all analyzed examples. More research could be done to examine how accurate this formula is in an actual situation.

Since the PRRFs are determined subjectively, if we exercise more care in obtaining them, they will more accurately reflect the expected request frequencies. It is important to specify the search in a manner that reflects these expected request frequencies as accurately as possible. Research could be done in the area of specifying fuzzy searches and subjectively defining "workable" membership functions. Also, techniques other than those examined here could be investigated for determining PRRFs.

A problem mentioned previously in using possibilistic search trees is that of constructing one tree to be used in several fuzzy searches and maybe other, more general, searches. Methods could be sought to incorporate possibilistic search trees relevant to fuzzy searches into one tree that is optimal relevant to all of the searches.

In conclusion, the use of possibility in place of probability to determine record-request frequencies, build a tree, and assess the average search-time provides a viable method for dealing with fuzzy searches.

7.7 FUZZY RELATIONAL DATA-BASE AS AN EXPERT SYSTEM TOOL

Primary concerns in the design of any information system are sound theoretical foundations, efficiency, and ease of use. An aspect often neglected is how well a given model satisfies information requirements of the system users. Much of human reasoning deals with imprecise, incomplete, or vague information. During the last decade, research in the fields of applied computer science such as information processing, artificial intelligence, knowledge processing in expert systems (e.g., medical diagnosis systems), and pattern recognition has established the need for formulation of models of imprecise information systems that would simulate human approximate reasoning. This section is based on the work by Zemankova-Leech and Kandel [1984].

The proposed Fuzzy Relational Data-Base (FRDB) model based on the research in the fields of relational data-bases and theories of fuzzy sets and possibility is designed to allow representation and manipulation of imprecise information. Furthermore, the system provides the means for "individualization" of data to reflect the user's perception of data. As such, the FRDB model is suitable for use in expert systems and other fields of imprecise information-processing that model human approximate reasoning.

The objective of the Fuzzy Relational Data-Base (FRDB) model is the capability to handle imprecise information. The FRDB should be able to retrieve information corresponding to natural language statements such as:

"Find men who are MUCH TALLER THAN Mark."

"Find CLOSE FRIENDS of John who are NOT SMALL AND are APPROXIMATELY AS OLD AS John."

"Find cities for which it is VERY POSSIBLE to have a PLEASANT CLIMATE."

Although problems of this kind cannot be solved within the framework of classical data-base management systems, they are illustrative of the types of problem that human beings are capable of solving through the use of approximate reasoning.

The proposed FRDB model retrieves the desired information by applying the rules of fuzzy linguistics to the fuzzy terms in the query. In the examples above, the terms in block capitals are treated as labels to fuzzy sets (OLD, PLEASANT CLIMATE), fuzzy modifiers (MUCH, APPROXIMATELY), fuzzy relationships (CLOSE FRIENDS, AS...AS), fuzzy connectives (AND), truth qualifiers (VERY POSSIBLE). In summary, "fuzzified" forms of the relational algebra operations: SET OPERATIONS, SELECT, and JOIN, with the regular form of PROJECT, are used in query evaluation.

The FRDB model is designed to take into consideration individual differences in data perception. For example, a definition of a fuzzy set TALL may differ from user to user although the actual height data in the data base are the same.

The FRDB model development was influenced by the need for easy-to-use systems with sound theoretical foundations as provided by the relational data-base model and theories of fuzzy sets and possibility. The FRDB model design addresses:

1. representation of imprecise information,
2. derivation of possibility/certainty measures of acceptance,
3. linguistic approximations of fuzzy terms in the query language,
4. development of fuzzy relational operators (IS, AS...AS, GREATER,...),
5. processing of queries with fuzzy connectors and truth quantifiers,
6. null-value handling using the concept of the possibilistic expected value,
7. modification of the fuzzy term definitions to suit the individual user.

A fuzzy relational data base is a collection of fuzzy, time-varying relations which may be characterized by tables or functions, and manipulated by recognition (retrieval) algorithms or translation rules. The organization of the FRDB can be divided into:

1. Value data-base (VDB),
2. Explanatory data-base (EDB), and
3. Translation rules.

The VDB is used to store actual data values, whereas the EDB consists of a collection of relations or functions (similarity, proximity, general fuzzy relations, and fuzzy-set definitions) that "explain" how to compute the degree of compliance of a given data value with a user's query. This part of the data-base definition can be used to reflect the subjective knowledge profile of a user, thus individualizing the entire FRDB without changing the actual VDB. It should be pointed out that both VDB and EDB have the same data structure, since the functions that may be used in the EDB could be represented in the tabular form.

Queries that may contain imprecise terms (i.e., labels to fuzzy sets, for example, LIGHT COLOR, OLD), linguistic modifiers (e.g., VERY, APPROXIMATELY), qualifiers (e.g., VERY POSSIBLE, CERTAIN), and may be composed (that is, using connectors AND, OR), are evaluated by applying relational algebra operations together with built-in linguistic translation rules. A user can define specific functions or rules during a query session that can be added to the system's "vocabulary." Therefore, the FRDB can evolve over a period of time, and reflect the user's sophistication level and specific needs.

Data Types

The *domains* in the FRDB can be of the following types:

1. discrete scalar set (e.g., COLOR = {red, white, blue});
2. discrete number sets, finite or infinite (limited by the maximum computer word size and precision);
3. the unit interval [0, 1].

The *attribute values* can be:

1. single scalars or numbers,
2. a sequence (list) of scalars or numbers,
3. a possibilistic distribution of scalar or numeric domain values,
4. a real number from the unit interval [0, 1] (membership or possibility distribution function value),
5. null value (denoted by a dash).

■ ***Example 7.7.1***

Attribute values for attributes NAME and AGE are singletons of scalar and integer sets, respectively (Type 1), with a null value (Type 5) in the attribute AGE. (See Table 7.7.1.) The attribute RESIDENCY allows a sequence of scalars (Type 2) with the interpretation that Ron lives in Miami OR in Tampa OR in Orlando, where OR is exclusive, i.e., it is equally possible that Ron lives in any *one* of the three cities. HAIR COLOR is specified by a possibilistic distribution over a domain {blond, red, brown, black} (Type 3). The interpretation of this attribute value is that, for example, Tom's hair color is between brown and black, but more brown than black. The real number in [0, 1] in the column SMART is an attribute value of Type 4 indicating degree of membership in a fuzzy set SMART, or $m_{\mathrm{SMART}}(x)$. ■

Table 7.7.1. Relation PERSON in the VDB

Name	Age	Residency	Hair Color	Smart
Tom	25	Seattle, Portland	0.8/brown + 0.6/black	0.6
Bob	30	Tallahassee	0.3/blond + 0.7/red	0.4
Al	35	Boston, New York	1/brown	0.9
Ron	—	Miami, Tampa, Orlando	0.4/brown + 0.9/red	0.2

Compare now, the following example and Table 7.7.2.

■ ***Example 7.7.2***

The relation in Table 7.7.2 is an "explanation" of the concept YOUNG, or a definition of a fuzzy set YOUNG defined over the domain of the attribute AGE (for example, [0, 100]) in the VDB. ■

Table 7.7.2. Relation YOUNG in the EDB

Age	m_{YOUNG}
10	1.0
20	0.9
25	0.8
30	0.6
40	0.4

It can be observed that Type 3 is the most expressive attribute value type. The other types are special cases and are classified separately for the sake of clarity.

Relations

Every relation must have a primary key attribute that can be only of Type 1.

It is desirable to maintain all relations in the third normal form to avoid redundancy problems and update/delete anomalies.

In general, if A_i is an imprecise attribute with a domain D_i, then an attribute value can be a possibilistic distribution specified on D_i (let it be denoted by Π_{A_i}). Since precise attributes are a special case of imprecise attributes, the following definition covers all relations in the FRDB.

Definition 7.7.1

Given domain sets $D_1, \ldots, D_n$ (not necessarily distinct) of scalar type, numeric type, or unit interval, R is a possibilistic relation on these n sets if it is a set of n-tuples $(\Pi_{A_1}, \ldots, \Pi_{A_n})$ where Π_{A_i} $(i = 1, \ldots, n)$ is a possibilistic distribution representing the possible values of the attribute A_i defined on the domain set D_i.

Relations defined in the EDB are used in translation of fuzzy propositions. In essence, they relax the dependence of relational algebra operators on the regular relational operators ($=$, $\neq$, $<$, $>$, $\leq$, $\geq$).

Similarity Relation

Let D_i be a scalar domain, $x, y \in D_i$. Then $s(x, y) \in [0, 1]$ is a similarity relation with the following properties:

Reflexivity: $s(x, x) = 1$;
Symmetry: $s(x, y) = s(y, x)$;
θ-transitivity: where θ is most commonly specified as max-min transitivity. If $x, y, z \in U$, then

$$s(x, z) \geq \max_{y \in D} \{\min(s(x, y), s(y, z))\}.$$

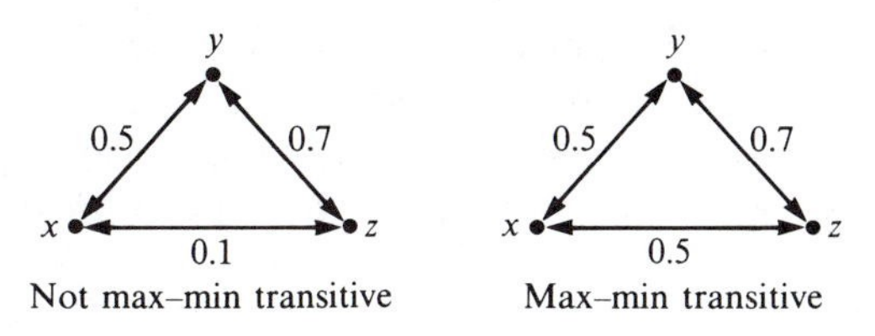

Figure 7.7.1.

■ ***Example** 7.7.3*

Max–min *transitivity*. Suppose now, that x, y, and z are elements from a set of colors, that is, x = blond, y = brown, z = black. The max–min transitivity forces the similarity of blond to black from 0.1 to 0.5 which is a contradiction to "accepted" similarity of colors. ■

Tversky* states:

> The theoretical analysis of similarity relations has been dominated by geometric models. These models represent objects as points in some coordinate space such that the observed dissimilarities between objects correspond to the metric distances between the respective points. However, . . . , metric assumptions are open to question.

Since a general similarity transitivity rule that would be appropriate for all possible data types has not been established, the requirement for transitivity is not enforced in the FRDB framework. In this respect, the role of similarity in the FRDB model differs from the max–min transitive similarity generally used in imprecise information systems based on the concepts of similarity.

Proximity Relation

Let D_i be a numerical domain and $x, y, z \in D_i$. Here $p(x, y) \in [0, 1]$ is a proximity relation that is reflexive, and symmetric with transitivity of the form

$$p(x, z) \geq \max_{y \in D_i} \{ p(x, y) * p(y, z) \}.$$

The generally used form of the proximity relations is

$$p(x, y) = e^{-\beta|x-y|}, \quad \text{where } \beta > 0.$$

This form assigns equal degrees of proximity to equally distant points. For this reason, it is referred to as *absolute proximity* in the FRDB model.

---

*A. Tversky, Features of similarity, *Psychological Review*, 84, 4, pp. 327–353, 1977.

In modelling proximities defined on numeric domains representing age, height, temperature, etc., the concept of absolute proximity does not reflect the increasing similarity (or diminishing difference) as numbers grow larger. The following definition was derived to express this type of proximity.

Definition 7.7.2

Let $x, y \in D$, where D is a numeric domain with a lower bound L. The relative proximity between x and y is defined as:

(i) $p(x, y) = e^{-\beta|(x-y)/(\max(x,y)-L)|}$ *for* $x, y \geq L$, *and* $p(L, L) \triangleq 1$, *where* $\beta > 0$.

(*ii*) $p(x, y) = e^{-\beta|(x-L)/(y-L)-(y-L)/(x-L)|}$ *for* $x, y > L$, *and* $p(L, L) \triangleq 1$, *where* $\beta > 0$.

Comment

The values of relative proximity increase faster for form (ii) for the same value of β. The form that is used depends on the user's perception of suitability to the particular attribute. Both the absolute and relative proximity satisfy product-transitivity.

Similarity and proximity are used in evaluation of queries of the general form:

"Find X such that $X.A \ominus d$"

where $X.A$ is an attribute of X, $d \in D$ is a value of attribute A defined on the domain D, and $\ominus$ is a fuzzy relational operator.

Definition 7.7.3

Let $s(x, y)$ and $p(x, y)$ be similarity and proximity relations defined for elements of scalar and numeric domains, respectively. Then the fuzzy relational operators IS, $\gtrsim$ (IS *greater than*), *and* $\lesssim$ (IS *less than*) *are defined by*

$$x \text{ IS } y \triangleq \begin{cases} s(x, y), & x, y \in \text{ scalar } D \\ p(x, y), & x, y \in \text{ numeric } D; \end{cases}$$

$$x \gtrsim y \triangleq \begin{cases} 1 - 0.5 * p(x, y), & x \geq y, \\ 0.5 * p(x, y), & x < y; \end{cases}$$

$$x \lesssim y \triangleq \begin{cases} 0.5 * p(x, y), & x \leq y, \\ 1 - 0.5 * p(x, y), & x > y. \end{cases}$$

Application of the above definition to the query translates into selecting such X where $\theta(x, d)$ satisfies user's criteria.

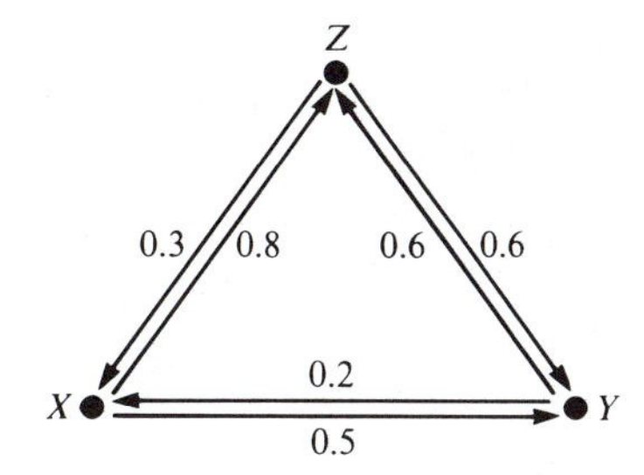

$a \to b$		$g(a, b)$
X	Y	0.5
Y	X	0.2
Y	Z	0.6
Z	Y	0.6
X	Z	0.8
Z	X	0.3

Figure 7.7.2.

General Fuzzy Relation (Link)

Let D_i be a scalar or numerical domain. A general fuzzy relation (link) can be defined in either VDB or EDB. Link may have any of the following properties:

Nonreflexive:	$g(x, x) = 0$;
ε-reflexive:	$g(x, x) = \varepsilon, \quad \varepsilon > 0$;
Reflexive:	$g(x, x) = 1$;
Nonsymmetric:	$g(x, y) \neq g(y, x)$;
$\hat{g}$-transitive:	generalized max-min transitivity.

Link can be used to express relationships that are not necessarily reflexive and symmetrical, but may obey a specific transitivity, or *transitivity improvement*. Typical relations that can be represented by a link are friendship or influence among the members of a group.

■ ***Example 7.7.4***

The influence pattern among X, Y, and Z can be represented by the directed graph of a table, as shown in Fig. 7.7.2. Suppose now that we want to find how strongly Y can influence X. In this case $g(Y, X) = 0.2$. However, it can be easily observed that X can be influenced through Z acting as an intermediary. Choosing the best of two-directional links leading from Y to X gives us a *transitivity improvement*, $\hat{g}(Y, X)$ as follows:

$$\hat{g}(Y, X) = \max[g(Y, X), \min(\min(g(Y, Z), g(Z, Y)), (\min(g(Z, X), g(X, Z))]$$

$$= \max[0.2, \min(\min(0.6, 0.6), \min(0.3, 0.8))]$$

$$= \max[0.2, \min(0.6, 0.3)]$$

$$= 0.3. \qquad ■$$

This observation can be stated formally.

Definition 7.7.4

Let $g(x,z)$ denote a link relation in D from x to z of strength $g \in [0,1]$. Then the transitivity improvement in D, $\hat{g}(x,z)$, is defined by

$$\hat{g}(x,z) \triangleq \max_{\substack{y \in D \\ y \neq x \neq z}} \{\min[\min(g(x,y), g(y,x)), \min(g(y,z), g(z,y))]\}$$

if $\hat{g}(x,z) > g(x,z)$.

Transitivity improvement $\hat{g}(x,z)$ replaces $g(x,z)$ in the query evaluation of the form:

"Find X such that X strongly influences Y"

or

"Find X such that X is strongly influenced by Y",

that is, select $\hat{g}(x,y)$, or $\hat{g}(y,x)$, respectively, with a degree of influence ($\hat{g} \in [0,1]$) that the user considers to be "strong."

Fuzzy Subsets

Let $A_1, \ldots, A_k$ be attributes in the VDB, with domains $D_1, \ldots, D_k$ (not necessarily distinct). A fuzzy set F can be defined as a mapping of elements of the Cartesian product $D_1 \times \cdots \times D_k$ onto a unit interval as follows:

(i) $F = \{(x_1, \ldots, x_k), \quad m_F(x_1, \ldots, x_k)\}$, where $(x_1, \ldots, x_k) \in D_1 \times \cdots \times D_k$, and $m_F \in [0,1]$.
(ii) $F = S(x; a, b, c)$, where x is a variable ranging over D, and S is the standard S-function with parameters a, b, and c.

Furthermore, new fuzzy sets may be defined in terms of previously defined fuzzy sets by the use of modifiers and connectives by applying basic fuzzy-set operations. The current FRDB model supports modifiers NOT, VERY/MUCH, and APPROXIMATELY, which may be combined to form modifier lists, e.g., NOT VERY, and connectives AND and OR.

Let fuzzy sets on the right-hand side denote previously defined fuzzy sets, m denote a modifier list, and let [ ] indicate optional use. Then a new fuzzy set F may be formed as follows:

(iii) $F = mF_i$; for example, SMART = NOT FOOLISH, BEAUTIFUL = VERY VERY PRETTY;
(iv) $F = [m]F_j(\{\text{AND/OR}\}\ [m]F_k)^k, \quad k = 0, 1, \ldots,$ where the clause in ( ) may be repeated k times with a choice of either AND or OR; for example, SUCCESSFUL = VERY INTELLIGENT AND DETERMINED OR RICH

Application of the weighted union of fuzzy sets can be used to express the level of importance of contributing factors in the formation of a new descriptor (fuzzy set):

(v) $F = \Sigma_{i=1}^{k} w_i * [m_i] F_i$, where $\Sigma_{i=1}^{k} w_i = 1$; for example, HEALTHY = 0.8 * NOT HEAVY + 0.2 * YOUNG.

Fuzzy sets defined in the EDB extend the expressive power of the query language by providing the means for representation of intrinsically imprecise concepts in the natural language in terms of the labels of fuzzy sets. In general, fuzzy sets are used in queries of the form:

"Find X such that $X.A$ is F,"

where $X.A$ is an attribute describing X, and F is a label of a fuzzy set defined over the attribute A.

Recall that relational algebra operators have relations as their arguments and produce a new relation.

Let us limit the present discussion to the "fuzzification" of the relational algebra operation

$$X = \text{SELECT } R \text{ WHERE } R.A = c,$$

where R is a relation, $R.A$ is an attribute name in R defined on a domain D, $c \in D$ is an attribute value, and X is the resulting relations whose tuples satisfy the comparison clause $R.A = c$.

For the sake of simplicity, let us first consider a simple query using a fuzzy set F in the comparison clause:

$$X = \text{SELECT } R \text{ SUCH THAT } A \text{ IS } F.$$

$R.A$ was simplified to A with the understanding that $A \in R$. In this case, X contains objects (tuples) whose attribute is "A IS F." The question is, what is the measure of agreement of each object in X with the query q? Recall that in general, attribute values may be possibilistic distributions, that is, the question translates into finding a measure of

"$\Pi_A(x)$ IS F."

Definition 7.7.5

Let X be a set of objects retrieved by the query $q \triangleq$ SELECT R SUCH THAT A IS F. Let R be a relation in the VDB *with attributes $A_1, \ldots, A_n$ with the corresponding domains $D_1, \ldots, D_n$ (not necessarily distinct). Let F be a fuzzy set defined in the* EDB *on k attribute(s) A of R, where $1 \leq k \leq n$. Let F be characterized by its membership function $m_F(a)$. Given the possibility distribution functions $p_A(x)$ restricting the possible values of the k attribute(s) A for all $x \in R$, the possibility measure, $p(q)$, associated with a query, q, for $x \in X$ is defined to be:*

$$p(q) = \text{Poss}\{A \text{ is } F\}$$

$$= \#_A(m_F(x) * p_A(x)).$$

Similarly, the certainty measure, $c(q)$, associated with a query, q, can be computed as follows:

$$c(q) \triangleq \mathrm{Cert}\{A \text{ is } F\}$$
$$\triangleq \max(0, \&_A(m_A(x) * p_A(x) > 0)).$$

These modified possibility/certainty measures are adopted in the FRDB query evaluation because of their ease of understanding and ease of use and because of their inherent consistency with logical intuition.

We can define fuzzy sets $P(X|q)$ and $C(X|q)$, that is, possibility and certainty sets of objects x satisfying the query q. The membership function values for $x \in X$ in these sets are given by the possibility and certainty measure values, that is, $p(q)$ and $c(q)$, respectively. Specification of a threshold of acceptance $t \in [0,1]$ results in the retrieval sets that are t-level sets $[P(X|q)]_t$ and $[C(X|q)]_t$, respectively. The interpretation of the possibility or certainty t-level set is that the objects in X can be described by the attribute A being F with a possibility or certainty measure of at least t. For example, a natural language query "Find persons who are tall with a certainty of 0.6" corresponds to a FRDB query "SELECT PERSON WHO IS TALL (CERT = 0.6)." The attribute HEIGHT is implied by the definition of TALL. Hence, the retrieval set is a t-level certainty set $[C(\text{NAME}|\text{HEIGHT IS TALL})]_{0.6}$.

The possibility/certainty threshold of acceptance is used in the conjunction with Reason's study*, which shows that human beings apply a "threshold" analog variable in coming to a binary decision.

Let us now define the general FRDB form of the SELECT operation.

Definition 7.7.6

Let R be a relation in the VDB *with attributes $A, B, \ldots$ Then the* FRDB SELECT *operation has the form:*

$$X = \text{SELECT } A, B, \ldots \text{ FROM } R \begin{Bmatrix} \text{such that} \\ \text{where} \\ \text{who} \\ \text{whose} \\ \text{whom} \\ \text{which} \end{Bmatrix} \langle \text{comparison clause} \rangle,$$

where any adjective may be chosen.

Note that the SELECT is SELECT-PROJECT($A, B, \ldots$), a combination similar to SQL "SELECT-FROM-WHERE" block.

Definition 7.7.7

Let $A, B, \ldots$ be attribute names in a VDB *relation R, c be an attribute value, F be a fuzzy set defined over the attributes of relation R, m be a modifier list, and θ be a*

*J. T. Reason, Motion Sickness — Some Theoretical Considerations, *Int. J. of Man-Machine Studies*, 1, 1, pp. 21–38, 1969.

fuzzy relational operator GREATER THAN, MORE, LESS THAN, or LESS. *A ⟨comparison clause⟩ has the forms*:

(i) $[A]$ IS mF
(ii) A IS $m\theta c$
(iii) A IS $m\theta B$
(iv) $[A]$ IS m AS F AS c
(v) IS m_1 AS F_1 AS $m_2 F_2$
(vi) X IS $m_1 \ominus m_2 F_1$ THAN $m_3 F_2$
(vii) LINK IS $[s]$ linkname {TO/FROM} c,

where linkname is the EDB *link relation defined over the attribute A, and $[s]$ is an optional modifier* STRONG (*equivalent to* VERY).

The more complex queries in the proposed FRDB system are evaluated by applying rules of:

1. fuzzy modifiers,
2. fuzzy relational operators,
3. composition,
4. qualified propositions.

1. *Modifiers* used in the FRDB queries are the following: NOT, VERY/MUCH/STRONG and APPROXIMATELY. They can be employed in conjunction with fuzzy relational operators, as fuzzy-set modifiers, and in qualification clauses.

Let f denote a possibility measure associated with a query having a comparison clause, C, with no modifiers. Let mC denote a comparison clause C with a modifier m. Then for the modifiers NOT, VERY/MUCH and APPROXIMATELY, we obtain:

- NOT $C \rightarrow 1 - f$
- VERY/MUCH $C \rightarrow f^2$
- APPROXIMATELY $C \rightarrow f^{1/2}$

2. The FRDB query language supports the following *fuzzy relational operators*:

- IS
- IS $[m]$ GREATER THAN
- IS $[m]$ LESS THAN
- IS $[m]$ AS...AS...
- IS $[m]$ MORE/LESS...THAN.

The fuzzy relational operators are evaluated by the application of the underlying relations of similarity, proximity, or fuzzy sets defined on the attributes specified in the query as introduced above.

3. *Composition* of comparison clauses using logical AND and OR is translated according to proposition composition rules. Namely, if $p(C_1)$ and $p(C_2)$ are possibility measures associated with comparison clauses C_1 and C_2, respectively, then the possibility measures associated with the composed comparison clauses are defined as:

- $p(C_1 \text{ AND } C_2) = \min(p(C_1), p(C_2))$
- $p(C_1 \text{ OR } C_2) = \max(p(C_1), p(C_2))$

4. *Qualified queries* are queries with a qualification clause:

$$([m]\text{POSS} = t) \qquad \text{or} \qquad ([m]\text{CERT} = t)$$

where $0 \leq t \leq 1$ is the *threshold of acceptance*. The default qualification clause is (POSS = 0.5), that is, the retrieval set is $[P(X|q)]_{0.5}$, or objects with a possibility measure associated with the query $p(q) \geq 0.5$.

The application of the above stated SELECT formats permits queries through "English-like" statements.

■ ***Example** 7.7.5*

Assume that all relations in the VDB and the underlying similarities, proximities, link relations, and fuzzy sets have been defined. Then we may form the following FRDB SELECT queries:

- SELECT NAME, IQ FROM PERSON WHO IS VERY SMART (POSS = .7)
- SELECT NAME FROM PERSON WHOSE IQ IS MUCH GREATER THAN 100 AND WHOSE RESIDENCY IS 'MIAMI'
- SELECT CITY FROM USA WHERE SUMMER-TEMPERATURE IS NOT MUCH GREATER THAN WINTER-TEMPERATURE AND IS NOT LARGE (VERY CERT = 0.8)
- SELECT NAME, HEIGHT FROM PERSON WHO IS APPROXIMATELY AS TALL AS "John"
- SELECT NAME FROM PERSON WHO IS MUCH MORE INTELLIGENT THAN ATHLETIC OR WHO IS NOT VERY YOUNG
- SELECT NAME FROM GROUP WHERE LINK IS STRONG INFLUENCE FROM/TO "Jim"
- SELECT NAME FROM PERSON WHOSE HAIR IS "BROWN" (CERT = 1.0) ■

Note that the last query is equivalent to using the Boolean relational operator = , or performing the regular relational algebra select:

SELECT PERSON WHERE HAIR = 'BROWN'

Table 7.7.3. Similarity (hair color)

Hair Color	*S*-Hair Color	Sim
Blond	Red	0.6
Blond	Brown	0.4
Blond	Black	0.0
Red	Brown	0.5
Red	Black	0.1
Brown	Black	0.8

■ ***Example 7.7.6***

Consider the relation PERSON given in Example 7.7.1. Let the similarity relation defined for HAIR COLOR be given by the similarity matrix shown below:

	Blond	Red	Brown	Black
Blond	1	0.6	0.4	0
Red	0.6	1	0.5	0.1
Brown	0.4	0.5	1	0.8
Black	0	0.1	0.8	1

This matrix can be expressed as the EDB relation shown in Table 7.7.3. Note that the reflexive and symmetric pairs do not have to be entered. ■

Let us evaluate a query q:

- SELECT NAME FROM PERSON WHOSE HAIR COLOR IS "BROWN" (POSS = 0.3)

Let us focus on Bob's hair color, which is represented by the possibility distribution

- $\Pi_{\text{HAIR COLOR}}(\text{Bob}) = 0.3/\text{blond} + 0.7/\text{red}.$

The possibility/certainty measures associated with the query q are:

$$\begin{aligned} p(q) &= \max(0.3 * s(\text{blond, brown}), 0.7 * s(\text{red, brown})) \\ &= \max(0.3 * 0.4, 0.7 * 0.5) = \max(0.12, 0.35) = 0.35 \\ c(q) &= \max(0, \min(0.3 * s(\text{blond, brown}) > 0, \\ &\qquad 0.7 * s(\text{red, brown}) > 0)) \\ &= \max(0, \min(0.12, 0.35)) = \max(0, 0.12) = 0.12. \end{aligned}$$

Note that the possibility and certainty measures are nonzero although the value of "brown" is not explicitly represented in the possibility distribution of Bob's hair color. Since the threshold of acceptance was set to possibility measure of at least 0.3, then Bob would be a member of the retrieval set with a degree of membership of 0.35.

The Fuzzy Relational Data-Base model presented here was designed to satisfy the requirements for sound formal foundations, real-world information models, individualization, and user's convenience. An experimental FRDB system was developed to test the feasibility of an imprecise information system. A commercial relational data-base management system RIM was enhanced to incorporate the FRDB model features. The system is in experimental use at the Florida State University, Tallahassee.

The relational data-base structure combined with the theory of fuzzy sets and possibility provide the solid theoretical foundation. The query language permits "natural-language-like" expressions that are easily understood by users, and can be further developed to incorporate fuzzy inferences or production rules.

The Value Data-Base appears to be an adequate schema for imprecise data representation. The Explanation Data-Base provides the means of individualization, and may be used to extend the query vocabulary by defining new fuzzy sets in terms of previously defined sets. This feature can become very useful in compounding knowledge and it can be projected that the underlying structure can be utilized in knowledge extrapolation, or learning. Hence, it can be concluded that the FRDB system can serve as the ideal data-base in applications related to decision making in "soft" expert systems.

Bibliography

A COMPILATION OF APPROXIMATELY 1000 IMPORTANT REFERENCES (FUZZY 1000) ON FUZZY SET THEORY AND ITS APPLICATIONS.

Abu Osman, M. T. (1984). On the direct product of fuzzy subgroups. *Fuzzy Sets and Systems*, Vol. 12, No. 1, Jan., pp. 87–92.

Achache, A. (1982). Galois connexion of a fuzzy subset. *Fuzzy Sets and Systems*, Vol. 8, No. 2, pp. 215–218.

Adamo, J. M. (1980a). L. P. L.—A fuzzy programming language—1. Syntactic aspects. *Fuzzy Sets and Systems*, Vol. 3, No. 2, pp. 151–180.

Adamo, J. M. (1980b). Fuzzy decision trees. *Fuzzy Sets and Systems*, Vol. 4, No. 3, Nov., pp. 207–220.

Adamo, J. M. (1980c). L. P. L.—A fuzzy programming language—2. Semantic aspects. *Fuzzy Sets and Systems*, Vol. 3, No. 3. pp. 261–290.

Adamo, J. M. (1981). Some applications of the L. P. L. language to combinatorial programming. *Fuzzy Sets and Systems*, Vol. 6, No. 1, July, pp. 43–60.

Adlassnig, K. P. (1980). A fuzzy logical model of computer-assisted medical diagnosis. *Methods of Information in Medicine*, Vol. 19, No. 3, July, pp. 141–148.

Aizermann, M. A. (1977). Some unsolved problems in the theory of automatic control and fuzzy proofs. *IEEE Trans. on Auto. Cont.*, pp. 116–118.

Albert, P. (1978). The algebra of fuzzy logic. *Fuzzy Sets and Systems*, Vol. 1, pp. 203–230.

Albrycht, J., and M. Matloka (1984). On fuzzy multi-valued functions. Part 1: Introduction and general properties. *Fuzzy Sets and Systems*, Vol. 12, No. 1, Jan., pp. 61–70.

Ali. S. T., and H. Doebner (1976). On the equivalence of nonrelativistic quantum mechanics based upon sharp and fuzzy measurements. *J. Math. Phys.*, Vol. 17, pp. 1105–1111.

Ali, S. T., and H. Doebner (1977). Systems of imprimitivity and representation of quantum mechanics of fuzzy phase spaces. *J. Math. Phys.*, Vol. 18, No. 2.

Allen, A. D. (1973). A method of evaluating technical journals on the basis of published comments through fuzzy implications—A survey of the major IEEE transactions. *IEEE Trans. Syst. Man. Cybern.*, SMC-3, pp. 422–425.

Allen, A. D. (1974). Measuring the empirical properties of sets. *IEEE Trans. Syst. Man. Cybern.*, SMC-4, pp. 66–73.

Alsina, C. (1985). On a family of connectives for fuzzy sets. *Fuzzy Sets and Systems,* Vol. 16, No. 3, pp. 231–235.

Alsina, C., E. Trillas, and L. Valverde (1983). On some logical connectives for fuzzy-set theory. *Journal of Mathematical Analysis and Applications*, Vol. 93, No. 1, Apr., pp. 15–26.

Ambrosio, R., and G. B. Martini (1984). Maximum and minimum between fuzzy symbols in noninteractive and weakly noninteractive situations. *Fuzzy Sets and Systems*, Vol. 12, No. 1, Jan., pp. 27–36.

Anthony, J. M., and H. Sherwood (1979). Fuzzy groups redefined. *J. Math. Anal. Appl.*, Vol. 69, No. 1, May, pp. 124–131.

Anthony, J. M., and H. Sherwood (1982). A characterization of fuzzy subgroups. *Fuzzy Sets and Systems*, Vol. 7, No. 3, May, pp. 297–306.

Arbib, M. A. (1977). Review of three books on fuzzy sets. *Bull. Am. Math. Soc.*, Vol. 83, No. 5, Sept., pp. 946–951.

Arbib, M. A., and E. G. Manes (1975a). A category-theoretic approach to systems in a fuzzy world. *Synthèse*, Vol. 30, pp. 381–406.

Arbib, M. A., and E. G. Manes (1975b). Fuzzy machines in a category. *Bulletin Australian Math. Soc.*, Vol. 13, pp. 169–210.

Arigoni, A. O. (1980). Mathematical developments arising from "Semantical implication" and the evaluation of membership characteristic functions. *Fuzzy Sets and Systems*, Vol. 4, No. 2, Sept., pp. 167–184.

Arigoni, A. O. (1982). Transformational-generative grammar for description of formal properties. *Fuzzy Sets and Systems*, Vol. 8, No. 3, Sept., pp. 311–322.

Aronson, A. R., B. E. Jacobs, and J. Minka (1980). A note on fuzzy deduction. *JACM*, Vol. 27, No. 4, Oct., pp. 599–603.

Artico, G., and R. Moresco (1984). Fuzzy proximities and totally bounded fuzzy uniformities. *Journal of Mathematical Analysis and Applications*, Vol. 99, No. 2, pp. 320–337.

Asai, K., and S. Kitajima (1971). A method for optimizing control of multimodal systems using fuzzy automata. *Inform. Sci.*, Vol. 3, pp. 343–353.

Asai, K., and S. Kitajima (1972). Optimizing control using fuzzy automata. *Automatica*, Vol. 8, pp. 101–104.

Asai, K., K. Tanaka, and T. Okuda (1977). On the discrimination of fuzzy states in probability space. *Kybernetes*, Vol. 6, pp. 185–192.

Asai, K., H. Tanaka, C. V. Negoita, and D. A. Ralescu (1978). *Introduction to Fuzzy Systems Theory*. Ohm-Sha Co. Ltd., Tokyo (in Japanese).

Aubin, J. P. (1976). Fuzzy core and equilibria of games defined in strategic form, in *Directions in Large-scale Systems*. (Ho, Y. C., and S. K. Mitter, Eds.). Plenum Press, New York, pp. 371–388.

Azad, K. K. (1981). Fuzzy grills and a characterization of fuzzy proximity. *Journal of Mathematical Analysis and Applications*, No. 1, pp. 13–18.

Baaklini, N., and E. H. Mamdani (1975). Prescriptive methods for deriving control policy in a fuzzy logic controller. *Electronics Letters*, Vol. 11.

Baas, S. M., and H. Kwakernaak (1977). Rating and ranking of multiple-aspect alternatives using fuzzy sets. *Automatica*, Vol. 13, No. 1, pp. 47–58.

Baldwin, J. F. (1979a). A model of fuzzy reasoning through multi-valued logic and set theory. *Int. J. Man-Machine Studies*, Vol. 11, pp. 351–380.

Baldwin, J. F. (1979b). A new approach to approximate reasoning using a fuzzy logic. *Fuzzy Sets and Systems*, Vol. 2, No. 4, pp. 309–326.

Baldwin, J. F. (1979c). Fuzzy logic and approximate reasoning for mixed-input arguments. *Int. J. Man-Machine Studies*, Vol. 11, pp. 381–396.

Baldwin, J. F., and N. C. F. Guild (1979a). Comment on the "fuzzy max" operator of Dubois and Prade. *Int. Jour. Syst. Sc.*, Vol. 10, No. 9, pp. 1063–1064.

Baldwin, J. F., and N. C. F. Guild (1979b). Comparison of fuzzy sets on the same decision space. *Fuzzy Sets and Systems*, Vol. 2, No. 3, pp. 213–232.

Baldwin, J. F., and N. C. F. Guild (1980). Feasibility algorithms for approximate reasoning using a fuzzy logic. *Fuzzy Sets and Systems*, Vol. 3, No. 3, pp. 225–252.

Baldwin, J. F., and B. Pilsworth (1980). Axiomatic approach to implication for approximate reasoning using a fuzzy logic. *Fuzzy Sets and Sys.*, Vol. 3, No. 2, pp. 193–220.

Bandler, W., and L. Kohout (1980). Fuzzy power sets and fuzzy implication operators. *Fuzzy Sets and Sys.*, Vol. 4, No. 1, July, pp. 13–30.

Bandler, W., and L. Kohout (1980). Semantics of implication operators and fuzzy relational products. *Int. J. Man-Machine Studies*, Vol. 12, No. 1.

Banon, G. (1979). Use of possibility measures for modelling an *a priori* knowledge in parametric estimation. *Advances in Control*, Vol. 2.

Banon, G. (1981). Distinction between several subsets of fuzzy measures. *Fuzzy Sets and Systems*, Vol. 5, No. 3, pp. 291–306.

Baptistella, L. F. B., and A. Ollero (1980). Fuzzy methodologies for interactive multicriteria optimization. *IEEE Trans. Syst. Man. Cybern*, SMC-10, No. 7, pp. 355–365.

Battle, N., and E. Trillas (1979). Entropy and fuzzy integrals. *Jour. Math. Anal. and Appl.*, Vol. 69, pp. 469–474.

Bellacicco, A. (1976). Fuzzy classification. *Synthèse*, Vol. 33, pp. 273–281.

Bellman, R. E., and M. Giertz (1973). On the analytic formalism of the theory of fuzzy sets. *Inform. Sci.*, Vol. 5, pp. 149–156.

Bellman, R. E., R. Kalaba, and L. A. Zadeh (1966). Abstraction and pattern classification. *J. Math. Anal. and Appl.*, Vol. 13, pp. 1–7.

Bellman, R. E., and L. A. Zadeh (1970). Decision-making in a fuzzy environment. *Management Sci.*, Vol. 17, pp. 141–164.

Bellman, R. E., and L. A. Zadeh (1977). Local and fuzzy logics, in *Modern Uses of Multiple-valued Logic* (Dunn, J. C., and G. Epstein, Eds.). D. Reidel, Dordrecht, pp. 103–165.

Berman, J. and M. Mukaidono (1983). Enumerating fuzzy switching functions and free Kleene algebras. *Computers and Mathematics with Applications*, Vol. 10, No. 3, pp. 179–194.

Bezdek, J. C. (1974a). Cluster validity with fuzzy sets. *J. Cybernetics*, Vol. 3, pp. 58–73.

Bezdek, J. C. (1974b). Numerical taxonomy with fuzzy sets. *J. Math. Biology*, Vol. 1, pp. 57–71.

Bezdek, J. C. (1976). A physical interpretation of fuzzy isodata. *IEEE Trans. Syst. Man. Cybern*, Vol. 6, pp. 387–389.

Bezdek, J. C. (1980a). A convergence theorem for the fuzzy isodata clustering algorithms. *IEEE Transactions on Pattern Analysis and Machine Intelligence*, Jan., Vol. 2, No. 1, pp. 1–8.

Bezdek, J. C. (1980b). Fuzzy imbeddings that are convex hulls. *J. Cybernet.*

Bezdek, J. C. (1981). *Pattern Recognition with Fuzzy Objective Function Algorithms*. Plenum Press, New York.

Bezdek, J. C., and P. F. Castelaz (1977). Prototype classification and feature selection with fuzzy sets. *IEEE Trans. Syst. Man. Cybernet*, Vol. 7, No. 2, pp. 87–92.

Bezdek, J. C., and J. C. Dunn (1975). Optimal fuzzy partitions—A heuristic for estimating the parameters in a mixture of normal distributions. *IEEE Trans. Comp.*, C-24, pp. 835–838.

Bezdek, J. C., and J. D. Harris (1978). Fuzzy relations and partitions — An axiomatic basis for clustering. *Fuzzy Sets and Sys.*, Vol. 1, pp. 111–126.

Bezdek, J. C., B. Spillman, and R. Spillman (1979). Fuzzy relation spaces for group decision theory—An application. *Fuzzy Sets and Sys.*, Vol. 2, No. 1, Jan., pp. 5–14.

Bhat, K. V. S. (1982). On the notion of fuzzy consensus. *Fuzzy Sets and Systems*, Vol. 8, No. 3, Sept., pp. 285–290.

Black, M. (1937). Vagueness — An exercise in logical analysis. *Philos. Sci.*, Vol. 4, pp. 427–455.

Black, M. (1963). Reasoning with loose concepts. *Dialogue*, Vol. 2, pp. 325–373.

Black, M. (1968). *The Labyrinth of Language*. Mentor Books, New York.

Black, M. (1970). *Margins of Precision*. Cornell University Press, Ithaca, New York.

Blackburn, S., Ed. (1975). *Meaning, Reference and Necessity*. Cambridge University Press, Cambridge.

Blin, J. M. (1974). Fuzzy relations in group decision theory. *J. Cybernetics*, Vol. 4, pp. 17–22.

Blin, J. M. (1977). Fuzzy sets in multiple-criteria decision-making, in *Multiple-Criteria Decision-Making*. (Slarr, M. K., and M. Zeleny, Eds.) Academic Press, New York.

Blin, J. M., and A. B. Whinston (1973). Fuzzy sets and social choice. *J. Cybernetics*, Vol. 3, pp. 28–33.

Blockley, D. I. (1979). The role of fuzzy sets in civil engineering. *Fuzzy Sets and Sys.*, Vol. 2, No. 4, Oct., pp. 267–278.

Bogardi, I., A. Tiszadata, and L. Duckstein (1983). Regional management of an aquifer for mining under fuzzy environmental objectives. *Water Resources Research*, Vol. 19, No. 6, pp. 1394–1402.

Borisov, A. N., and O. Krumberg (1983). A theory of possibility for decision-making. *Fuzzy Sets and Systems*, Vol. 9, No. 1, Jan., pp. 13–24.

Borisov, A. N., and E. A. Kokle (1970). Recognition of fuzzy patterns (Kristinkov, D. S., J. J. Osis, and L. A. Rastrigin, Eds.), *Kibernetika I Diagnostika*, Zinatne, Riga, U.S.S.R. Vol. 4, pp. 135–247 (In Russian).

Borisov, A. N., and J. J. Osis (1970). Search for the greatest divisibility of fuzzy sets (Kristinkov, D. S., J. J. Osis, and L. A. Rastrigin, Eds.), *Kibernetika I Diagnostika*, Zinatne, Riga, U.S.S.R. Vol. 3, pp. 79–88. (In Russian).

Borodkin, L. I. (1979). Aggregated structures of graphs with fuzzy blocks. *Automation and Remote Control*, Vol. 39, No. 8, pp. 1219–1229.

Bortolan, G., and R. Degani (1985). A review of some methods for ranking fuzzy subsets. *Fuzzy Sets and Systems*, Vol. 15, No. 1, pp. 1–19.

Bouchon, B. (1981). Fuzzy questionnaires. *Fuzzy Sets and Systems*, Vol. 6, No. 1, July, pp. 1–10.

Bouza, C. N. (1983). Estimation of the mean in populations with a fuzzy connected structure. *Rev. Investigacion Operational* (Cuba), Vol. 4, No. 3, pp. 151–157.

Braae, M., and D. A. Rutherford (1978). Fuzzy relations in a control setting. *Kybernetes*, Vol. 7, pp. 185–188.

Braae, M., and D. A. Rutherford (1979a). Theoretical and linguistic aspect of the fuzzy logic controller. *Automatica*, Vol. 15, pp. 553–557.

Braae, M., and D. A. Rutherford (1979b). Selection of parameters for a fuzzy logic controller. *Fuzzy Sets and Sys.*, Vol. 2, No. 3, July, pp. 185–200.

Brown, C. B. (1979). A fuzzy safety measure. *Journal of the Engineering Mechanics Division*, Vol. 105, No. EM5, Oct., pp. 855–872.

Brown, J. G. (1971). A note on fuzzy sets. *Inform. and Control*, Vol. 18, pp. 32–39.

Brunner, J., and W. Wechler (1976). The behaviour of *R*-fuzzy automata (pp. 210–215 in *Lecture Notes in Computer Science*, 45, A. Mazurkiewicz, (Ed.)), Springer-Verlag, Berlin.

Buckles, B. P., and F. E. Petry (1982). A fuzzy representation of data for relational databases. *Fuzzy Sets and Systems*, Vol. 7, No. 3, May, pp. 213–226.

Buckles, B. P., and F. E. Petry (1983). Information-theoretical characterization of fuzzy relational databases. *IEEE Transactions on Systems, Man, and Cybernetics*, Vol. SMC-13, No. 1, pp. 74–77.

Buckley, J. J. (1983). Fuzzy programming and the Pareto optimal set. *Fuzzy Sets and Systems*, Vol. 10, No. 1, Apr., pp. 57–64.

Buckley, J. J. (1984a). The multiple-judge, multiple-criteria ranking problem: A fuzzy-set approach. *Fuzzy Sets and Systems*, Vol. 13, No. 1, May, pp. 25–38.

Buckley, J. J. (1984b). Multiple-goal noncooperative conflicts under uncertainty: A fuzzy-set approach. *Fuzzy Sets and Systems*, Vol. 13, No. 2, July, pp. 107–127.

Buckley, J. J. (1985a). Ranking alternatives using fuzzy numbers. *Fuzzy Sets and Systems*, Vol. 15, No. 1, pp. 21–31.

Buckley, J. J. (1985b). Fuzzy decision making with data: Applications to statistics. *Fuzzy Sets and Systems*, Vol. 16, No. 2, pp. 139–147.

Buell, D. A. (1982). An analysis of some fuzzy-subset applications to information retrieval systems. *Fuzzy Sets and Systems*, Vol. 7, No. 1, Jan., pp. 35–42.

Buoncristiani, J. F. (1983). Probability on fuzzy sets. *J. Math. Anal. and Appl.*, Vol. 96, pp. 24–41.

Burdzy, K., and J. B. Kiszka (1983). The reproducibility property of fuzzy control systems. *Fuzzy Sets and Systems*, Vol. 9, No. 2, Feb., pp. 161–178.

Butnariu, D. (1976). Fuzzy games and their minimax theorem (In Romanian). *St. Cerc. Math.*, Vol. 28, No. 2, pp. 142–160.

Butnariu, D. (1977). Fuzzy games — A description of the concept. *Fuzzy Sets and Systems*, Vol. 1, pp. 181–192.

Butnariu, D. (1980). Stability and Shapley value for *N*-person fuzzy games. *Fuzzy Sets and Systems*, Vol. 4, No. 1, pp. 63–72.

Butnariu, D. (1982). Fixed points for fuzzy mappings. *Fuzzy Sets and Systems*, Vol. 7, No. 2, Mar., pp. 191–208.

Butnariu, D. (1983). Decompositions and range for additive fuzzy measures. *Fuzzy Sets and Systems*, Vol. 10, No. 2, June, pp. 135–156.

Cakraborty, M. K., and M. Das (1983). Studies in fuzzy relations over fuzzy subsets. *Fuzzy Sets and Systems*, Vol. 9, No. 1, Jan., pp. 79–90.

Cao, H., and G. Chen (1983). Some applications of fuzzy sets to meteorological forecasting. *Fuzzy Sets and Systems*, Vol. 9, No. 1, Jan., pp. 1–12.

Capocelli, R. M., and A. De Luca (1973). Fuzzy sets and decision theory. *Inform. and Control*, Vol. 23, pp. 446–473.

Carlsson, C. (1978). *Fuzzy Automata as Cybernetic Control Functions*. Instit. of Management Sci., ABO Swedish Univ. School of Economics, Henriksgatan 7, SF-20500 ABO 50, Finland.

Carlsson, C. (1983). An approach to handling fuzzy problem structures. *Cybernetics and Systems*, Vol. 14, No. 1, pp. 33–54.

Carlstrom, I. F. (1975). Truth and entailment for a vague quantifier. *Synthèse*, Vol. 30, pp. 461–495.

Carrega, J. C. (1983a). The categories Set H and Fuz H. *Fuzzy Sets and Systems*, Vol. 9, No. 3, pp. 327–332.

Cattaneo, G. (1980). Fuzzy events and fuzzy logics in classical information systems. *Journal of Mathematical Analysis and Applications*, Vol. 75, No. 2, pp. 523–548.

Cattaneo, G. (1983). Canonical embedding of an abstract quantum logic into the partial Baer-ring of complex fuzzy events. *Fuzzy Sets and Systems*, Vol. 9, No. 2, Feb., pp. 179–198.

Cavallo, R. E., and G. J. Klir (1982). Reconstruction of possibilistic behavior systems. *Fuzzy Sets and Systems*, Vol. 8, No. 2, Aug., pp. 175–198.

Cella, C., A. Fadina, and R. Sarno (1981). A sharpening operator for a class of fuzzy preference relations. *Bulletin pour les Sous-ensembles Flous et Leurs Applications*, Vol. 6, pp. 76–86.

Cerruti, U. (1981). The Stone-Cech compactification in the category of fuzzy topological spaces. *Fuzzy Sets and Systems*, Vol. 6, No. 2, Sept., pp. 197–204.

Cerruti, U. (1983). Completion of L-fuzzy relations. *Journal of Mathematical Analysis and Applications*, Vol. 94, No. 2, pp. 312–327.

Chakraborty, M. K., and M. Das (1983a). On fuzzy equivalence — I. *Fuzzy Sets and Systems*, Vol. 11, No. 2, Oct., pp. 195–198.

Chakraborty, M. K., and M. Das (1983b). On fuzzy equivalence — II. *Fuzzy Sets and Systems*, Vol. 11, No. 3, Nov., pp. 299–308.

Chanas, S. (1983). The use of parametric programming in fuzzy linear programming, *Fuzzy Sets and Systems*, Vol. 11, No. 3, Nov., pp. 243–252.

Chanas, S., and J. Kamburowski (1981). The use of fuzzy variables in P.E.R.T., *Fuzzy Sets and Systems*, Vol. 5, No. 1, pp. 11–20.

Chanas, S., and W. Kolodziejckyk (1982). Maximum flow in a network with fuzzy arc capacities. *Fuzzy Sets and Systems*, Vol. 8, No. 2, Aug., pp. 165–174.

Chanas, S., and W. Kolodziejczyk (1984). Real-valued flows in a network with fuzzy arc capacities. *Fuzzy Sets and Systems*, Vol. 13, No. 2, July, pp. 139–152.

Chang, C.L. (1968). Fuzzy topological spaces, *J. Math. Anal. and Appl.*, Vol. 24, pp. 182–190.

Chang, R. L. P., and T. Pavlidis (1977). Fuzzy decision-tree algorithms. *IEEE Trans. Syst., Man, and Cybern.*, Vol. 7, pp. 28–34.

Chang, R. L. P., and T. Pavlidis (1979). Applications of fuzzy sets in curve fitting. *Fuzzy Sets and Systems*, Vol. 2, No. 1, Jan., pp. 67–74.

Chang, S. K. (1971a). Automated interpretation and editing of fuzzy line drawings. *SJCC*, Vol. 38, pp. 393–399.

Chang, S. K. (1971b). Picture processing grammar and its applications. *Inform. Sci.*, pp. 121–148.

Chang, S. K. (1972). On the execution of fuzzy programs using finite-state machines. *IEEE Trans. Comp.*, C-21, pp. 241–253.

Chang, S. K., and J. S. Ke (1979). Translation of fuzzy queries for relational database system, *IEEE Trans. in Pattern Analysis and Machine Intelligence*, Vol. 1., No. 3, pp. 281–294.

Chang, S. S. L. (1972). Fuzzy mathematics, man, and his environment. *IEEE Trans. Syst. Man Cybern.*, SMC-2, p. 93.

Chang, S. S. L. (1977). Application of fuzzy-set theory to economics. *Kybernetes*, Vol. 6, pp. 203–207.

Chang, S. S. L. (1978). On a fuzzy algorithm and its implementation. *IEEE Trans. on Syst., Man, and Cybern.*, Vol. 8, No. 1, p. 31.

Chang, S. S. L., and L. A. Zadeh (1972). On fuzzy mapping and control. *IEEE Trans. on Systems, Man and Cybernetics*, Vol. 2, pp. 30–34.

Chang, S. Y., E. D. Brill, and L. D. Hopkins (1983). Modeling to generate alternatives — A fuzzy approach. *Fuzzy Sets and Systems*, Vol. 9, No. 2, Feb., pp. 137–152.

Chen, C. (1974). Realizability of communication nets — An application of the Zadeh criterion. *IEEE Trans. Circuits and Syst.*, CAS-21, pp. 150–151.

Chorayan, O. G. (1982). Identifying elements of the probabilistic neuronal ensembles from the standpoint of fuzzy-set theory. *Fuzzy Sets and Systems*, Vol. 8, No. 2, Aug., pp. 141–148.

Christopher, F. T. (1977). Quotient fuzzy topology and local compactness. *J. of Math. Anal. and Appl.*, Vol. 53, No. 3.

Chu, A. T. W., R. E. Kalaba, and K. Spingarn (1979). A comparison of two methods for determining the weights of belonging to fuzzy sets. *J. Optimization Theory and Applications*, Vol. 27, No. 4.

Clark, C. M., M. Ben-David, and A. Kandel (1981). On the enumeration of distinct fuzzy switching functions. *Fuzzy Sets and Systems*, Vol. 5, No. 1, Jan., pp. 69–82.

Conrad, F. (1980). Fuzzy topological concepts. *J. Math. Anal. and Appl.*, Vol. 74, No. 2.

Czogala, E. (1983). On a distribution function description of probabilistic sets and its application in decision-making. *Fuzzy Sets and Systems*, Vol. 10, No. 1, Apr., pp. 21–30.

Czogala, E. (1983). A generalized concept of a fuzzy probabilistic controller. *Fuzzy Sets and Systems*, Vol. 11, No. 3, Nov.

Czogala, E. (1984). An introduction to probabilistic *L*-valued logic. *Fuzzy Sets and Systems*, Vol. 13, No. 2, July, pp. 179–186.

Czogala, E. and J. Drewniak (1984). Associative monotonic operations in fuzzy-set theory. *Fuzzy Sets and Systems*, Vol. 12, No. 3, pp. 249–290.

Czogala, E., J. Drewniak, and W. Pedrycz (1982). Fuzzy relation equations on a finite set. *Fuzzy Sets and Systems*, Vol. 7, No. 1, Jan., pp. 89–102.

Czogala, E., S. Gottwald, and W. Pedrycz (1982). Contribution to application of energy measure of fuzzy sets. *Fuzzy Sets and Systems*, Vol. 8, No. 2, Aug., pp. 205–214.

Czogala, E., S. Gottwald, and W. Pedrycz (1983). Logical connectives of probabilistic sets, *Fuzzy Sets and Systems*, Vol. 10, No. 3, July, pp. 299–308.

Czogala, E., and W. Pedrycz (1981). On identification in fuzzy systems and its applications in control problems. *Fuzzy Sets and Systems*, Vol. 6, No. 1, July, pp. 73–84.

Czogala, E., and W. Pedrycz (1982a). Fuzzy rules generation for fuzzy control. *Cybernetics and Systems*, Vol. 13, No. 3, pp. 275–294.

Czogala, E., and W. Pedrycz, (1982b). Control problems in fuzzy systems. *Fuzzy Sets and Systems*, Vol. 7, No. 3, May, pp. 257–274.

Czogala, E., and W. Pedrycz (1983). On the concept of fuzzy probabilistic controllers. *Fuzzy Sets and Systems*, Vol. 10, No. 2, June, pp. 109–122.

Czogala, E., and H. J. Zimmermann (1984). Some aspects of synthesis of probabilistic fuzzy controllers, *Fuzzy Sets and Systems*, Vol. 13, No. 2, July, pp. 169–178.

Dalcin, M. (1975a). Fuzzy-state automata, their stability and fault-tolerance. *Int. J. Comp. Inf. Sciences*, Vol. 4, p. 63–80.

Dalcin, M. (1975b). Modification tolerance of fuzzy-state automata, *Int. J. Comp. Inf. Sciences*, Vol. 4, pp. 81–93.

Dale, A. I. (1980). Probability, vague statements, and fuzzy sets. *Philosophy of Science*, Vol. 47.

De Glas, M. (1983). Theory of fuzzy systems, *Fuzzy Sets and Systems*, Vol. 10, No. 1, Apr., pp. 65–78.

De Glas, M. (1984). Invariance and stability of fuzzy systems, *Journal of Mathematical Analysis and Applications*, Vol. 99, No. 2, pp. 299–319.

De Kerf, J. (1975). A bibliography on fuzzy sets. *J. Computational and Applied Mathematics*, Vol. 1, pp. 205–212.

De Mitri, C., and E. Pascali (1983). Characterization of fuzzy topologies from neighborhoods of fuzzy points, *Journal of Mathematical Analysis and Applications*, Vol. 93, No. 1, pp. 1–14.

De Mori, R. (1983). Computer Models of Speech Using Fuzzy Algorithms, Plenum Press, New York.

De Mori, R., and P. LaFace (1980). Use of fuzzy algorithms for phonetic and phonemic labeling of continuous speech. *IEEE Trans. on Pattern Anal. and Machine Intelligence*, Mar., Vol. 2, No. 2.

De Mori, R., and L. Saitta (1980). Automatic learning of fuzzy naming relations over finite languages. *Information Sciences*, Vol. 20.

DeFays, D. (1975). Relations floues et analyse hierarchique des questionnaires. *Math. Sci. Hum.*, No. 55, p. 45–60.

DeLuca, A., and R. M. Capocelli (1973). Fuzzy sets and decision theory. *Inf. and Control*, Vol. 23, pp. 446–473.

DeLuca, A., and S. Termini (1972a). A definition of a nonprobabilistic entropy in the setting of fuzzy sets, *Inform. and Control*, Vol. 20, pp. 301–312.

DeLuca, A., and S. Termini (1972b). Algebraic properties of fuzzy sets. *J. Math. Anal. and Appl.*, Vol. 40, pp. 373–386.

DeLuca, A., and S. Termini (1974). Entropy of *L*-fuzzy sets. *Inform. and Control*, Vol. 24, pp. 55–73.

DeLuca, A., and S. Termini (1977a). Measures of ambiguity in the analysis of complex systems. In *Mathematical Foundations of Computer Science*, pub. Springer-Verlag, pp. 382–389.

DeLuca, A., and S. Termini (1977b). On the convergence of entropy measures of a fuzzy set. *Kybernetes*, Vol. 6, pp. 219–227.

Deutsch, S. J., and C. J. Malborg (1985). A fuzzy set approach to data set evaluation for decision support. *IEEE Trans. SMC*, Vol. SMC-15, No. 6, Nov.–Dec., pp. 777–783.

Dhar, S. B. (1979). Power-system long-range decision analysis under fuzzy environment. *IEEE Transactions on Power Apparatus and Systems*, Vol. PAS-89, No. 2, March–April.

Di Concilio, A., and G. Gerla (1984). Almost compactness in fuzzy topological spaces, *Fuzzy Sets and Systems*, Vol. 13, No. 2, July, pp. 187–192.

Dijkman, J., H. van Haeringen, and S. J. De Lange (1983). Fuzzy numbers, *Journal of Mathematical Analysis and Applications*, Vol. 92, No. 2, pp. 302–341.

Dimitrov, V. D. (1970). G. M. D. H. algorithms on fuzzy sets of Zadeh. *Soviet Automatic Control*, No. 3, pp. 40–45.

Dimitrov, V. D. (1976). Learning decision-making with fuzzy automata. In *Computer-Oriented Learning Processes* (J. C. Simon, Ed.). Noordhoff, pp. 149–154.

Dimitrov, V. D. (1977). Social choice and self-organization under fuzzy management. *Kybernetes*, Vol. 6, p. 153.

Dimitrov, V. D. (1983). Group choice under fuzzy information. *Fuzzy Sets and Systems*, Vol. 9, No. 1, Jan., pp. 25–40.

Dimitrov, V. D., and F. Luban (1982). Membership functions, some mathematical programming models, and production scheduling. *Fuzzy Sets and Systems*, Vol. 8, No. 1, June, pp. 19–34.

Di Nola, A., and A. Fadina (1982). A hyperspatial representation of a particular set of fuzzy preference relations. *Fuzzy Sets and Systems*, Vol. 7, No. 1, Jan., pp. 79–88.

Di Nola, A., and S. Sessa (1983a). On the set of solutions of composite fuzzy relation equations. *Fuzzy Sets and Systems*, Vol. 9, No. 3, pp. 275–285.

Di Nola, A. and S. Sessa (1983b). On the fuzziness of solutions of *o*-fuzzy relation equations on finite spaces. *Fuzzy Sets and Systems*, Vol. 11, No. 1, Aug., pp. 65–78.

Di Nola, A., and A. G. S. Ventre (1979). On some sequences on fuzzy sets, R. A. I. R. O. *Informatique*, Paris, Vol. 13, pp. 199–204.

Di Nola, A., and A. G. S. Ventre (1980). On some chains of fuzzy sets. *Fuzzy Sets and Systems*, Vol. 4, No. 2, pp. 185–192.

Dishkant, H. (1981). About membership-function estimation. *Fuzzy Sets and Systems*, Vol. 5, No. 2, pp. 141–148.

Dodson, C. T. J. (1974). Hazy spaces and fuzzy spaces. *Bull. London Math. Soc.*, Vol. 6, pp. 191–197.

Dodson, C. T. J. (1975). Tangent structure for hazy spaces. *Jour. London Math. Soc.*, Vol. 2, No. 11, pp. 465–473.

Dodson, C. T. J. (1981). A new generalization of graph theory. *Fuzzy Sets and Systems*, Vol. 6, No. 3, Nov., pp. 293–308.

Dombi, J. (1982). A general class of fuzzy operators, the De Morgan class of fuzzy operators, and fuzziness measures induced by fuzzy operators. *Fuzzy Sets and Systems*, Vol. 8, No. 2, Aug., pp. 149–164.

Dreyfess, G. R., M. Kochen, J. Robinson, and A. N. Badre (1975). On the psycholinguistic reality of fuzzy sets, In *Functionalism* (Grossman, R. E., L. J. San, and T. J. Vance, Eds.). University of Chicago Press, pp. 135–149.

Dubois, D., and H. Prade (1978a). A procedure for multiple-aspect decision-making. *Int. J. Sys. Sci.*, Vol. 9, No. 3, pp. 357–360.

Dubois, D., and H. Prade (1978b). Comment on tolerance analysis using fuzzy sets and a procedure for multiple-aspect decision-making. *Int. J. Sys. Sci.*, Vol. 9, pp. 357–360.

Dubois, D., and H. Prade (1978c). Operations on fuzzy numbers. *Int. J. Sys. Sci.*, Vol. 9, p. 613–626.

Dubois, D., and H. Prade (1979a). Fuzzy real algebra — Some results. *Fuzzy Sets and Systems*, Vol. 2, No. 4, pp. 327–348.

Dubois, D., and H. Prade (1979b). *Fuzzy Sets and Systems: Theory and Applications.* Academic Press, New York.

Dubois, D., and H. Prade (1980a). Fuzzy logics and fuzzy control. *Int. Jour. on Man-Machine Studies.*

Dubois, D., and H. Prade (1980b). Systems of linear fuzzy constraints. *Fuzzy Sets and Systems*, Vol. 3, No. 1, pp. 37–48.

Dubois, D., and H. Prade (1982a). A class of fuzzy measures based on triangular norms — A general framework for the combination of uncertain information. *Int. J. of General Systems*, Vol. 8, No. 1.

Dubois, D., and H. Prade (1982b). Towards fuzzy differential calculus, Part 1 — Integration of fuzzy mappings. *Fuzzy Sets and Systems*, Vol. 8, No. 1, June, pp. 1–18.

Dubois, D., and H. Prade (1982c). Towards fuzzy differential calculus, Part 2 — Integration on fuzzy intervals. *Fuzzy Sets and Systems*, Vol. 8, No. 2, Aug., pp. 105–116.

Dubois, D., and H. Prade (1982d). Towards fuzzy differential calculus, Part 3 — Differentiation. *Fuzzy Sets and Systems*, Vol. 8, No. 3, Sept., pp. 225–234.

Dubois, D., and H. Prade (1983). Unfair coins and necessity measures: Towards a possibilistic interpretation of histograms. *Fuzzy Sets and Systems*, Vol. 10, No. 1, Apr., pp. 15–20.

Dubois, D., and H. Prade (1985a). *Theorie des possibilities.* Masson, Paris.

Dubois, D., and H. Prade (1985b). Fuzzy cardinality and the modeling of imprecise quantification. *Fuzzy Sets and Systems,* Vol. 16, No. 3, pp. 199–230.

Dubois, T. (1977). A teaching system using fuzzy subsets and multi-criteria analysis. *Int. J. Math. Ed. Sci. Tech.*, Vol. 8, No. 2, pp. 203–217.

Dubois, T., A. Jones, M. Peteau, and A. N. Huynen (1977). Toward continuous learning management. In CAI, *Int. J. Math. Ed. Sci. Tech.*, Vol. 8, No. 3, pp. 335–350.

Dunn, J. C. (1973). A fuzzy relative of the isodata process and its use in detecting compact well-separated clusters. *J. Cybernetics*, Vol. 3, pp. 32–57.

Dunn, J. C. (1974a). A graph-theoretic analysis of pattern classification via Tamura's fuzzy relation. *IEEE Transactions on Systems, Man, and Cybernetics*, Vol. SMC-3, pp. 310–313.

Dunn, J. C. (1974b). Some recent investigations of a new fuzzy partitioning algorithm and its application to pattern classification problems. *J. Cybernetics*, Vol. 4, pp. 1–15.

Dunn, J. C. (1974c). Well-separated clusters and optimal fuzzy partitions. *J. Cybernetics*, Vol. 4, pp. 95–104.

Dussauchoy, A. E. (1982). Generalized information theory and decomposability of systems. *International Journal of General Systems*, Vol. 9, No. 1, pp. 13–36.

Economakos, E. (1979). Application of fuzzy concepts to power-demand forecasting, *IEEE Transactions on Systems, Man, and Cybernetics*, Vol. SMC-9, No. 10, Oct., pp. 651–657.

Efstathiou, J., and V. Rajkovic (1979). Multi-attribute decision-making using a fuzzy heuristic approach. *IEEE Trans. on Syst., Man and Cyber.*, Vol. 9, No. 6, pp. 326–333.

Emptoz, H. (1981). Nonprobabilistic entropies and indetermination measures in the setting of fuzzy-set theory. *Fuzzy Sets and Systems*, Vol. 5, No. 3, pp. 307–318.

Erceg, M. A. (1979). Metric spaces in fuzzy-set theory. *J. Math. Anal. Appl.*, Vol. 69, No. 1, May, pp. 205–231.

Erceg, M. A. (1980). Functions, equivalence relations, quotient spaces and subsets in fuzzy-set theory. *Fuzzy Sets and Sys.*, Vol. 3, No. 1, Jan., pp. 75–92.

Eshragh, F., and E. H. Mamdani (1979). A general approach to linguistic approximation. *Int. Jour. on Man-Machine Studies*, Vol. 11, pp. 501–519.

Esogbue, A. O. (1981). A fuzzy-set approach to public participation effectiveness measurement in water-quality planning. *Appl. Sys. and Cyber*, Vol. 6, pp. 3076–3081.

Esogbue, A. O. (1983). Dynamic programming, fuzzy sets, and the modelling of research and development management-control systems. *IEEE Transactions on Systems, Man and Cybernetics*, Vol. SMC-13, No. 1, pp. 18–30.

Esogbue, A. O., and R. C. Elder (1980). Fuzzy sets and modeling of physician decision processes, Part 1 — Initial interview information-gathering session. *Fuzzy Sets and Systems*, Vol. 2, No. 4, pp. 279–292.

Esogbue, A. O., and R. C. Elder (1980). Fuzzy sets and the modeling of physician decision processes. Part 2 — Diagnosis decision models. *Fuzzy Sets and Systems*, Vol. 3, No. 1, pp. 1–10.

Esogbue, A. O., and R. C. Elder (1983). Measurement and valuation of a fuzzy mathematical model for medical diagnosis. *Fuzzy Sets and Systems*, Vol. 10, No. 3, July, pp. 223–242.

Estera, F. (1981). On a representation theorem oF De Morgan algebras by fuzzy sets. *Stochastica*, Vol. 5, No. 2, pp. 109–115.

Etschmaier, M. M. (1980). Fuzzy controls for maintenance scheduling in transportation systems. *Automatica*, Vol. 16, pp. 255–264.

Eytan, M. (1981). Fuzzy sets: a topos-logical point of view. *Fuzzy Sets and Systems*, Vol. 5, No. 1, pp. 47–68.

Ezhkova, I. V., and D. A. Pospelov (1978). Decision-making on fuzzy premises, II — Deduction schemes, engineering cybernetics. *Soviet Jour. of Comput. Syst.*, Vol. 16, No. 2, pp. 1–6.

Feng, Y. J. (1983). A method using fuzzy mathematics to solve the vectormaximum problem. *Fuzzy Sets and Systems*, Vol. 9, No. 2, Feb., pp. 129–136.

Feron, R. (1976a). Ensembles flous, ensembles aléatoires flous, et économie aléatoire floue. *Publ. Économetriques*, Vol. IX, Fasc. 1.

Feron, R. (1976b). Ensembles flous attachés à un ensemble aléatoire flou. *Publications Économetriques*, Vol. IX, Fasc. 2.

Feron, R. (1979). Sur les ensembles aléatoires flous dont la fonction d'apparentance prend ses valeurs dans un treillis distributif fermé. *Publications Économetriques*, Vol. 12, Fasc. 1, pp. 81–118.

Fine, K. (1975). Vagueness, truth, and logic. *Synthèse*, Vol. 30, pp. 265–300.

Fishburn, P. C. (1975). A theory of subjective expected utility with vague preference. *Theory and Decision*, Vol. 6, pp. 287–310.

Flachs, J., and M. A. Pollatschek (1978). Further results on fuzzy-mathematical programming. *Inf. and Control*, Vol. 38, pp. 241–257.

Flonder, P. (1977). An example of a fuzzy system. *Kybernetes*, Vol. 6, pp. 229–230.

Flowers, P. L., and A. Kandel (1985). Possibilistic search trees. *Fuzzy Sets and Systems*, Vol. 16, No. 1, pp. 1–24.

Foster, D. H. (1979). Fuzzy topological groups. *J. Math. Anal. Appl.*, Vol. 67, No. 2, Feb., pp. 549–565.

Francioni, J. M., and A. Kandel (1983). Decomposable fuzzy-valued switching functions. *Fuzzy Sets and Systems*, Vol. 9, No. 1, Jan., pp. 41–68.

Franksen, O. I. (1979). On fuzzy sets, subjective measurements, and utility. *Int. Jour. of Man-Machine Studies*, Vol. 11, pp. 521–545.

Freeling, A. N. S. (1980a). Fuzzy sets and decision analysis. *IEEE Trans. on Systems, Man, and Cybernetics*, Vol. SMC-10, No. 7.

Freeling, A. N. S. (1980b). Possibilities versus fuzzy probabilities — Two alternative decision aids. In *Decision Analysis Through Fuzzy Sets*. (Zimmerman, H. J., and others, Eds.). North-Holland, New York.

Fukami, S., M. Mizumoto, and K. Tanaka (1980). Some considerations on fuzzy conditional inference. *Fuzzy Sets and Systems*, Vol. 4, No. 3, Nov., pp. 243–274.

Fung, L. W., and K. S. Fu (1974). The K″th optimal policy algorithm for decision-making in fuzzy environments. In *Identification and System Parameter Estimation* (P. Eykhoff, Ed.). North-Holland, pp. 1025–1059.

Furuta, H., and N. Shiraishi (1984). Fuzzy importance in fault tree analysis. *Fuzzy Sets and Systems*, Vol. 12, No. 3, Apr., pp. 205–217.

Gaines, B. R. (1975). Stochastic and fuzzy logics. *Electronics Lett.*, Vol. 11, pp. 188–189.

Gaines, B. R. (1976). Behaviour-structure transformations under uncertainty. *Int. J. Man-Machine Studies*, Vol. 8, pp. 337–365.

Gaines, B. R. (1978). Fuzzy and probability uncertainty logics. *Inform. and Control*, Vol. 38, pp. 154–169.

Gaines, B. R., and L. J. Kohout (1975). The logic of automata. *Int. J. General Syst.*, Vol. 2, pp. 191–208.

Gaines, B. R., and L. J. Kohout (1977). The fuzzy decade — A bibliography of fuzzy systems and closely related topics. *Int. J. Man-Machine Studies*, Vol. 9, pp. 1–68.

Gale, S. (1972). Inexactness, fuzzy sets, and the foundations of behavioral geography. *Geographical Analysis*, Vol. 4, pp. 337–349.

Ganter, T. E., R. C. Steinlage, and R. H. Warren (1978). Compactness in fuzzy topological spaces. *J. Math. Anal. Appl.*, Vol. 62, pp. 547–562.

Gerla, G. (1983). Representations of fuzzy topologies. *Fuzzy Sets and Systems*, Vol. 1, No. 2, Oct., pp. 103–114.

Ghanim, M. H., and A. S. Mashhour (1983). Characterization of fuzzy topologies by *S*-neighbourhood systems (Short communication). *Fuzzy Sets and Systems*, Vol. 9, No. 2, Feb., pp. 211–214.

Giering, E. W., and A. Kandel (1983). The application of fuzzy-set theory to the modeling of competition in ecological systems. *Fuzzy Sets and Systems*, Vol. 9, No. 2, Feb., pp. 103–128.

Giles, R. (1976a). A logic for subjective belief, In *Foundations of Probability Theory, Statistical Inference, and Statistical Theories of Science*, 1 (Harper, W., and C. A. Hooker, Eds.). D. Reidel, Dordrecht, Holland, pp. 41–72.

Giles, R. (1976b). Lukasiewicz logic and fuzzy set theory. *Int. J. Man-Machine Studies*, Vol. 8, pp. 313–327.

Giles, R. (1980a). A formal system for fuzzy reasoning. *Fuzzy Sets and Systems*, Vol. 2, No. 3, July, pp. 233–258.

Giles, R. (1980b). A computer program for fuzzy reasoning. *Fuzzy Sets and Systems*, Vol. 4, No. 3, Nov., pp. 221–234.

Gitman, I., and M. D. Levine (1970). An algorithm for detecting unimodal fuzzy sets and its application as a clustering technique. *IEEE Trans. Comp.*, C-19, pp. 583–593.

Gluss, B. (1973). Fuzzy multistage decision-making. *Int. J. Control*, Vol. 17, pp. 177–192.

Godal, R. C., and T. J. Goodman (1980). Fuzzy sets and borel. *IEEE Trans. on Sys. Man. and Cybernetics*, Oct., Vol. SMC-10, No. 10.

Goetschel, R., and W. Voxman (1983). Topological properties of fuzzy numbers. *Fuzzy Sets and Systems*, Vol. 10, No. 1, April, pp. 87–100.

Goguen, J. A. (1967). *L*-fuzzy sets. *J. Math. Anal. and Appl.*, Vol. 18, pp. 145–174.

Goguen, J. A. (1969a). Categories of *L*-fuzzy sets. *Bull. Amer. Soc.*, Vol. 75, No. 3, May, pp. 622–624.

Goguen, J. A. (1969b). The logic of inexact concepts. *Synthèse*, Vol. 19, pp. 325–373.

Goguen, J. A. (1974a). Concept representation in natural and artificial languages — Axioms extensions and applications for fuzzy sets. *Int. J. Man-Machine Studies*, Vol. 6, pp. 513–561.

Goguen, J. A. (1974b). The fuzzy Tychonoff theorem. *J. Math. Anal. and Appl.*, Vol. 43, pp. 734–742.

Goguen, J. A., J. L. Weiner, and C. Linde (1983). Reasoning and natural explanation. *Int. J. Man-Machine Studies*, Vol. 19, No. 6, pp. 521–559.

Gottinger, H. W. (1973). Competitive processes — Application to urban structure. *Cybernetica*, Vol. 16, pp. 177–197.

Gottinger, H. W. (1973). Towards a fuzzy reasoning in the behavioural sciences. *Cybernetica*, pp. 113–135.

Gottinger, H. W. (1976). Some basic issues connected with fuzzy analysis. In *Systems Theory in the Social Sciences* (Bossel, H., S. Klaczko, and N. Muller, Eds.). Birkhauser Verlag, Basel, pp. 323–325.

Gottwald, S. (1974). Fuzzy topology, product, and quotient theorems. *J. of Math. Anal. Appl.*, Vol. 45, pp. 512–521.

Gottwald, S. (1976). Fuzzy propositional logics. *Fuzzy Sets and Systems*, Vol. 3, No. 2, pp. 181–192.

Gottwald, S. (1978). Theoretische betrachtungen über fuzzy logik. *Schriftenreihe Weiterbildungszentrum Math. Kybernetik Rechentechnik* (Tu Dresden), Heff, pp. 3–22.

Gottwald, S. (1979). A note on measures of fuzziness. *Elek. Informationsverarbeitung und Kyber.*, Eik, Vol. 15, No. 4, pp. 221–223.

Gottwald, S. (1980). Fuzzy uniqueness of fuzzy mappings. *Fuzzy Sets and Systems*, Vol. 3, No. 1, pp. 49–74.

Gottwald, S. (1981). Fuzzy points and local properties of fuzzy topological spaces. *Fuzzy Sets and Systems*, Vol. 5, No. 2, pp. 199–202.

Gottwald, S. (1984). On the existence of solutions of systems of fuzzy equations (short communication). *Fuzzy Sets and Systems*, Vol. 12, No. 3, Apr., pp. 301–302.

Gottwald, S., E. Czogala, and W. Pedrycz (1983). Measures of fuzziness and operations with fuzzy sets. *Stochastica*, Vol. 6, No. 3, pp. 187–205.

Gratten-Guiness (1976). Fuzzy membership mapped onto interval and many-valued quantities. *Z. Math. Logik Grundlagen Math.*, Vol. 22, pp. 149–160.

Grofman, B., and G. Hyman (1973). Probability and logic in belief systems. *Theory and Decision*, Vol. 4, pp. 179–195.

Guo-Quan, C. (1980). An approach to fuzzy controller algorithms. *Inf. and Cont.*, Vol. 9, No. 3.

Guo-Quan, C. (1981). *Fuzzy Sets, Fuzzy Linguistic Variables, and Fuzzy Logic.* Book translation and edition supported by Chinese Academic Press.

Gupta, M. M. (1979). Guest editorial to the special issue on fuzzy sets and applications. *Fuzzy Sets and Systems*, Vol. 2, No. 1, Jan., pp. 1–4.

Gupta, M. M., A. Kandel, W. Bandler, and J. B. Kiszka (Eds). (1985). *Approximate Reasoning in Expert Systems.* North-Holland, Amsterdam.

Gupta, M. M., and E. H. Mamdani (1976). Second IFAC round table on fuzzy automata and decision processes. *Automatica*, Vol. 12, pp. 291–296.

Gupta, M. M. (Ed.), G. N. Saridis, and B. R. Gaines, (Assoc. Eds.) (1977). *Fuzzy Automata and Decision Processes.* North-Holland, Amsterdam.

Gupta, M. M., R. K. Ragade, and R. R. Yager (1979a). Report on the IEEE Symp. on fuzzy-set theory and applications (Short communication). *Fuzzy Sets and Systems*, Vol. 2, No. 1, Jan., pp. 105–111.

Gupta, M. M., Ed., Ragade, R. K., and R. R. Yager, (Assoc. Eds.) (1979b). *Advances in Fuzzy-Set Theory and Applications.* North-Holland, Amsterdam.

Gupta, M. M., and E. Sanchez, (Eds.) (1982a). *Fuzzy Information and Decision Processes.* North-Holland, Amsterdam.

Gupta, M. M., and E. Sanchez, (Eds.) (1982b). *Approximate Reasoning in Decision Analysis.* North-Holland, Amsterdam.

Gusev, L. A., and I. M. Smirnova (1973). Fuzzy sets — Theory and applications (A survey). *Automation and Remote Control*, No. 5, May, pp. 66–85.

Haack, S. (1974). *Deviant Logic*, Cambridge University Press, Cambridge.

Haack, S. (1975). “Alternative” in “Alternative Logic.” In *Meaning, Reference and Necessity* (Blackburn, S. (Ed.)). Cambridge University Press, Cambridge, pp. 32–55.

Haack, S. (1976). The justification of deduction. *MIND*, Vol. 85, pp. 112–119.

Haack, S. (1979). Do we need “fuzzy logic”?. *Int. Jour. Man-Machine Studies*, Vol. 11, pp. 437–445.

Hacking, I. (1975). All kinds of possibility. *Philosophical Review*, Vol. 84, pp. 319–337.

Hagg, C. (1977). Possibility and cost in decision analysis. *Fuzzy Sets and Systems*, Vol. 1, pp. 81–86.

Hall, L. O., and A. Kandel (1985). Studies in possibilistic recognition. *Fuzzy Sets and Systems*, Vol. 17, No. 2, pp. 167–179.

Halpern, J. (1975). Set-adjacency measures in fuzzy graphs. *J. of Cybern.*, Vol. 5, No. 4, pp. 77–87.

Hamacher, H., H. Leberling, and H. J. Zimmermann (1978). Sensitivity analysis in fuzzy linear programming. *Fuzzy Sets and Systems*, Vol. 1, pp. 269–281.

Hammerbach, I. M., and R. R. Yager (1981). The personalization of security selection — An application of fuzzy-set theory, *Fuzzy Sets and Systems*, Vol. 5, No. 1, Jan., pp. 1–10.

Hannan, E. L. (1979). On the efficiency of the product operator in fuzzy programming with multiple objectives. *Fuzzy Sets and Sys.*, Vol. 2, No. 3, July, pp. 252–263.

Hannan, E L. (1981). Linear programming with multiple fuzzy goals. *Fuzzy Sets and Systems*, Vol. 6, No. 3, Nov., pp. 235–248.

Hart, W. D. (1972). Probability as a degree of possibility. *Notre Dame J. Formal Logic*, Vol. 13, pp. 286–288.

Hashimoto, H. (1983a). Convergence of powers of a fuzzy transitive matrix. *Fuzzy Sets and Systems*, Vol. 9, No. 2, Feb., pp. 153–160.

Hashimoto, H. (1983b). Szpilrajn's theory on fuzzy orderings. *Fuzzy Sets and Systems*, Vol. 10, No. 1, Apr., pp. 101–108.

Hashimoto, H. (1983c). Canonical form of a transitive fuzzy matrix. *Fuzzy Sets and Systems*, Vol. 11, No. 2, Oct., pp. 157–162.

Heshmaty, B., and A. Kandel (1985). Fuzzy linear regression and its applications to forecasting in uncertain environment. *Fuzzy Sets and Systems*, Vol. 15, No. 3, pp. 159–191.

Hersh, H. M., and A. Caramazza (1976). A fuzzy-set approach to modifiers and vagueness in natural language. *J. Experimental Psychology*, Vol. 105, pp. 254–276.

Higashi, M., and G. J. Klir (1982). Measures of uncertainty of general systems. *International Journal of General Systems*, Vol. 9, No. 1, pp. 43–52.

Higashi, M., and G. J. Klir (1983). On the notion of distance representing information closeness: Possibility and probability distributions. *Int. J. General Systems*, Vol. 9, pp. 103–115.

Higashi, M., and G. J. Klir (1984a). Identification of fuzzy-relation systems, *IEEE Trans. Syst., Man, Cybern.*, Vol. SMC-14, No. 2, pp. 349–355.

Higashi, M., and G. J. Klir (1984b). Resolution of finite fuzzy-relation equations. *Fuzzy Sets and Systems*, Vol. 13, No. 1, May, pp. 65–82.

Higashi, M., G. J. Klir, and M. A. Pittarelli (1984). Reconstruction families of possibilistic structure systems. *Fuzzy Sets and Systems*, Vol. 12, No. 1, Jan., pp. 37–60.

Hinde, C. J. (1983). Inference of fuzzy-relational tableaux from fuzzy exemplications. *Fuzzy Sets and Systems*, Vol. 11, No. 1, Aug., pp. 91–102.

Hiramatsu, K., K. Kabasawa, and S. Kaibara, (1974). Fuzzy logic applied to the medical diagnosis. *Medical Electronics and Bioscience*, Vol. 12, pp. 34–41.

Hirota, K. (1981). Concepts of probabilistic sets. *Fuzzy Sets and Systems*, Vol. 5, No. 1, pp. 31–46.

Hirota, K., and W. Pedrycz (1982a). Probabilistic sets in identification of fuzzy systems. *Archiwum Automatyki i Tilemechaniki*, Poland, Vol. 27, pp. 119–130.

Hirota, K., and W. Pedrycz (1982b). Fuzzy-systems identification via probability sets. *Information Science*, Vol. 28, No. 1, pp. 21–44.

Hirota, K., and W. Pedrycz (1983). Analysis and synthesis of fuzzy systems by the use of probabilistic sets. *Fuzzy Sets and Systems*, Vol. 10, No. 1, Apr., pp. 1–14.

Hirsch, G., M. Lamotte, M. T. Mas, and M. J. Virneron (1981). Phonemic classification using a fuzzy dissimilitude relation. *Fuzzy Sets and Systems*, Vol. 5, No. 3, pp. 267–276.

Hisdal, E. (1978). Conditional possibilities, independence, and noninteraction. *Fuzzy Sets and Systems*, Vol. 1, pp. 283–297.

Hisdal, E. (1980). Generalized fuzzy-set systems and particularization. *Fuzzy Sets and Systems*, Vol. 4, No. 3, Nov., pp. 275–292.

Hohle, U. (1977). Probabilistic uniformization of fuzzy topologies. *Fuzzy Sets and Systems*, Vol. 1, pp. 311–332.

Hohle, U. (1981). Representation theorems for *L*-fuzzy quantities. *Fuzzy Sets and Systems*, Vol. 5, No. 1, Jan., pp. 83–108.

Hohle, U. (1982). Probabilistic metrization of fuzzy uniformities. *Fuzzy Sets and Systems*, Vol. 8, No. 1, June, pp. 63–70.

Hohle, U. (1983). Fuzzy measures as extensions of stochastic measures. *Journal of Mathematical Analysis and Applications*, Vol. 92, No. 2, pp. 372–380.

Hohle, U. (1984). Compact *G*-fuzzy topological spaces. *Fuzzy Sets and Systems*, Vol. 13, No. 1, May, pp. 39–67.

Honda, N. (1971). Fuzzy sets. *J. Inst. Electron. Comm. Eng.* (Japan), Vol. 54, pp. 1359–1363.

Honda, N. (1975). Applications of fuzzy-set theory to automata and linguistics. *J. JAACE*, Vol. 19, pp. 249–254.

Hughes, J. S., and A. Kandel (1977). Applications of fuzzy algebra to hazard detection in combinational switching circuits. *Int. J. of Comp. and Inf. Sci.*, Vol. 6, pp. 71–82.

Hunt, R. M., and W. B. Rouse (1984). A fuzzy rule-based model of human problem-solving. *IEEE Trans. Syst., Man, Cybern.*, Vol. SMC-14, No. 1, pp. 112–120.

Hutton, B. (1975). Normality in fuzzy topological spaces. *J. Math. Anal. and Appl.*, Vol. 50, pp. 74–79.

Hutton, B. (1977). Uniformities on fuzzy topological spaces. *J. Math. Anal. Appl.*, Vol. 58, pp. 559–571.

Hutton, B., and J. L. Reilly (1980). Separation axioms in fuzzy topological spaces. *Fuzzy Sets and Systems*, Vol. 3, No. 1, pp. 93–104.

Ignizio, J. P., and S. C. Daniels (1983). Fuzzy multicriteria integer programming via fuzzy generalized networks. *Fuzzy Sets and Systems*, Vol. 10, No. 2, July, pp. 261–270

Ishikawa, A., and H. Mieno (1979). The fuzzy entropy concept and its application. *Fuzzy Sets and Systems*, Vol. 2, No. 2, Apr.

Jacobson, D. H. (1976). On fuzzy goals and maximizing decisions in stochastic optimal control. *J. Math. Anal. and Appl.*, Vol. 55, pp. 434–440.

Jahn, K. U. (1977). Grundfragen einer mehrwertigen (fuzzy) analysis. *Vortrage zu Grundlagen der Informatik* (Tu Dresden, D.D.R.), pp. 36–45.

Jain, R. (1976a). Convolution of fuzzy variables. *JIETE*, Vol. 22.

Jain, R. (1976b). Decision-making in the presence of fuzzy variables. *IEEE Trans. Syst., Man, Cybern.*, Vol. SMC-6, pp. 698–703.

Jain, R. (1977a). Tolerance analysis using fuzzy sets. *Int. Jour. System Sci.*, Vol. 7, No. 12, pp. 1393–1401.

Jain, R. (1977b). A procedure for multiple-aspect decision-making using fuzzy sets. *Int. Jour. System Sci.*, Vol. 8, pp. 1–7.

Jain, R., and W. Stallings (1978). Comments on fuzzy-set theory versus Bayesian statistics. *IEEE Trans. Syst., Man, Cybern.*, Vol. SMC-8, pp. 332–333.

Jarvis, R. A. (1975). Optimization strategies in adaptive control — A selective survey. *IEEE Trans. Syst., Man, Cybern.*, Vol. SMC-5, pp. 83–94.

Jinwen, Z. (1980). A kind of nonstandard model of the axiomatic set theory with urelements — The normal fuzzy-set structure. Published in the *Journal of Huazhong Institute of Technology* [English Edition], Vol. 2, No. 1.

Jones, W. T. (1976). A fuzzy-set characterization of interaction in scientific research. *J. Am. Soc. Inf. Sci.* (Sept.–Oct).

Joyce, J. (1976). Fuzzy sets and the study of linguistics. *Pac. Coast Philol.*, Vol. 11, pp. 39–42.

Jumarie, G. (1977). Some technical applications of relativistic information theory, Shannon information, fuzzy sets, linguistics, relativistic sets, and communication. *Cybernetica*, Vol. 20, No. 2, pp. 91–128.

Jumarie, G. (1983). Entropy of fuzzy events revisited. *Cybernetica*, Vol. 16, No. 2.

Kacprzyk, J. (1977a). Decision-making in a fuzzy environment with fuzzy termination time. *Fuzzy Sets and Systems*, Vol. 1, pp. 169–179.

Kacprzyk, J. (1977b). Control of a nonfuzzy system in a fuzzy environment with fuzzy termination time. *Systems Science*, Vol. 3.

Kacprzyk, J. (1979a). A branch-and-bound algorithm for the multistage control of a nonfuzzy system in a fuzzy environment. *Contr. Cybernet.*, Vol. 8, No. 2, pp. 139–147.

Kacprzyk, J. (1979b). On some multistage decision-making problems in a fuzzy environment. In *Proceedings of the Seminar on "Nonconventional Problems of Optimization*," Part 1, (J. Gutenbaum, Ed.). Warzawa.

Kacprzyk, J. (1983). *Multistage Decision-Making under Fuzziness*. Verlag TÜV, Rheinland, Köln, Germany.

Kacprzyk, J., and P. Staniewski (1982). Long-term inventory policy-making through fuzzy decision-making models. *Fuzzy Sets and Systems*, Vol. 8, No. 2, Aug., pp. 117–132.

Kacprzyk, J., and P. Staniewski (1983). Control of a deterministic system in a fuzzy environment over infinite planning horizon. *Fuzzy Sets and Systems*, Vol. 10, No. 3, July, pp. 291–298.

Kacprzyk, J., and A. Straszak (1984). Determination of "stable" trajectories of integrated regional development using fuzzy decision models. *IEEE Trans. Syst., Man, Cybern.*, Vol. SMC-14, No. 2, pp. 310–313.

Kacprzyk, J., and R. R. Yager (Eds.) (1985). *Management Decision Support Systems Using Fuzzy Sets and Possibility Theory.* Verlag TÜV, Rheinland, Köln, Germany.

Kalaba, R. E., and K. Springarn (1978). Numerical approaches to the eigenvalues of Saaty's matrices for fuzzy sets. *J. Comput. Math. Appl.*, Vol. 4, pp. 369–375.

Kaleva, O., and S. Seikkala (1984). On fuzzy metric spaces. *Fuzzy Sets and Systems*, Vol. 12, No. 3, Apr., pp. 215–230.

Kalmanson, D., and H. E. Stegall (1975). Cardiovascular investigations and fuzzy-set theory. *Amer. Jour. of Cardiology*, Vol. 35, pp. 30–34.

Kalsaras, A. K., and D. B. Liu (1977). Fuzzy vector spaces and fuzzy topological vector spaces. *J. of Math Anal. and Appl.*, Vol. 58, pp. 135–146.

Kameda, T., and E. Sadeh (1977). Bounds on the number of fuzzy functions. *Information and Control.*, Vol. 35, pp. 139–145.

Kandel, A. (1973a). Comment on an algorithm that generates fuzzy prime implicants, by Lee and Chang. *Inform. and Control*, Vol. 22, pp. 279–282.

Kandel, A. (1973b). Comments on "minimization of fuzzy functions." *IEEE Trans. Comp.*, Vol. C-22, p. 217.

Kandel, A. (1973c). On minimization of fuzzy functions. *IEEE Trans. Comp.*, Vol. C-22, pp. 826–832.

Kandel, A. (1974a). Codes over languages. *IEEE Trans. Syst., Man, Cybern.*, Vol. SMC-4, pp. 135–138.

Kandel, A. (1974b). On the minimization of incompletely specified fuzzy functions. *Inform. and Control*, Vol. 26, pp. 141–153.

Kandel, A. (1974c). On the properties of fuzzy switching functions. *J. Cybernetics*, Vol. 4, pp. 119–126.

Kandel, A. (1974d). Synthesis of fuzzy logic with analog modules — preliminary developments. *Computers in Education Transaction* (ASEE Div.), Vol. 6, pp. 71–79.

Kandel, A. (1975). Fuzzy hierarchical classifications of dynamic patterns. *NATO ASI Pattern Recognition and Classification*, France, September.

Kandel, A. (1976a). Inexact switching logic. *IEEE Trans. Syst., Man, Cybern.*, Vol. SMC-6, pp. 215–219.

Kandel, A. (1976b). On the decomposition of fuzzy functions. *IEEE Trans. Comp.*, Vol. C-25, pp. 1124–1130.

Kandel, A. (1977a). A note on the simplification of fuzzy functions. *Information Sci.*, Vol. 13, pp. 91–94.

Kandel, A. (1977b). Comments on comments by Lee. *Information and Control*, Vol. 35, pp. 109–113.

Kandel, A. (1978). Fuzzy statistics and forecast evaluation. *IEEE Trans. Syst., Man, Cybern.*, Vol. SMC-8, No. 5, pp. 396–401.

Kandel, A. (1979). Reply to Tribus' comments. *Proc. IEEE*, Vol. 67, No. 8, pp. 1168–1169.

Kandel, A. (1981). Fuzzy expectation and energy states in fuzzy media. *Fuzzy Sets and Systems*, Vol. 6, No. 2, Sept., pp. 145–160.

Kandel, A. (1982). *Fuzzy Techniques in Pattern Recognition*. Wiley-Interscience, New York.

Kandel, A., and W. J. Byatt (1978). Fuzzy sets, fuzzy algebra, and fuzzy statistics. *Proceedings of the IEEE*, Vol. 66, No. 12, Dec., pp. 1619–1639.

Kandel, A., and W. J. Byatt (1980). Fuzzy processes. *Fuzzy Sets and Systems*, Vol. 4, No. 2, Sept., pp. 117–152.

Kandel, A., and C. M. Clark (1982). The enumeration of distinct fuzzy-valued switching functions. *Fuzzy Sets and Systems*, Vol. 8, No. 3, Sept., pp. 291–310.

Kandel, A., and J. Francioni (1980). On the properties and applications of fuzzy-valued switching functions. *IEEE Transactions on Computers*, Vol. C-29, No. 11, November.

Kandel, A., and S. C. Lee (1979). *Fuzzy Switching and Automata — Theory and Applications*. Crane, Russak and Co., New York, and Edward Arnold, London.

Kandel, A., and L. Yelowitz (1974). Fuzzy chains. *IEEE Trans. Syst., Man, Cybern.*, Vol. SMC-4, pp. 472–475.

Kania, A. A., J. B. Kiszka, M. B. Gorzalczany, and J. R. Maj (1980). On stability of formal fuzziness systems. *Information Sciences, Internat. J.*, Vol. 21.

Karwowski, W., M. M. Ayoub, L. R. Alley, and J. L. Smith (1984). Fuzzy approach in psychophysical modeling of human operator-manual lifting system. *Fuzzy Sets and Systems*, Vol. 14, No. 1, pp. 65–66.

Katsaras, A. K. (1979). Fuzzy proximity spaces. *J. Math. Anal. Appl.*, Vol. 68, No. 1, Mar., pp. 100–111.

Katsaras, A. K. (1981). Fuzzy topological vector spaces I. *Fuzzy Sets and Systems*, Vol. 6, No. 1, July, pp. 85–95.

Katsaras, A. K. (1984). Fuzzy topological vector spaces II. *Fuzzy Sets and Systems*, Vol. 12, No. 2, Feb., pp. 143–151.

Katsaras, A. K., and D. B. Liu (1977). Fuzzy vector spaces and fuzzy topological vector spaces. *J. Math. Anal. Appl.*, Vol. 58, pp. 135–146.

Katz, M. (1980). Inexact geometry. *Notre Dame J. of Formal Logics*, Vols. 21–23, pp. 521–535.

Kaufmann, A. (1973). *Introduction à la Théorie des Sous-Ensembles Flous*, 1: *Elements Théoretiques de Base*. Masson et Cie, Paris, France.

Kaufmann, A. (1975a). *Introduction à la Théorie des Sous-Ensembles Flous*, 2: *Applications à la Linguistique et à la Sémantique*. Masson et Cie, Paris, France.

Kaufmann, A. (1975b). *Introduction à la Théorie des Sous-Ensembles Flous*, 3: *Applications à la Classification et la Reconnaissance des Formes, aux Automates et aux Systemes, aux Choix des Critares*. Masson et Cie, Paris, France.

Kaufmann, A. (1975c). *Theory of Fuzzy Subsets*, Vol. 1. Academic Press, New York.

Kaufmann, A. (1976). La théorie des sous-ensembles flous et ses applications dans les sciences humains, in *Economie Appliqués*. Librairie Droz, Genève, Tome 29, No. 3, pp. 469–478.

Kaufmann, A. (1977). *Introduction à la Théorie des Sous-Ensembles Flous*, 4: *Complément et Nouvelles Applications*. Masson et Cie, Paris, France.

Kaufmann, A. (1979a). *Modèles Mathématiques pour la Stimulation Inventive* (Albin-Michel, Ed.). Paris, Nov.

Kaufmann, A. (1979b). *Compléments à la Théorie des Sous-Ensembles Flous* (unpublished volumes).

Kaufmann, A. (1980). Bibliography on fuzzy sets and their applications. Busefal, No. 1–3, *LSI Lab, Univ. Paul Sabatier*, Toulouse, France.

Kaufmann, A. and M. Gupta (1985). *Introduction to Fuzzy Arithmetic*. Van Nostrand, New York.

Keller, J. M., and D. J. Hunt (1985). Incorporating fuzzy membership function into the perceptron algorithm. *IEEE Trans. on Pattern Analysis and Machine Intelligence*, Vol. PAMI-7, No. 6, Nov., pp. 693–699.

Kempton, W. (1984). Interview methods for eliciting fuzzy categories. *Fuzzy Sets and Systems*, Vol. 14, No. 1, Sept., pp. 43–64.

Kerre, E. E. (1980). Fuzzy Sierpinski space and its generalizations. *J. Math. Anal. and Appl.*, Vol. 74, No. 1.

Khalili, S. (1979a). Fuzzy measures and mappings. *J. Math. Anal. and Appl.*, Vol. 68, pp. 92–99.

Khalili, S. (1979b). Independent fuzzy events. *J. Math. Anal. and Appl.*, Vol. 67, No. 2, Feb., pp. 412–421.

Kickert, W. J. M. (1978). *Fuzzy Theories on Decision-Making*. Martinns Mijhoff, Leiden, The Netherlands.

Kickert, W. J. M. (1979). Towards an analysis of linguistic modelling. *Fuzzy Sets and Systems*, Vol. 2, No. 4, pp. 293–308.

Kickert, W. J. M., and H. Koppelaar (1976). Applications of fuzzy-set theory to syntactic pattern recognition of handwritten capitals. *IEEE Trans. Syst., Man, Cybern.*, Vol. 6, pp. 148–151.

Kickert, W. J. M., and E. H. Mamdani (1978). Analysis of a fuzzy logic controller. *Fuzzy Sets and Systems*, pp. 29–44.

Kickert, W. J. M., and H. R. Van Nauta-Lemke (1976). Application of a fuzzy controller in a warm water plant. *Automatica*, Vol. 12, pp. 301–308.

Kim, H. H., M. Mizumoto, J. Toyoda, and K. Tanaka (1974). Automated editing of fuzzy line drawings for picture description. *Trans. IECE*, Vol. 57-A, pp. 216–223.

Kim, H. H., M. Mizumoto, J. Toyoda, and K. Tanaka (1975). *L*-fuzzy grammars. *Inform. Sci.*, Vol. 8, pp. 123–140.

Kim, J. B. (1983). Fuzzy rational-choice functions. *Fuzzy Sets and Systems*, Vol. 10, No. 1, Apr., pp. 37–44.

Kim, K. H., and F. W. Roush (1980). Generalized fuzzy matrices. *Fuzzy Sets and Systems*, Vol. 4, No. 3, pp. 293–316.

Kim, K. H., and F. W. Roush (1982). Fuzzy flows on networks. *Fuzzy Sets and Systems*, Vol. 8, No. 1, June, pp. 35–38.

Kiszka, J. B. (1980). On stability of formal fuzziness systems. *Information Sciences*, Vol. 22, pp. 51–68.

Kiszka, J. B., M. M. Gupta, and P. N. Nikiforuk (1985). Energetistic stability of fuzzy dynamic systems. *IEEE Trans. SMC*, Vol. SMC-15, No. 6, Nov.–Dec., pp. 783–792.

Kiszka, J. B., M. E. Kochańska, and D. S. Sliwińska (1985). The influence of some fuzzy implication operators on the accuracy of a fuzzy model. *Fuzzy Sets and Systems*. Part I — Vol. 15, No. 2, pp. 111–128. Part II — Vol. 15, No. 3, pp. 223–240.

Kitajima, S., and K. Asai (1970). Learning controls by fuzzy automata. *Journal of JAACE*, Vol. 14, pp. 551–559.

Kitajima, S., and K. Asai (1974). A method of learning control varying search domain by fuzzy automata. In *Learning Systems and Intelligent Robots* (Fu, K. S., and J. T. Tou, Eds.). Plenum Press, New York, pp. 249–262.

Klein, A. J. (1983). Generating fuzzy topologies with semi-closure operators. *Fuzzy Sets and Systems*, Vol. 9, No. 3, Mar., pp. 267–274.

Klein, A. J. (1984). Generalizing the *L*-fuzzy unit interval. *Fuzzy Sets and Systems*, Vol. 12, No. 3, Apr., pp. 271–286.

Klement, E. P. (1980). Fuzzy sigma-algebras and fuzzy measurable functions. *Fuzzy Sets and Systems*, Vol. 4, No. 1, pp. 83–95.

Klement, E. P., and D. Ralescu (1983). Nonlinearity of fuzzy integral (short communication). *Fuzzy Sets and Systems*, Vol. 11, No. 3, Nov., pp. 309–316.

Klement, E. P., and W. Schwyhla (1982). Correspondence between fuzzy measures and classical measures. *Fuzzy Sets and Systems*, Vol. 7, No. 1, pp. 57–70.

Klement, E. P., W. Schwyhla, and R. Lowen (1981). Fuzzy probability measures. *Fuzzy Sets and Systems*, Vol. 5, No. 1, Jan., pp. 21–30.

Kling, R. (1974). Fuzzy-planner — Reasoning with inexact concepts in a procedural problem-solving language. *J. Cybernetics*, Vol. 4, pp. 105–122.

Klir, G. J. (1971). On universal logic primitives. *IEEE Trans. on Computers*, Vol. C20, No. 4, Apr., pp. 457–467.

Klir, G. J. (1975). Processing of fuzzy activities of neural systems. In *Progress in Cybernetics and Systems Research* (Trappel, R., and F. R. Pichler, Eds.), Vol. 1, pp. 21–24.

Kloeden, P. E. (1980). Compact supported endographs and fuzzy sets. *Fuzzy Sets and Systems*, Vol. 4, No. 2, Sept., pp. 193–202.

Kloeden, P. E. (1982). Fuzzy dynamical systems. *Fuzzy Sets and Systems*, Vol. 7, No. 3, May, pp. 275–296.

Knopfmacher, K. (1975). On measures of fuzziness. *J. Math. Anal. and Appl.*, Vol. 49, pp. 529–534.

Kochen, M. (1979). Enhancement of coping through blurring. *Fuzzy Sets and Systems*, Vol. 2, No. 1, Jan., pp. 37–52.

Kochen, M., and A. N. Badre (1974). On the precision of adjectives which denote fuzzy sets. *J. Cybernetics*, Vol. 4, pp. 49–59.

Koczy, L.T. (1977). On some basic theoretical problems of fuzzy mathematics. *Acta Cybernetica*, Vol. 3, pp. 225–237.

Koczy, L. T. (1978). Interactive sigma-algebras and fuzzy objects of type *N*. *Jour. of Cyber.*, Vol. 8.

Koczy, L. T. (1979). Some questions of *B*-Algebras of fuzzy objects of type *N*. *IEEE Trans. Syst., Man, Cybern.*, Sept., Vol. SMC-9, No. 9.

Kohout, L. J. (1975). Generalized topologies and their relevance to general systems. *Int. J. General Syst.*, Vol. 2, pp. 25–34.

Kohout, L. J. (1976). Representation of functional hierarchies of movement in the brain. *Int. J. Man-Machine Studies*, Vol. 8, pp. 699–709.

Kokawa, M., K. Nakamura, and M. Oda (1974a). Fuzzy process of memory behaviour. *Trans. Soc. Instrum. Control Engrs.*, Vol. 10, pp. 385–386.

Kokawa, M., K. Nakamura, and M. Oda (1974b). Fuzzy-theoretical and concept-formational approaches to memory and inference experiments. *Trans. Inst. Electron. Comm. Eng.* (Japan), Vol. 57-D, pp. 487–493.

Kokawa, M., K. Nakamura, and M. Oda (1979). Fuzzy-theoretic and concept-formational approaches to hint-effect experiments in human decision processes. *Fuzzy Sets and Systems*, Vol. 2, No. 1, Jan., pp. 25–36.

Kokawa, M., M. Oda, and K. Nakamura (1975). Fuzzy-theoretical one-dimensionalizing method of multidimensional quantity. *Trans. Society of Instrument and Control Engineers*, Vol. 11, No. 5, pp. 8–14.

Kokawa, M., M. Oda, and K. Nakamura (1979). A study of handling method of fuzzy data. *Trans. Inst. of Elec. and Comm. Eng. (Japan)*, Vol. 62-A, No. 1, pp. 97–102.

Kotoh, K., and K. Hiramatsu (1973). A representation of pattern classes using the fuzzy sets. *Systems, Computers, Controls*, Vol. 1-8 (Orig. T.I.E.C.E. 56-D, 275–282).

Kramosil, I., and J. Michalek (1975). Fuzzy matrices and statistical metric spaces. *Kybernetika* (Prague), Vol. 11, pp. 336–344.

Kruse, R. (1982a). A note on lambda-additive fuzzy measures (short communication). *Fuzzy Sets and Systems*, Vol. 8, No. 2, Aug., pp. 219–222.

Kruse, R. (1982b). On the construction of fuzzy measures (short communication). *Fuzzy Sets and Systems*, Vol. 8, No. 3, Sept., pp. 323–328.

Kruse, R. (1983a). Fuzzy integrals and conditional measures. *Fuzzy Sets and Systems*, Vol. 10, No. 3, July, pp. 309–314.

Kruse, R. (1983b). On the entropy of fuzzy events. *Kybernetes*, Vol. 12, No. 1, pp. 53–58.

Kumar, A. (1977). A real-time system for pattern recognition of human sleep stages by fuzzy systems analysis. *Pattern Recognition*, Pergamon Press, Vol. 9, pp. 43–46.

Kuroki, N. (1981). On fuzzy ideals and fuzzy bi-ideals in semigroups. *Fuzzy Sets and Systems*, Vol. 5, No. 2, pp. 203–215.

Kuroki, N. (1982). Fuzzy semiprime ideals in semigroups. *Fuzzy Sets and Systems*, Vol. 8, No. 1, June, pp. 71–80.

Kuzmin, V. B. (1981a). A parametric approach to the description of linguistic values of variables and hedges. *Fuzzy Sets and Systems*, Vol. 6, pp. 27–41.

Kuzmin, V. B. (1981b). Corrections to "A parametric approach to description of linguistic values of variables and hedges" (Erratum). *Fuzzy Sets and Systems*, Vol. 6, No. 2, Sept., pp. 205.

Kuzmin, V. B., and S. V. Ovchinnikov (1980). Group decisions in arbitrary spaces of fuzzy binary relations. *Fuzzy Sets and Systems*, Vol. 4, No. 1, pp. 53–62.

Kwakernaak, H. (1979a). An algorithm for rating multiple-aspect alternatives using fuzzy sets. *Automatica*, Vol. 15, pp. 615–616.

Kwakernaak, H. (1979b). Fuzzy random variables, algorithms and examples for the discrete case. *Inform. Science*, Vol. 17, pp. 253–278.

Lakeoff, G. (1973a). Hedges — A study in meaning criteria and the logic of fuzzy concepts. *J. Philos. Logic*, Vol. 2, pp. 458–508.

Lakeoff, G. (1973b). Notes on what it would take to understand how one adverb works. *Monist*, Vol. 57, pp. 328–343.

Lakeoff, G. (1973c). Pragmatics in natural logic. In *Formal Semantics of Natural Language* (Keenan, E. L., Ed.). Cambridge University Press. Cambridge, pp. 253–286.

Lakov, D. R., and N. Naplatanoff (1977). Decision-making in vague conditions. *Kybernetes*, Vol. 6, pp. 91–93.

Lakshmivaran, S., and K. S. Rajasethupathy (1978). Considerations for fuzzifying formal languages and synthesis for fuzzy games. *J. of Cybern.*, Vol. 8, pp. 83–100.

Larsen, P. M. (1980). Industrial applications of fuzzy logic control. *Int. Man-Machine Studies*, Vol. 12, No. 1.

Lasker, G. E. (Ed.) (1981). *Applied Systems and Cybernetics*, Vol. VI. Pergamon Press. New York.

Leal, A., and J. Pearl (1977). An interactive program for conversational elicitation of decision structures. *IEEE Trans. Syst., Man, Cybern.*, Vol. SMC-7, No. 5.

Leberling, H. (1981). On finding compromise solutions in multicriteria problems using the fuzzy min-operator. *Fuzzy Sets and Systems*, Vol. 6, No. 2, Sept., pp. 105–118.

Lee, E. T. (1972). Proximity measures for the classification of geometric figures. *J. Cybernetics*, Vol. 2, pp. 43–59.

Lee, E. T. (1975). Shape-oriented chromosome classification. *IEEE Trans. Syst., Man, Cybern.*, Vol. SMC-5, pp. 629–632.

Lee, E. T. (1977a). Application of fuzzy languages to pattern recognition. *Kybernetes*, Vol. 6, pp. 167–173.

Lee, E. T. (1977b). A similarity-directed picture database. *Policy Analysis and Information Systems*, Vol. 1, No. 2, pp. 113–125.

Lee, E. T. (1980). Applications of fuzzy-set theory to image sciences. *J. of Cybernetics*, Vol. 10.

Lee, E. T. (1983). Algorithms for finding Chomsky and Greibach normal forms for a fuzzy context-free grammar using an algebraic approach. *Kybernetes*, Vol. 12, No. 2, pp. 125–134.

Lee, E. T., and L. A. Zadeh (1969). Notes on fuzzy languages. *Inform. Sci.*, Vol. 1, pp. 421–434.

Lee, R. C. T. (1972). Fuzzy logic and the resolution principle. *J. Assn. Comp. Mach.*, Vol. 19, pp. 109–119.

Lee, R. C. T., and C. L. Chang (1971). Some properties of fuzzy logic. *Inform. and Control*, Vol. 19, pp. 417–431.

Lee, S. C., and E. T. Lee (1974). Fuzzy sets and neural networks. *J. Cybernetics*, Vol. 4, pp. 83–103.

Lee, S. C., and E. T. Lee (1975). Fuzzy neural networks. *Mathematical Biosciences*, Vol. 23, pp. 151–177.

LeFaivre, R. (1974a). Fuzzy — A Programming language for fuzzy problem-solving. *Report* PB-231813/7, University of Wisconsin.

LeFaivre, R. A. (1974b). The representation of fuzzy knowledge. *J. Cybernetics*, Vol. 4, pp. 57–66.

Levi, I. (1967). *Gambling with Truth*. M.I.T. Press, Cambridge, Mass.

Lientz, B. P. (1972). On time-dependent fuzzy sets. *Inform. Sci.*, Vol. 4, pp. 367–376.

Liu, W. J. (1982). Fuzzy invariant subgroups and fuzzy ideals. *Fuzzy Sets and Systems*, Vol. 8, No. 2, Aug., pp. 133–140.

Liu, W. (1983). Operations on fuzzy ideals. *Fuzzy Sets and Systems*, Vol. 11, No. 1, Aug., pp. 31–42.

Loo, S. G. (1977). Measures of fuzziness. *Cybernetica*, Vol. 3, pp. 201–207.

Loo, S. G. (1978). Fuzzy relations in social and behavioral sciences. *J. of Cybern.*, Vol. 8, pp. 1–16.

Lou, S. P., and S. H. Pan (1980). Fuzzy structure. *Journal of Mathematical Analysis and Applications*, Vol. 76, No. 2, pp. 631–642.

Lowen, R. (1974). *Topologies Flous*. C. R. Acad. Des Sciences (Paris), Vol. 278A, pp. 925–928.

Lowen, R. (1976a). Fuzzy topological spaces and fuzzy compactness. *J. Math. Anal. Appl.*, Vol. 57.

Lowen, R. (1976b). Initial and final fuzziness topologies and the fuzzy Tychonoff theorem. *J. Math. Anal. Appl.*, Vol. 57.

Lowen, R. (1977a). A comparison of different compactness notions in fuzzy topological spaces — 1. Notices of the AMS. Oct., 1976. *J. Math. Anal. Appl.*, Vol. 64, pp. 446–454.

Lowen, R. (1977b). Initial and fuzzy topologies and the fuzzy Tychonoff theorem. *J. Math. Anal. Appl.*, Vol. 58, pp. 11–21.

Lowen, R. (1978). On fuzzy complements. *Infor. Sci.*, Vol. 14, pp. 107–113.

Lowen, R. (1979). Convergence in fuzzy topological spaces. *General Topology Appl.*, Vol. 10, No. 2, pp. 147–160.

Lowen, R. (1980). Convex fuzzy sets. *Fuzzy Sets and Systems*, Vol. 3, No. 3, May, pp. 291–310.

Lowen, R. (1981a). Compact Hausdorff fuzzy topological spaces are topological. *Topological Appl.*, Vol. 12, No. 1, pp. 65–74.

Lowen, R. (1981b). Fuzzy uniform spaces. *J. Math. Anal. Appl.*, Vol. 82, No. 2, pp. 370–385.

Lowen, R. (1982). Fuzzy neighborhood spaces. *Fuzzy Sets and Systems*, Vol. 7, No. 2, Mar., pp. 165–190.

Lowen, R. (1983). Hyperspaces of fuzzy sets. *Fuzzy Sets and Systems*, Vol. 9, No. 3, pp. 287–311.

Lowen, R., and P. Wuyts (1983). Completeness, compactness and precompactness in fuzzy uniform spaces. *Journal of Mathematical Analysis and Applications*, Vol. 92, No. 2, pp. 342–371.

Luhandjula, M. K. (1982). Compensatory operators in fuzzy linear programming with multiple objectives. *Fuzzy Sets and Systems*, Vol. 8, No. 3, Sept., pp. 245–252.

Luhandjula, M. K. (1983). Linear programming under randomness and fuzziness. *Fuzzy Sets and Systems*, Vol. 10, No. 2, June, pp. 123–134.

Lusk, E. J. (1981). Evaluating performance statistics used to monitor performance: A fuzzy approach. *Fuzzy Sets and Systems*, Vol. 5, No. 2, pp. 149–158.

Lusk, E. J. (1982). Priority assignment — A conditioned sets approach. *Fuzzy Sets and Systems*, Vol. 7, No. 1, Jan., pp. 43–56.

Lysvag, B. (1975). Verbs of hedging. In *Syntax and Semantics*, Vol. 4 (Kimball, J. P., Ed.). Academic Press, New York, pp. 125–154.

MacCormac, E. R. (1982). Metaphors and fuzzy sets. *Fuzzy Sets and Systems*, Vol. 7, No. 3, May, pp. 243–256.

Machina, K. F. (1972). Vague predicates. *Amer. Philos. Quart.*, Vol. 9, pp. 225–233.

Machina, K. F. (1976). Truth, belief and vagueness. *J. Philos. Logic*, Vol. 5, pp. 47–77.

MacVicar-Whelan, P. J. (1976). Fuzzy sets for man-machine interaction. *Int. J. Man-Machine Studies*, Vol. 8.

Ma, Ji Liang, and Chun Hai Yu (1984). Fuzzy topological groups. *Fuzzy Sets and Systems*, Vol. 12, No. 3, Apr., pp. 289–300.

Majumder, D. D., and S. K. Pal (1977). On some applications of fuzzy algorithm in man-machine communication research. *J. Inst. Telecom. Electron Eng.*, Vol. 23, pp. 117–120.

Makarovitsch, A. (1976). How to build fuzzy visual symbols. *Computer Graphics and Art*, February.

Makarovitsch, A. (1977). Visual fuzziness. *Comput. Graphics and Art*, November.

Malghen, S. R., and S. S. Benchalli (1984). Open maps, closed maps, and local compactness in fuzzy topological spaces. *Journal of Mathematical Analysis and Applications*, Vol. 99, No. 2, pp. 338–349.

Mamdani, E. H. (1974). Applications of fuzzy algorithms for control of simple dynamic plant. *Proc. IEEE*, Vol. 121, pp. 1585–1588.

Mamdani, E. H. (1976). Advances in the linguistic synthesis of fuzzy controllers. *Int. J. Man-Machine Studies*, Vol. 8, pp. 669–678.

Mamdani, E. H., and S. Assilian (1975). An experiment in linguistic synthesis with a fuzzy logic controller. *Int. J. Man-Machine Studies*. Vol. 7, pp. 1–13.

Mamdani, E. H., and N. Baaklini (1975). Prescriptive method for deriving control policy in a fuzzy-logic controller. *Electronics Lett.*, Vol. 11. pp. 625–626.

Mamdani, E. H., and B. R. Gaines (Eds.) (1981). *Fuzzy Reasoning and Its Applications*. Academic Press, London.

Mares, M. (1977a). How to handle fuzzy quantities. *Kybernetica*, Vol. 13, No. 1, pp. 22–40.

Mares, M. (1977b). On fuzzy quantities with real and integer values. *Kybernetica*, Vol. 13, No. 1, pp. 41–56.

Mares, M., and J. Horak (1983). Fuzzy quantities in networks. *Fuzzy Sets and Systems*, Vol. 10, No. 2, June, pp. 123–134.

Marinos, P. N. (1969). Fuzzy logic and its application to switching systems. *IEEE Trans. Comp.*, Vol. C-18, pp. 343–348.

Martin, T. (1966). Fuzzy algorithmische schemata, in *Vortrage aus Dem Problemseminar Automaten- und Algorithmentheorie*. April, Weissig, pp. 44–51.

Materna, P. (1972). Intentional semantics of vague constants — An application of Tichy's concept of semantics. *Theory and Decision*, Vol. 2, pp. 267–273.

McCain, R. A. (1983). Fuzzy confidence intervals. *Fuzzy Sets and Systems*, Vol. 10, No. 3, July, pp. 281–290.

McCloskey, M. E., and S. Glucksberg (1978). Natural categories well-defined or fuzzy sets? *Memory and Cognition*, Vol. 6, pp. 462–472.

Menges, G., and E. Kofler (1976). Linear partial information as fuzziness. In *Systems Theory in the Social Sciences* (Bossel, H., S. Klaczko, and N. Muller, Eds.). Birkhauser Verlag, Basel, pp. 307–322.

Menges, G., and H. J. Skala (1974). On the problem of vagueness in the social sciences. In *Information, Inference, and Decision* (Menges, G., Ed.). D. Reidel, Dordrecht, Holland, pp. 51–61.

Michalek, J. (1975). Fuzzy topologies. *Kybernetika* (Prague), Vol. 11, pp. 345–354.

Mizumoto, M. (1981). Fuzzy sets and their operations. *Information and Control*, Vol. 48, pp. 30–48.

Mizumoto, M., and K. Tanaka (1975). Algebraic structures of fuzzy-fuzzy sets. *Trans. IECE(D)*. Vol. 58-D, pp. 421–428.

Mizumoto, M., and K. Tanaka (1976a). Bounded-sum and bounded-difference for fuzzy sets. *Trans. IECE(D)*, Vol. 59-D, pp. 905–912.

Mizumoto, M., and K. Tanaka (1976b). Four arithmetic operations of fuzzy numbers. *Trans. IECE(D)*, Vol. 59-D, pp. 703–710.

Mizumoto, M., and K. Tanaka (1976c). Fuzzy-fuzzy automata. *Kybernetes*, Vol. 5, pp. 107–112.

Mizumoto, M., and K. Tanaka (1976d). Some properties of fuzzy sets of type 2. *Inform. and Control*, Vol. 31, pp. 312–340.

Mizumoto, M., and K. Tanaka (1976e). Various kinds of automata with weights. *J. Comp. Syst. Sci.*

Mizumoto, M., and K. Tanaka (1981). Fuzzy sets of type 2 under algebraic product and algebraic sum. *Fuzzy Sets and Systems*, Vol. 5, No. 3, pp. 277–290.

Mizumoto, M., J. Toyoda, and K. Tanaka (1969). Some considerations on fuzzy automata. *J. Comp. Syst. Sci.*, Vol. 3, pp. 409–422.

Mizumoto, M., J. Toyoda, and K. Tanaka (1970). Fuzzy languages. *Systems, Computers, Controls*, Vol. 1, No. 36 (Orig. *T.I.E.C.E.*, Vol. 53-C, pp. 333–340).

Mizumoto, M., J. Toyoda, and K. Tanaka (1972). Normal grammars with weights. *Trans. IECE* (Japan), Vol. 55D, pp. 292–293.

Mizumoto, M., J. Toyoda, and K. Tanaka (1973). *N*-fold fuzzy grammars. *Inform. Sci.*, Vol. 5, pp. 25–43.

Mizumoto, M., J. Toyoda, and K. Tanaka (1975). *B*-fuzzy grammars. *Comp. Math.*, Vol. 4, pp. 343–368.

Mizumoto, M., and H. J. Zimmermann (1982). Comparison of fuzzy reasoning methods. *Fuzzy Sets and Systems*, Vol. 8, No. 3, Sept., pp. 253–284.

Moisil, G. C. (1975). *Lectures on the Logic of Fuzzy Reasoning*. Scientific Editions, Bucharest.

Morgan, C., and F. Pelletier (1977). Some notes concerning fuzzy logics. *Linguist. Philos*, Vol. 1, No. 1.

Muir, A. (1981). Fuzzy sets and probability. *Kybernetes*, Vol. 10, pp. 197–200.

Mukaidono, M. (1975a). An algebraic structure of fuzzy logical functions and its minimal and irredundant form. *Trans. IECE* (D), Vol. 58-D, pp. 748–755.

Mukaidono, M. (1975b). Some properties of fuzzy logics. *Trans. IECE* (D), Vol. 58-D, pp. 150–157.

Nahmias, S. (1977). Fuzzy variables. *Fuzzy Sets and Syst.*, Vol. 1, No. 2 pp. 97–110.

Nakamura, K. (1982a). Quantifications of social utilities for multi-aspect decision-making. *Policy and Information*, Vol. 6, pp. 59–69.

Nakamura, K. (1982b). A procedure for social utility assessment: Consumer preference for residential air-coolers. *Behaviormetrika*, Vol. 15, pp. 33-54.

Nakata, H., M. Mizumoto, J. Toyoda, and K. Tanaka (1972). Some characteristics of *N*-fold fuzzy grammars. *Trans. Inst. Electron. Comm. Eng.* (Japan), Vol. 55-D, pp. 287–288.

Narasimhan, R. (1982). A geometric averaging procedure for constructing supertransitive approximation to binary comparison matrices. *Fuzzy Sets and Systems*, Vol. 8, No. 1, June, pp. 53–62.

Nasu, M., and N. Honda (1968). Fuzzy events realized by finite probabilistic automata. *Inform. and Control*, Vol. 12, pp. 284–303.

Nasu, M., and N. Honda (1969). Mapping induced by PGSM — Mapping and some recursively unsolvable problems of finite probabilistic automata. *Inform. and Control*, Vol. 15, pp. 250–273.

Nath, A. K., and T. T. Lee (1983). On the design of a classifier with linguistic variables as inputs. *Fuzzy Sets and Systems*. Vol. 11, No. 3, Nov., pp. 265–286.

Natvig, B. (1983). Possibility versus probability. *Fuzzy Sets and Systems*, Vol. 10, No. 1, Apr., pp. 31–36.

Nazaroff, G. J. (1973). Fuzzy topological polysystems. *J. Math. Anal. and Appl.*, Vol 41, pp. 478–485.

Neff, T. P., and A. Kandel (1977). Simplification of fuzzy switching functions. *Int. J. Comput. Information Sci.*, Vol. 6. pp. 55–70.

Negoita, C. V. (1976). Fuzzy models for social processes. In *Systems Theory in the Social Sciences* (Bossel, H., S. Klaczko, and N. Muller Eds.). Birkhauser Verlag, Basel, pp. 283–291.

Negoita, C. V. (1977a). On dynamics and fuzziness in management systems. In *Modern Trends in Cybernetics and Systems* (Rose, J. and C. Belciu Eds.). Springer-Verlag, Berlin.

Negoita, C. V. (1977b). Review of fuzzy sets and their application to cognition and decision processes. *IEEE Trans. SMC*, Vol. SMC-7, No. 2.

Negoita, C. V. (1981a). The current interest in fuzzy optimization. *Fuzzy Sets and Systems*, Vol. 6, No. 3, pp. 261–270.

Negoita, C. V. (1981b). *Fuzzy Systems*. Tunbridge Wells, England.

Negoita, C. V. (1982). Fuzzy sets in topology. *Fuzzy Sets and Systems*, Vol. 8, No. 1, June, pp. 93–100.

Negoita, C. V. (1983). Fuzzy sets in decision-support systems. *Human Systems Management*, Vol. 4, pp. 27–33.

Negoita, C. V. (1985). *Expert Systems and Fuzzy Systems*. Benjamin/Cummings, Menlo Park, California.

Negoita, C. V., and P. Flondor (1976). On fuzziness in information retrieval. *Int. J. Man-Machine Studies*, Vol. 8, pp. 711–716.

Negoita, C. V., and P. Flondor (1979). Le concept fuzzy dans le processus de recherche des informations. *Problème de Informare si Documentare*, Vol. 13, pp. 136–141.

Negoita, C. V., P. Flondor, and M. Sularia (1977). On fuzzy environments in optimization problems. *Econ. Comput. Econ. Cybernet. Stud. Res.*, pp. 13–24.

Negoita, C. V., and D. A. Ralescu (1974). Fuzzy systems and artificial intelligence. *Kybernetes*, Vol. 3, pp. 173–178.

Negoita, C. V., and D. A. Ralescu (1974). Inexactness in dynamic systems. *Economic Computation and Economic Cybernetics Studies and Research*, Vol. 4, pp. 69–81.

Negoita, C. V., and D. A. Ralescu (1975a). *Applications of Fuzzy Sets to Systems Analysis*. Birkhäuser Verlag, Basel.

Negoita, C. V., and D. A. Ralescu (1975b). Representation theorems for fuzzy concepts. *Kybernetes*, Vol. 4, pp. 169–174.

Negoita, C. V., and D. A. Ralescu (1976). Comment on a comment on an algorithm that generates fuzzy prime implicants, by Lee and Chang. *Inform. and Control*, Vol. 30, pp. 199–201.

Negoita, C. V., and D. A. Ralescu (1977a). On fuzzy optimization. *Kybernetics*, Vol. 6, pp. 193–195.

Negoita, C. V., and D. A. Ralescu (1977b). Some results in fuzzy systems theory. In *Modern Trends in Cybernetics and Systems* (Rose, J. and C. Bilciu, Eds.). Springer-Verlag, Berlin.

Negoita, C. V., and A. C. Stefanescu (1975). On the state equation of fuzzy systems. *Kybernetes*, Vol. 4, pp. 213–214.

Negoita, C. V., and M. Sularia (1976). Fuzzy linear programming and tolerances in planning. *Econ. Comput. Econ. Cybernet. Stud. Res.*, Vol. 1, pp. 3–15.

Negoita, C. V., and M. Sularia (1978). A selection method of nondominated points in multi-criteria decision problems. *Econ. Comp., Econ. Cybern. Stud. Res.*, No. 1, pp. 19–23.

Nguyen, H. T. (1977). On fuzziness and linguistic probabilities. *J. Math. Anal. and Appl.*, Vol. 61, No. 3, pp. 658–671.

Nguyen, H. T. (1978). On conditional possibility distribution. *Fuzzy Sets and Systems*, Vol. 1, No. 4, pp. 299–309.

Niedenthal, P. M., and N. Cantor (1984). Making use of social prototypes: From fuzzy concepts to firm decisions. *Fuzzy Sets and Systems*, Vol. 14, No. 1, Sept., pp. 5–28.

Nieminen, J. (1977). On the algebraic structure of fuzzy sets of Type 2. *Kybernetika*, Vol. 13, pp. 261–273.

Nieminen, J. (1978). Fuzzy mappings and algebraic structures. *Fuzzy Sets and Systems*, Vol. 1, pp. 231–235.

Nishida, T., and E. Takeda (1978). *Fuzzy Sets and Its Applications*. Morikita, Tokyo (in Japanese).

Nojiri, H. (1980). On the fuzzy team decision in a changing environment. *Fuzzy Sets and Systems*, Vol. 3, No. 2, pp. 137–150.

Nojiri, H. (1982). A model of the executives' decision processes in new-product development. *Fuzzy Sets and Systems*, Vol. 7, No. 3, May, pp. 227–242.

Nola, A. D., and A. C. S. Ventre (1979). On some sequences of fuzzy sets. *R.A.I.R.O. Inform./Comp. Sci.*, Vol. 12, No. 2, pp. 199–204.

Norwich, A. M., and I. B. Turksen (1984). A model for the measurement of membership and the consequences of its empirical implementation. *Fuzzy Sets and Systems*, Vol. 12, No. 1, Jan., pp. 1–26.

Novak, V. (1980). An attempt at Gödel-Bernays-like axiomatization of fuzzy sets. *Fuzzy Sets and Systems*, Vol. 3, No. 3, pp. 323–326.

Nowakowska, M. (1977). Fuzzy concepts in the social sciences. *Behav. Sci.*, Vol. 22, pp. 107–115.

Nowakowska, M, (1979). New ideas in decision theory. *Int. Jour. on Man-Machine Studies*, Vol. 11, pp. 213–234.

Nurmi, H. (1981a). A fuzzy solution to a majority voting game. *Fuzzy Sets and Systems*, Vol. 6, No. 2, pp. 187–198.

Nurmi, H. (1981b). Approaches to collective decision-making with fuzzy preference relations. *Fuzzy Sets and Systems*, Vol. 6, No. 3, pp. 249–260.

Nurmi, H. (1982). Imprecise notions in individual and group decision theory — Resolutions of Allais paradox and related problems. *Stochastica*, Vol. 6, No. 3, pp. 283–303.

Oden, G. C. (1977a). Fuzziness in semantic memory — choosing exemplars of subjective categories. *Memory and Cognition*, Vol. 5, pp. 198–204.

Oden, G. C. (1977b). Integration of fuzzy logical information. *J. Exp. Psychol.*, Vol. 3, pp. 565–575.

Oden, G. C. (1980). A fuzzy logical model of letter identification. *J. of Exp. Psy. Human Perception and Performance*, Vol. 5.

Oden, G. C. (1984). Integration of fuzzy linguistic information in language comprehension. *Fuzzy Sets and Systems*, Vol. 14, No. 1, Sept., pp. 29–42.

Oden, G. C., and N. H. Erson (1974). Integration of semantic constraints. *J. Verbal Learning and Behaviour*, Vol. 13, pp. 138–148.

Oguntade, O. O. (1981a). Pragmatic aspect of fuzzy programming. *Int. J. of Sys. Sci.*, Vol. 12.

Oguntade, O. O. (1981b). Semantics and pragmatics of fuzzy sets and systems. *Fuzzy Sets and Systems*, Vol. 6, No. 2, pp. 119–143.

Oguntade, O. O., and J. S. Gero (1981c). Evaluation of architectural design profiles using fuzzy sets. *Fuzzy Sets and Systems*, Vol. 5, No. 3, pp. 221–234.

Oguntade, O. O., and P. E. Beaumont (1982). Ophthalmological prognosis via fuzzy subsets. *Fuzzy Sets and Systems*, Vol. 7, No. 2, Mar., pp. 123–138.

Oheigeartaigh, M. (1982). A fuzzy transportation algorithm. *Fuzzy Sets and Systems*, Vol. 8, No. 3, Sept., pp. 235–244.

Ohsato, A., T. Ohta, and T. Sekiguchi (1983). Solar-hydrogen energy system-model applied to isolated island. *International Journal of Hydrogen Energy*, Vol. 8, No. 3, pp. 163–174.

Okuda, T., H. Tanaka, and K. Asai (1976). Decision problems and quantity of information in fuzzy events. *Trans. S.I.C.E.*, Vol. 12, pp. 63–68.

Okuda, T., H. Tanaka, and K. Asai (1978). A formulation of fuzzy decision problems with fuzzy information using probability measure of fuzzy events. *Inf. and Control*, Vol. 38, pp. 135–147.

Ollero, A., and E. Freire (1981). The structure of relations in personnel management. *Fuzzy Sets and Systems*, Vol. 5, No. 2, pp. 115–126.

Orlovsky, S. A. (1977). On programming with fuzzy constraint sets. *Kybernetes*, Vol. 6, pp. 197–201.

Orlovsky, S. A. (1978). Decision-making with a fuzzy preference relation. *Fuzzy Sets and Systems*, Vol. 1, No. 3, pp. 155–167.

Orlovsky, S. A. (1980). On formalization of a general fuzzy mathematical problem. *Fuzzy Sets and Systems*, Vol. 3, No. 3, May, pp. 311–322.

Orlovsky, S. A., and D. I. Shapiro (1979). The first Soviet seminar on control in a fuzzy environment. *Fuzzy Sets and Systems*, Vol. 2, No. 4, Oct., pp. 349–351.

Osis, J. J. (1968). Fault detection in complex systems using the theory of fuzzy sets. In *Kibernetika i Diagnostika* 2 (Kristinkov, D. S., J. J. Osis, L. A. Ravtrigin, Eds.). Zinatne, Riga, U.S.S.R., pp. 13–18 (In Russian).

Ostasiewicz, W. (1982). A new approach to fuzzy programming. *Fuzzy Sets and Systems*, Vol. 7, No. 2, Mar. pp. 139–152.

Ovchinnikov, S. V. (1980a). Design of group decisions II. In spaces of partial-order fuzzy relations. *Fuzzy Sets and Systems*, Vol. 4, pp. 153–165.

Ovchinnikov, S. V. (1980b). In arbitrary spaces of fuzzy binary relations. *Fuzzy Sets and Systems*, Vol. 4, pp. 53–62.

Ovchinnikov, S. V. (1980c). Involutions in fuzzy set theory. *Stochastica*, Vol. 4, No. 3, pp. 227–231.

Ovchinnikov, S. V. (1981a). General negations in fuzzy-set theory. *J. of Math. Anal. and Appl.*

Ovchinnikov, S. V. (1981b). Structure of fuzzy binary relations. *Fuzzy Sets and Systems*, Vol. 6, No. 2, Sept., pp. 169–196.

Ovchinnikov, S. V., (1983). General negations in fuzzy-set theory. *Journal of Mathematical Analysis and Applications*, Vol. 92, No. 1.

Pal, S. K., A. K. Datta, and D. D. Majumder (1978). Adaptive learning algorithms in classification of fuzzy patterns — An application to vowels in CNC context. *Int. J. Systems Sci.*, Vol. 9, pp. 887–897.

Pal, S. K., and R. A. King (1983). On edge detection of X-ray images using fuzzy sets.

IEEE Transactions on Pattern Analysis and Machine Intelligence, Vol. PAMI-5, No. 1, pp. 69–77.

Pal, S. K., R. A King, and A. A. Hashim (1983). Image description and primitive extraction using fuzzy sets. *IEEE Trans. Syst., Man, Cybern.*, Vol. SMC-13, No. 1, pp. 94–100.

Pal, S. K., and D. D. Majumder (1977). Fuzzy sets and decision-making approaches in vowel and speaker recognition. *IEEE Trans. Syst., Man, Cybern.*, Vol. SMC-7, pp. 625–629.

Pal, S. K., and D. D. Majumder (1978a). On automatic plosive identification using fuzziness in property sets. *IEEE Trans. Syst., Man, Cybern.*, Vol. SMC-8, pp. 302–307.

Pal, S. K., and D. D. Majumder (1978b). Correction to "on automatic plosive identification using fuzziness in property sets." *IEEE Trans. Syst., Man, Cybern.*, Vol. SMC-8, No. 12, pp. 907.

Pal, S. K., and D. D. Majumder (1978c). Effect of fuzzification and the plosive cognition system. *Int. J. Systems. Sci.*, Vol. 9, pp. 873–886.

Pao-Ming, Pu, and L. Ying-Ming (1980). Fuzzy topology (1 and 2). 1. Neighborhood structure of a point and Moore-Smith convergence. *JMAA*, Vol. 76, pp. 571–599. 2. *Product and Quotient Spaces*, *JMAA*, Vol. 77, pp. 20–37.

Paun, G. (1983). An impossibility theorem for indicator aggregation (short communication). *Fuzzy Sets and Systems*, Vol. 9, No. 2, Feb., pp. 205–210.

Paz, A. (1967). Fuzzy star functions, probabilistic automata, and their approximation by nonprobabilistic automata. *J. Comp. Syst. Sci.*, Vol. 1, pp. 371–389.

Pearl, J. (1975). On the storage economy of inferential question-answering systems. *IEEE Trans. Syst., Man, Cybern.*, Vol. SMC-5, pp. 595–602.

Pedrycz, W. (1981). An approach to the analysis of fuzzy sets. *Int. J. Control*, Vol. 3, No. 3, pp. 403–429.

Pedrycz, W. (1983a). Some applicational aspects of fuzzy relational equations in systems analysis. *International Journal of General Systems,* Vol. 9, pp. 125–132.

Pedrycz, W. (1983b). Fuzzy relational equations with generalized connectives and their applications. *Fuzzy Sets and Systems*, Vol. 10, No. 2, June, pp. 185–202.

Pedrycz, W. (1983c). Numerical and applicational aspects of fuzzy relational equations. *Fuzzy Sets and Systems*, Vol. 11, No. 1, Aug., pp. 1–18.

Pedrycz, W. (1984a). An identification algorithm in fuzzy relational systems. *Fuzzy Sets and Systems*, Vol. 13, No. 2, July, pp. 153–168.

Pedrycz, W. (1984b). Identification in fuzzy systems. *IEEE Trans. Syst., Man, Cybern.*, Vol. SMC-14, No. 2, pp. 361–367.

Pedrycz, W., E. Czogala, and K. Hirota (1984). Some remarks on the identification problem in fuzzy systems (short communication). *Fuzzy Sets and Systems*, Vol. 12, No. 2, Feb., pp. 185–190.

Pfeilsticker, A. (1981). The systems approach and fuzzy-set theory bridging the gap between mathematical and language-oriented economists. *Fuzzy Sets and Systems*, Vol. 6, No. 3, Nov., pp. 209–234.

Pinkava, V. (1976). "Fuzzification" of binary and finite multivalued logical calculi. *Int. J. Man-Machine Studies*, Vol. 8, pp. 717–730.

Pinkava, V. (1976). On the nature of some logical paradoxes. *Int. J. Man-Machine Studies*.

Pitts, A. M. (1982). Fuzzy sets do not form a topos (short communication). *Fuzzy Sets and Systems*, Vol. 8, No. 1, June, pp. 101–104.

Ponasse, D. (1983). Some remarks on the category FUZ(H) of M. Eytan (short communication). *Fuzzy Sets and Systems*, Vol. 9, No. 2, Feb., pp. 199–204.

Ponsard, C. (1977). Hierarchies des places centrales et graphes phi-flous. *Environment. Planning*, Vol. 9, pp. 233–252.

Ponsard, C. (1980). An application of fuzzy-set theory to the analysis of the consumer's spatial preferences. *Fuzzy Sets and Systems*, Vol. 5, No. 3, pp. 235–244.

Potoczny, H. B. (1984). On similarity relations in fuzzy relational databases. *Fuzzy Sets and Systems*, Vol. 12, No. 3, Apr., pp. 231–236.

Prade, H. (1979). Using fuzzy-set theory in a scheduling problem — A case study. *J. of Fuzzy Sets and Systems*, Vol. 2, No. 2, pp. 153–165.

Prade, H. (1980). Fuzzy programming — Why and how? — Some hints and examples. *4th IEEE Int. Comp. Software and Appl. Conf.*, Chicago, October.

Prade, H. (1980). Possibilité et logique trivalente de Lukasiewicz — Une Remarque. *Bulletin for Studies and Exchanges on Fuzziness and Its Appl.*, Printemps (internal publ.).

Preparata, F. P., and R. T. Yeh (1972). Continuously valued logic. *J. Comp. Syst. Sci.*, Vol. 6, pp. 397–418.

Prevot, M. (1981). Algorithm for the solution of fuzzy relations (short communication). *Fuzzy Sets and Systems*, Vol. 5, No. 3, pp. 319–322.

Prugovecki, E. (1974). Fuzzy sets in the theory of measurement of incompatible observables. *Foundations of Physics*, Vol. 4, pp. 9–18.

Prugovecki, E. (1975). Measurement in quantum mechanics as a stochastic process on spaces of fuzzy events. *Foundations of Physics*, Vol. 5, pp. 557–571.

Prugovecki, E. (1976a). Localizability of relativistic particles in fuzzy phase space. *J. Phys. Math.*, Vol. 9, No. 11.

Prugovecki, E. (1976b). Probability measures on fuzzy events in phase space. *J. Math. Physics*, Vol. 17, pp. 517–523.

Prugovecki, E. (1977). On fuzzy spin spaces. *J. Phys. and Math.*, Vol. 10, No. 4.

Puri, M., and D. Ralescu (1983). Differentials of fuzzy functions. *Journal of Mathematical Analysis and Applications*, Vol. 91, No. 2, pp. 552–558.

Puri, M. L., and D. Ralescu (1982). A possibility measure is not a fuzzy measure (short communication). *Fuzzy Sets and Systems*, Vol. 7, No. 3, May, pp. 311–314.

Qu, Y. (1983). Measures of fuzzy sets. *Fuzzy Sets and Systems*, Vol. 9, No. 3, pp. 219–227.

Raad, M. B. (1978). Fuzzy relations in a control setting. *Kybernetes*, Vol. 7, pp. 185–188.

Ragade, R. K. (1976). Fuzzy interpretive structural modeling. *J. Cybernetics*. Vol. 6, pp. 189–211.

Ragade, R. K. (1977). A mathematical model of approximate communication in information systems, in *The General Syst. Paradigm* (*J. White, Ed.*), pp. 334–346.

Ralescu, D. A. (1976). *L*-fuzzy sets and *L*-flou sets. *Elektronische Informat. und Kybernetik*, Vol. 12, pp. 599–605.

Ralescu, D. A. (1978). Fuzzy subobjects in a category and the theory of *C*-sets, *Fuzzy Sets and Systems*, Vol. 1, No. 3, pp. 193–202.

Ralescu, D., and G. Adams (1980). The fuzzy integral. *J. of Math. Anal. and Appl.*

Ramik, J. (1983). Extension principle and fuzzy-mathematical programming. *Kybernetika*, Vol. 19, No. 6, pp. 516–525.

Rao, M. B., and A. Rashed (1981). Some comments on fuzzy variables. *Fuzzy Sets and Systems*, Vol. 6, No. 3, Nov., pp. 285–292.

Ray, K. S., and D. D. Majumder (1984). Application of circle criteria for stability analysis of linear SISO and MIMO systems associated with fuzzy logic controller. *IEEE Trans. SMC*, Vol. SMC-14, No. 2, pp. 345–349.

Rescher, N. (1968). *Topics in Philosophical Logic*, D. Reidel, Holland.

Rescher, N. (1969). *Many-Valued Logic*, McGraw-Hill, New York.

Restian, A. (1977). Shannon information, fuzzy sets, linguistics, relativistic sets and communication. *Cybernetica*, Vol. 20.

Rickman, S. M., and A. Kandel (1976). Tabular minimization of fuzzy switching functions. *IEEE Trans. Syst., Man, Cybern.*, Vol. SMC-6, pp. 761–769.

Rickman, S. M., and A. Kandel (1977). Column table approach for the minimization of fuzzy functions. *Inf. Sci.*, Vol. 12, No. 2, pp. 111–128.

Riera, T. (1978). How similarity matrices are. *Stochastics*, Vol. 11, No. 4, pp. 77–80.

Rocha, A. F. (1981). Neural fuzzy point processes. *Fuzzy Sets and Systems*, Vol. 5, No. 2, pp. 127–140.

Rocha, A. F. (1982). Basic properties of neural circuits. *Fuzzy Sets and Systems*, Vol. 7, No. 2, Mar., pp. 109–122.

Rocha, A. F., E., Francozo, M. I., Hadler, and M. A. Balduino (1980). Neural languages. *Fuzzy Sets and Systems*, Vol. 3, No. 1, Jan., pp. 11–35.

Rodabaugh, S. E. (1979). A lattice of continuities for fuzzy topological spaces. *Journal of Mathematical Analysis and Applications*, No. 1, pp. 244–256.

Rodabaugh, S. E. (1982). Fuzzy addition in the *L*-fuzzy real line. *Fuzzy Sets and Systems*, Vol. 8, No. 1, June, pp. 39–52.

Rodabaugh, S. E. (1983a). A categorical accommodation of various notions of fuzzy topology. *Fuzzy Sets and Systems*, Vol. 9, No. 3, pp. 241–265.

Rodabaugh, S. E. (1983b). Separation axioms and the fuzzy real lines. *Fuzzy Sets and Systems*, Vol. 11, No. 2, Oct., pp. 163–184.

Rosenfeld, A. (1971). Fuzzy groups. *J. Math. Anal. and Appl.*, Vol. 35, pp. 512–517.

Roubens, M. (1977). Pattern classification problems and fuzzy sets. *Fuzzy Sets and Systems*, Vol. 1, No. 4, pp. 239–253.

Roubens, M., and P. Vincke (1983). Linear fuzzy graphs. *Fuzzy Sets and Systems*, Vol. 10, No. 1, Apr., pp. 79–86.

Rubin, P. A. (1982). A note on the geometry of reciprocal fuzzy relations (short communication). *Fuzzy Sets and Systems*, Vol. 7, No. 3, May, pp. 307–310.

Ruspini, E. H. (1969). A new approach to clustering. *Inform. and Control*, Vol. 15, pp. 22–32.

Ruspini, E. H. (1970). Numerical methods for fuzzy clustering. *Inform. Sci.*, Vol. 2, pp. 319–350.

Ruspini, E. H. (1972). Optimization in sample descriptions — Data reduction and pattern recognition using fuzzy clustering. *IEEE Trans. Syst., Man, Cybern.*, Vol. SMC-2, p. 541.

Ruspini, E. H. (1973). New experimental results in fuzzy clustering. *Inform. Sci.*, Vol. 6, pp. 273–284.

Saaty, T. L. (1974). Measuring the fuzziness of sets. *J. Cybernet.*, Vol. 4, No. 4, pp. 53–61.

Saaty, T. L. (1978). Exploring the interface between hierarchies, multiple objectives, and fuzzy sets. *Fuzzy Sets and Systems*, Vol. 1, No. 1, pp. 57–68.

Saitta, L., and P. Torasso (1981). Fuzzy characterization of coronary disease. *Fuzzy Sets and Systems*, Vol. 5, No. 3, pp. 245–250.

Sakawa, M., and H. Yano (1985). An interactive fuzzy satisfying method using augmented minimax problems and its application to environmental systems, *IEEE Trans. SMC*, Vol. SMC-15, No. 6, Nov.–Dec., pp. 720–729.

Sales, T. (1982). Fuzzy sets and set classes. *Stochastica*, Vol. 6, No. 3, pp. 249–264.

Salomaa, A. (1959). On many-valued systems of logic. *Ajatus*, Vol. 22, pp. 115–119.

Sanchez, E. (1976). Resolution of composite fuzzy relation equations. *Inform. and Control*, Vol. 30, pp. 38–47.

Sanchez, E. (1977). Resolution of eigen fuzzy-set equations. *Fuzzy Sets and Systems*, Vol. 1, No. 1, pp. 69–74.

Sanchez, E. (1981). Eigen fuzzy sets and fuzzy relations. *J. Math. Anal. Appl.*, Vol. 81, pp. 399–421.

Sanchez, E. (1984a). Solution of fuzzy equations with extended operations. *Fuzzy Sets and Systems*, Vol. 12, No. 3, Apr., pp. 237–248.

Sanchez, E. (Ed.) (1984b). Fuzzy Information, Knowledge Representation and Decision Analysis, *Proceedings of the IFAC Symposium* (Marseille, France July 1983), Pergamon Press.

Sanford, D. H. (1975). Infinity and vagueness. *Philosophical Review*, Vol. 84, pp. 520–535.

Santos, E. S. (1968a). Maximin automata. *Inform. and Control*. Vol. 13, pp. 363–377.

Santos, E. S. (1968b). Maximin, minimax and composite sequential machines. *J. Math. Anal. and Appl.*, Vol. 24, pp. 246–259.

Santos, E. S. (1969a). Maximin sequential chains. *J. Math. Anals. and Appl.*, Vol. 26, pp. 28–38.

Santos, E. S. (1969b). Maximin sequential-like machines and chains. *Mathematical Systems Theory*, Vol. 3, pp. 300–309.

Santos, E. S. (1970). Fuzzy algorithms. *Inform. and Control*, Vol. 17, pp. 326–339.

Santos, E. S. (1972a). Max-product machines. *J. Math. Anal. and Appl.*, Vol. 37, pp. 677–686.

Santos, E. S. (1972b). On reductions of maximin machines. *J. Math. Anal. and Appl.*, Vol. 40, pp. 60–78.

Santos, E. S. (1973). Fuzzy sequential functions. *J. Cybernetics*, Vol. 3, pp. 15–31.

Santos, E. S. (1974). Context-free fuzzy languages. *Inform. and Control*, Vol. 26, pp. 1–11.

Santos, E. S. (1975a) Max-product grammars and languages. *Inform. Sci.*, Vol. 9, pp. 1–23.

Santos, E. S. (1975b). Realization of fuzzy languages by probabilistic max-product and maximin automata. *Inform. Sci.*, Vol. 8, pp. 39–53.

Santos, E. S. (1976). Fuzzy automata and languages. *Inform. Sci.*, Vol. 10, pp. 193–197.

Santos, E. S., and W. G. Wee (1968). General formulation of sequential machines. *Inform. and Control*, Vol. 12, pp. 5–10.

Saradis, G. N., and H. E. Stephanou (1977). Fuzzy decision-making of a prosthetic arm. *IEEE Trans. Syst., Man, Cybern.*, Vol. SMC-7, No. 6, pp. 407–420.

Saradis, G. N. (1974). Fuzzy notions in nonlinear system classification. *J. Cybernetics*, Vol. 4, pp. 67–82.

Schmucker, K. J. (1984). *Fuzzy Sets, Natural Languages, Computations, and Risk Analysis*, Computer Science Press, Maryland.

Schek, H. J. (1977). Tolerating fuzziness in keywords by similarity searches. *Kybernetes*, Vol. 6, pp. 175–184.

Schotch, P. K. (1975). Fuzzy modal logic. *Proc. 1975 Int. Symp. Multiple-valued Logic*, IEEE 75CH0959-7C, May, pp. 176–182.

Schwartz, D. G. (1985). The case for an interval-based representation of linguistic truth. *Fuzzy Sets and Systems*, Vol. 17, No. 2, pp. 153–165.

Schwede, G. W., and A. Kandel (1977). Fuzzy maps. *IEEE Trans. Syst., Man, Cybern.*, Vol. SMC-7, pp. 619–674.

Seif, A., and J. Aguilar-Martin (1980). Multi-group classification using fuzzy correlation. *Fuzzy Sets and Systems*, Vol. 3, No. 2, pp. 109–122.

Sessa, S. (1984). On fuzzy subgroups and fuzzy ideals under triangular norms (short communication). *Fuzzy Sets and Systems*, Vol. 13, No. 1, May, pp. 95–97.

Sherwood, H. (1983). Products of fuzzy subgroups. *Fuzzy Sets and Systems*, Vol. 11, No. 1, Aug., pp. 79–90.

Shimura, M. (1973). Fuzzy-set concept in rank-ordering objects. *J. Math. Anal. and Appl.*, Vol. 43, pp. 717–733.

Shimura, M. (1975). Applications of fuzzy-set theory to pattern recognition. *J. JAACE*, Vol. 19, pp. 243–248.

Silvert, W. (1979). Symmetric summation: A class of operations on fuzzy sets. *IEEE Trans. Syst., Man, Cybern.*, pp. 657–659.

Siy, P., and C. S. Chen (1971). Some properties of fuzzy logic. *Inform. and Control*, Vol. 19, pp. 417–431.

Siy, P., and C. S. Chen (1972). Minimization of fuzzy functions. *IEEE Trans. Comp.*, Vol. C-21, pp. 100–102.

Siy, P., and C. S. Chen (1974). Fuzzy logic for handwritten numerical character recognition. *IEEE Trans. Syst., Man, Cybern.*, Vol. SMC-4, pp. 570–575.

Skala, H. J. (1976). Fuzzy concepts — logic, motivation application. In *Systems Theory in the Social Sciences* (Bossel, H., S. Klaczko, and N. Muller, Eds.). Birkhauser Verlag, Basel, pp. 292–306.

Skala, H. J. (1977). On many-valued logics, fuzzy sets, fuzzy logics and their applications. *Fuzzy Sets and Systems*, Vol. 1, No. 1, pp. 129–149.

Skala, H. J., S. Termini, and E. Trillas (Eds.) (1984). *Aspects of Vagueness*. D. Reidel, Dordnecht.

Smets, P. (1980). Medical diagnosis, fuzzy sets, and degree of belief. *Fuzzy Sets and Systems*, Vol. 5, No. 3, pp. 259–266.

Smets, P. (1982). Probability of a fuzzy event — An axiomatic approach. *Fuzzy Sets and Systems*, Vol. 7, No. 2, Mar., pp. 153–164.

Smets, P., H. Vainsel, R. Bernard, and F. Kornreich (1977). Bayesian probability of fuzzy diagnosis. *Proc. Medinfo* 77 (Shires Wolf, Ed.), pp. 121–122.

Spillman, B., J. Bezdek, and R. Spillman (1979). Coalition analysis with fuzzy sets. *Kybernetes*, Vol. 8, No. 3, pp. 203–211.

Spillman, B., R. Spillman, and J. Bezdek (1982). A dynamic perspective on leadership — Development of a fuzzy measurement procedure. *Fuzzy Sets and Systems*, Vol. 7, No. 1, Jan., pp. 19–34.

Srini, V. P. (1975). Realization of fuzzy forms. *IEEE Trans. on Computers*, Sept., pp. 941–943.

Srivastava, P., and R. L. Gupta (1980). Fuzzy proximity bases and subbases. *Journal of Mathematical Analysis and Applications*, No. 2, pp. 588–597.

Srivastava, P. and R. L. Gupta (1983). Fuzzy proximity structures and fuzzy ultrafilters. *Journal of Mathematical Analysis and Applications*, Vol. 94, No. 2, pp. 297–311.

Stalling, V. (1977). Fuzzy-set theory versus Bayesian statistics. *IEEE Trans. Syst., Man, Cybern.*, Vol. SMC-7, pp. 216–219.

Stein, W. E. (1980). Optimal stopping in a fuzzy environment. *Fuzzy Sets and Systems*, Vol. 3, No. 3, pp. 253–260.

Stein, W. E. (1985). Fuzzy probability vectors. *Fuzzy Sets and Systems*, Vol. 15, No. 3, pp. 263–267.

Stein, W. E., and K. Talati (1981). Convex fuzzy random variables. *Fuzzy Sets and Systems*, Vol. 6, No. 3, Nov., pp. 271–284.

Stocia, M., I. Slancu-Minasian, and E. Scarlat (1977) On large-scale classification problems using fuzzy sets. *Econ. Comp. and Cyber. Studies and Research*, Vol. 1, pp. 93–100.

Stoica, M., and E. Scarlat (1977). Some fuzzy concepts in the management of production systems. *Modern Trends in Cybern. and Sys.*, Vol. 2, pp. 175–181.

Sugeno, M. (1973). Constructing fuzzy measure and grading similarity of patterns by fuzzy integrals. *Transaction SICE*, Vol. 9, pp. 359–367.

Sugeno, M. (1974). Theory of fuzzy integrals and its applications. Ph.D. Thesis, Tokyo Institute of Technology, Tokyo, Japan.

Sugeno, M. (1975a). Fuzzy decision-making problems. *Trans SICE*, Vol. 11, pp. 709–714.

Sugeno, M. (1975b). Inverse operation of fuzzy integrals and conditional fuzzy measures. *Trans. SICE*, Vol. 11, pp. 32–37.

Sugeno, M. (1975c). Theoretical developments of fuzzy sets. *J. JAACE*, Vol. 19, pp. 229–234.

Sugeno, M. (1979). Application of fuzzy sets and logic to control — A survey. *J. of the Soc. of Instrument and Cont. Eng.*, Japan, Vol. 18, No. 2.

Sugeno, M. and M. Nishida (1985). Fuzzy control of model car. *Fuzzy Sets and Systems*, Vol. 16, No. 2, pp. 103–113.

Sugeno, M. and M. Sasaki (1983). *L*-fuzzy category. *Fuzzy Sets and Systems*, Vol. 11, No. 1, Aug., pp. 43–64.

Sugeno, M., and T. Takagi (1983). Multi-dimensional fuzzy reasoning. *Fuzzy Sets and Systems*, Vol. 9, No. 3, pp. 313–325.

Sugeno, M., and T. Terano (1977). A model of learning based on fuzzy information. *Kybernetes*, Vol. 6, pp. 157–166.

Sularia, M. (1977). On fuzzy programming in planning. *Kybernetes*, Vol. 6, p. 230.

Tahani, V. A. (1976). A fuzzy model of document retrieval systems. *Inf. Proc. and Man.*, Vol. 12, pp. 177–187.

Tahani, V. A. (1977). A conceptual framework for fuzzy query processing — A step towards very intelligent data systems. *Inf. Proc. and Man.*, Vol. 13, pp. 289–303.

Takeda, E., and T. Nishida (1976). An application of fuzzy graphs to the problem concerning group structure. *J. Op. Res. Soc.* (Japan), Vol. 19, No. 3.

Takeda, E., and T. Nishida (1980). Multiple-criteria decision problems with fuzzy domination structures. *Fuzzy Sets and Systems*, Vol. 3, No. 2, pp. 123–136.

Takeguchi, T., and H. Akashi (1984). Analysis of decisions under risk with incomplete knowledge. *IEEE Trans. Syst., Man, Cybern.*, Vol. SMC-14, No. 4, pp. 618–670.

Tamura, S., S. Higuchi, and K. Tanaka (1971). Pattern classification based on fuzzy relations. *IEEE Trans. Syst., Man, Cybern.*, Vol. SMC-1, pp. 61–66.

Tamura, S., and K. Tanaka (1973). Learning of fuzzy formal language. *IEEE Trans. Syst., Man, Cybern*, Vol. SMC-3, pp. 98–102.

Tanaka, H. (1975). Fuzzy-set theory and its application. *Journal of JAACE*, Vol. 19, pp. 227–228.

Tanaka, H., and K. Asai (1973). Fuzzy mathematical programming. *Transactions of SICE*, Vol. 9, pp. 109–115.

Tanaka, H. and K. Asai (1984a). Fuzzy solution in fuzzy linear programming problems. *IEEE Trans. Syst., Man, Cybern.*, Vol. SMC-14, No. 2, pp. 325–328.

Tanaka, H. and K. Asai (1984b). Fuzzy linear-programming problems with fuzzy numbers. *Fuzzy Sets and Systems*, Vol. 13, No. 1, May, pp. 1–10.

Tanaka, H., and S. Kaneku (1974). On a fuzzy decoding procedure for cyclic codes. *Trans. IEECE* (A), Vol. 57-A, pp. 505–510.

Tanaka, H., T. Okuda, and K. Asai (1974). On fuzzy mathematical programming. *J. Cybernet.*, Vol. 3, pp. 37–46.

Tanaka, H., T. Okuda, and K. Asai (1976). A formulation of fuzzy decision problems and its application to an investment problem. *Kybernetics*, Vol. 5, pp. 25–30.

Tanaka, H., T. Okuda, and K. Asai (1977). On decision-making in fuzzy environment. *Fuzzy Inf. and Dec.-Making, Int. J. of Prod. Res.*, Vol. 15, pp. 623–635.

Tanaka, H., S. Uejima, and K. Asai (1982). Linear-regression analysis with fuzzy model. *IEEE Trans. Syst., Man, Cybern.*, Vol. SMC-12, No. 6, pp. 903–907.

Tanino, T. (1984). Fuzzy preference orderings in group decision making. *Fuzzy Sets and Systems*, Vol. 12, No. 2, Feb., pp. 117–132.

Tazaki, E., and M. Amagasa (1979). Heuristic structure synthesis in a class of systems using a fuzzy automaton. *IEEE Trans. Syst., Man, Cybern.*, Vol. SMC-9, pp. 73-79.

Terano, T. (1972). Fuzziness of systems. *Nikka-Giren Engineers*, pp. 21–25.

Terano, T. (1975). Methodology of fuzzy systems. *J. of IECE*, Vol. 58, pp. 875–876.

Terano, T., Y. Murayama, and N. Akiyama (1983). Human reliability and safety evaluation of man-machine systems. *Automatica*, Vol. 19, No. 6, pp. 719–722.

Thole, U., H. J. Zimmermann, and P. Zysno (1979). On the suitability of minimum and product operators for the intersection of fuzzy sets. *Fuzzy Sets and Systems*, Vol. 2, No. 2, pp. 167–180.

Thomason, M. G. (1974a). Fuzzy syntax-directed translations. *J. Cybernetics*, Vol. 4, pp. 87–94.

Thomason, M. G. (1974b). The effect of logic operations on fuzzy logic distributions. *IEEE Trans. Syst., Man, Cybern.*, Vol. SMC-4, pp. 309–310.

Thomason, M. G. (1975). Finite fuzzy automata, regular languages and pattern recognition. *Pattern Recognition*, Vol. 5, pp. 383–390.

Thomason, M. G. (1977). Convergence of powers of a fuzzy matrix. *J. Math. Anal. and Appl.*, Vol. 57, pp. 476–480.

Thomason, M. G., and P. N. Marionos (1974). Deterministic acceptors of regular fuzzy languages. *IEEE Trans. Syst., Man, Cybern.*, Vol. SMC-4, pp. 228–230.

Thum, M. and A. Kandel (1984). On the complexity of growth of the number of distinct fuzzy switching functions. *Fuzzy Sets and Systems*, Vol. 13, No. 2, July, pp. 125–138.

Togai, M. (1985). A fuzzy inverse relation based on Gödelian logic and its applications. *Fuzzy Sets and Systems*, Vol. 17, No. 2, pp. 211–219.

Tong, R. M. (1977a). A control engineering review of fuzzy systems. *Automatica*, Vol. 13, pp. 559–569.

Tong, R. M. (1977b). Analysis and control of fuzzy systems using finite discrete relations. *Int. J. Control*, Vol. 27, pp. 431–440.

Tong, R. M. (1980a). Some properties of fuzzy feedback systems. *IEEE Trans. Syst., Man, Cybern.*, Vol. SMC-10, No. 6, June.

Tong, R. M. (1980b). The evaluation of fuzzy models derived from experimental data. *Fuzzy Sets and Systems*, Vol. 4, No. 1, July, pp. 1–12.

Tong, R. M., and J. Efstathiou (1982). A critical assessment of truth functional modification and its use in approximate reasoning (short communication). *Fuzzy Sets and Systems*, Vol. 7, No. 1, Jan., pp. 103–109.

Tribus, M. (1979). Comments on "fuzzy sets, fuzzy algebra and fuzzy statistics" (and reply by A. Kandel). *Proceedings of the IEEE*, Vol. 67, No. 8.

Tribus, M. (1980). Fuzzy sets and Bayesian methods applied to the problem of literature search. *IEEE Trans. Syst., Man, Cybern.*, Vol. SMC-10, No. 8, August.

Trillas, E. (1979). Sobre functiones de negacion in la teoria de conjuntos difusos. *Stochastica* (Barcelona). Vol. 111, No. 1, pp. 47–59.

Trillas, E., and C. Alsina (1979). Sur les mesures de degré de flou. *Stochastica* (Barcelona), Vol. 111, No. 1, pp. 81–84.

Trillas, E., and T. Riera (1978). Entropies in finite fuzzy sets. *Infor. Sciences*, Vol. 15, pp. 159–168.

Tsukamoto, T., and T. Tashiro (1979). Method of solution to fuzzy inverse problem. *Tran. of SICE*, Vol. 15, No. 1.

Turksen, I. B., and D. D. W. Yao (1984). Representation of connectives in fuzzy reasoning: The view through normal forms. *IEEE Trans. Syst., Man, Cybern.*, Vol. SMC-14, No. 1, pp. 146–151.

Twareque, Ali, S., and E. Prugovecki (1977). Systems of imprim. and representation of quantum mech. on fuzzy phase spaces. *J. of Math. Physics*, Vol. 18, No. 2.

Umano, M., M. Mizumoto, and K. Tanaka (1978). FSTDS — A fuzzy-set manipulation system. *Information Science*, Vol. 14, pp. 115–159.

Uno, K., H. Itakura, N. Sannomiya, and Y. Nishikawa (1976). Learning controls that use a fuzzy controller. *Systems and Control*, Vol. 20, pp. 262–268.

Vagin, V. N., D. A. Pospelov, and W. Papke (1977). Application of fuzzy logic in control systems. *Foundation of Control Eng.*, Vol. 2, pp. 153–160.

Van Gigch, J. P., and L. L. Pipino (1980). From absolute to probable and fuzzy in decision-making. *Kybernetes*, Vol. 9, No. 1.

Van Laarhoven, P. J. M., and W. Pedrycz (1983). A fuzzy extension of Saaty's priority theory. *Fuzzy Sets and Systems*, Vol. 11, No. 3, Nov., pp. 229–242.

Van Nauta-Lemke, and W. J. M. Kickert (1976). The application of fuzzy-set theory to control a warm water. *Jour. A.*, Vol. 17, No. 1, Delft Univ. of Technology.

Vari, A. (1983). *Insiemi Sfocati e Decisioni* (*Fuzzy Sets and Decisions*), Edizioni Scientifiche Italiane.

Vila, M. A., and M. Delgado (1983). Problems of classification in a fuzzy environment. *Fuzzy Sets and Systems*, Vol. 9, No. 3, pp. 229–239.

Vila, M. A., and M. Delgado (1983). On medical diagnosis using possibility measures. *Fuzzy Sets and Systems*, Vol. 10, No. 3, July, pp. 211–222.

Vira, J. (1981). Fuzzy expectation values in multistage optimization problems. *Fuzzy Sets and Systems*, Vol. 6, No. 2, Sept., pp. 161–168.

Von Kaenel, P. A. (1982). Fuzzy codes and distance properties. *Fuzzy Sets and Systems*, Vol. 8, No. 2, Aug., pp. 199–204.

Voxman, W., and R. Goetschel (1983). A note on the characterization of max and min operators. *Information Sciences*, Vol. 30, pp. 5–10.

Wagenknecht, M., and K. Hartmann (1983). On fuzzy rank ordering in polyoptimization. *Fuzzy Sets and Systems*, Vol. 11, No. 3, Nov., pp. 253–264.

Wagner, W. (1981). A fuzzy model of concept representation in memory. *Fuzzy Sets and Systems*, Vol. 6, No. 1, July, pp. 11–26.

Wang, Guo-Jan (1984). Order-homomorphisms on fuzzes. *Fuzzy Sets and Systems*, Vol. 12, No. 3, Apr., pp. 281–288.

Wang, P. P. (ed.) (1983). *Advances in Fuzzy Sets, Possibility Theory, and Applications*, Plenum Press, New York.

Wang, P. P., and S. K. Chang, (Eds.) (1980). *Fuzzy Sets*, Plenum Press, New York.

Wang, P. Z. (1982). Fuzzy contactibility and fuzzy variables. *Fuzzy Sets and Systems*, Vol. 8, No. 1, June, pp. 81–92.

Warren, R. H. (1976). Optimality in fuzzy topological polysystems. *J. Math. Anal. Appl.*, Vol. 54, pp. 309–315.

Warren, R. H. (1977). Boundary of fuzzy set. *Indiana Univ. Math. J.*, Vol. 26, No. 2, pp. 191–197.

Warren, R. H. (1978). Neighborhoods, bases and continuity in fuzzy topological spaces. *Rocky Mountain J. Math*, Vol. 8, pp. 459–470.

Warren, R. H. (1981). Equivalent fuzzy sets (short communication). *Fuzzy Sets and Systems*, Vol. 6, No. 3, Nov., pp. 309–312.

Watanabe, S. (1969). Modified concepts of logic, probability and information based on generalized continuous characteristic function. *Inform. and Control*, Vol. 15, pp. 1–21.

Watanabe, S. (1978). A generalized fuzzy-set theory. *IEEE Trans. Syst., Man, Cybern.*, Vol. SMC-8, No. 10, pp. 756–759.

Watson, S. R., J. J. Weiss, and M. L. Donnell (1976). Fuzzy decision analysis. *IEEE Trans. Syst., Man, Cybern.*, Vol. SMC-9, No. 1, pp. 1–9.

Weber, S. (1983). A general concept of fuzzy connectives, negations and implications based on *t*-norms and *t*-conorms. *Fuzzy Sets and Systems*, Vol. 11, No. 2, Oct., pp. 115–134.

Wechsler, W. (1975a). Gesteuerte *R*-fuzzy automaten. *ZKI-Informationen* (Akad. D. Wiss. Der D.D.R.). Vol. 1, pp. 9–13.

Wechsler, W. (1975b). *R*-fuzzy automata with a time structure. In *Mathematical Foundations of Computer Science*, Vol. 28 (Blikle, A. Ed.) Springer-Verlag, Berlin, pp. 73–76.

Wechsler, W. (1976). A fuzzy approach to medical diagnosis. *Int. J. Bio-Medical Comput.*, Vol. 7, pp. 191–203.

Wechsler, W. (1977). Families of *R*-fuzzy languages. *Lecture Notes in Comp. Sci.*, Vol. 56, pp. 117–186.

Wee, W. G., and K. S. Fu (1969). A formulation of fuzzy automata and its application as a model of learning systems. *IEEE Trans. Syst., Man, Cybern.*, Vol. SMC-5, pp. 215–223.

Weidner, A. J. (1981). Fuzzy sets and Boolean-valued universes. *Fuzzy Sets and Systems*, Vol. 6, No. 1, July, pp. 61–72.

Weiss, M. D. (1975). Fixed points separation and induced topologies for fuzzy sets. *J. Math. Anal. and Appl.*, Vol. 50, pp. 142–150.

Wenstop, F. (1976). Fuzzy-set simulation models in a systems dynamic perspective. *Kybernetes*, Vol. 6, pp. 209–218.

Wenstop, F. (1977). Fuzzy sets and decision-making. *California Engineer*, Vol. 16, pp. 20–24.

Wenstop, F. (1980). Quantitative analysis with linguistic values. *Fuzzy Sets and Systems*, Vol. 4, No. 2, Sept., pp. 99–116.

Wiedey, G., and H. J. Zimmermann (1978). Media selection and fuzzy linear programming. *J. Op. Res. Soc.*, Vol. 29, No. 11, pp. 1071–1084.

Wierzchon, S. T. (1982). Applications of fuzzy decision-making theory to coping with ill-defined problems. *Fuzzy Sets and Systems*, Vol. 7, No. 1, Jan., pp. 1–18.

Wierzchon, S. T. (1983). An algorithm for identification of fuzzy measure. *Fuzzy Sets and Systems*, Vol. 9, No. 1, Jan., pp. 69–78.

Wilkinson, J. (1973). Archetypes, language, dynamic programming and fuzzy sets. In *The Dynamic Programming of Human Systems* (Wilkinson, J., R. Bellman, and R. Garaudy, Eds.). MSS Information Corp., New York, pp. 44–53.

Willaeys, D., and N. Malvache (1981). The use of fuzzy sets for the treatment of fuzzy information by computer (short communication). *Fuzzy Sets and Systems*, Vol. 5, No. 3, pp. 323–328.

Willmott, R. C. (1978). Two fuzzier implication operators in the theory of fuzzy power sets. *Fuzzy Sets and Systems*, Vol. 4, No. 1, pp. 31–36.

Windham, M. P. (1981). Cluster validity for fuzzy clustering algorithms. *Fuzzy Sets and Systems*, Vol. 5, No. 2, pp. 177–186.

Windham, M. P. (1983). Geometric fuzzy clustering algorithms. *Fuzzy Sets and Systems*, Vol. 10, No. 3, July, pp. 271–280.

Wong, C. K. (1973). Covering properties of fuzzy topological spaces. *J. Math. Anal. and Appl.*, Vol. 43, pp. 697–704.

Wong, C. K. (1974a). Fuzzy points and logical properties of fuzzy topology. *J. Math. Anal. and Appl.*, Vol. 46, pp. 316–328.

Wong, C. K. (1974b). Fuzzy topology — Product and quotient theorems. *J. Math. Anal. and Appl.*, Vol. 45, pp. 512–521.

Wong, C. K. (1976). Categories of fuzzy sets and fuzzy topological spaces. *J. Math. Anal. and Appl.*, Vol. 53, pp. 704–714.

Wong, G. A., and D. C. Sheng (1975). On the learning behaviour of fuzzy automata. In *Advances in Cybernetics and Systems,* Vol. 2 (Rose, J., Ed.). Cordon and Breach, London, pp. 885–896.

Woodbury, M. A., and J. Clive (1974). Clinical pure types as a fuzzy partition. *J. Cybernetics*, Vol. 4, No. 3, pp. 111–121.

Wuyts, P. (1984). On the determination of fuzzy topological spaces and fuzzy neighborhood spaces by their level-topologies. *Fuzzy Sets and Systems*, Vol. 12, No. 1, Jan., pp. 71–86.

Wuyts, P., and R. Lowen (1983). On separation axioms in fuzzy topological spaces, fuzzy neighborhood spaces, and fuzzy uniform spaces. *Journal of Mathematical Analysis and Applications*, Vol. 93, No. 1, pp. 27–41.

Wygralak, M. (1983). Fuzzy inclusion and fuzzy equality of two fuzzy operations for fuzzy subsets. *Fuzzy Sets and Systems*, Vol. 10, No. 2, June, pp. 157–168.

Xie, W. X., and S. D. Bedrosian (1984). An information measure for fuzzy sets. *IEEE Trans. Syst., Man, Cybern.*, Vol. SMC-14, No. 1, pp. 151–156.

Yager, R. R. (1977). Fuzzy decision-making including unequal objectives. *Fuzzy Sets and Systems*, Vol. 1, pp. 87–95.

Yager, R. R. (1977a). Multiple-objective decision-making using fuzzy sets. *Int. J. Man-Machine Studies*, Vol. 9, pp. 375–382.

Yager, R. R. (1977b). On validity and building fuzzy models, Report on the *IEEE Symp. on Fuzzy-Set Theory and Appl.*, held at the 1977 IEEE Control and Decision Conference.

Yager, R. R. (1978a). Building fuzzy systems models, in *Applied General Systems Research — Recent Developments and Trends* (George J. Klir, Ed.) (NATA Conf. Series, Series 11 Sys. Sci., Vol. 5), pp. 313–321.

Yager, R. R. (1978b). Linguistic models and fuzzy truths. *Int. J. of Man-Machine Studies.*

Yager, R. R. (1978c). Validation of fuzzy linguistic models. *J. of Cybernetics*, Vol. 8, pp. 17–30.

Yager, R. R. (1979a). Fuzzy-set probabilities and decision. *J. of Cybernetics.*

Yager, R. R. (1979b). Fuzzy subsets of Type II in decisions. *J. of Cybernetics.*

Yager, R. R. (1979c). Mathematical programming with fuzzy constraints and a preference ordering on the object. *Kybernetes.*

Yager, R. R. (1979d). On choosing between fuzzy subsets. *Kybernetes.*

Yager, R. R. (1979e). On solving fuzzy mathematical relationships. *Info. and Control.*

Yager, R. R. (1979f). On the lack of inverses in fuzzy arithmetic. *Fuzzy Sets and Systems*, Vol 4, No. 1, pp. 73–82.

Yager, R. R. (1979g). Properties of connectives useful in local logics. In *General Systems Research — A Science, A Methodology, A Technology* (B. R. Gaines, Ed.). SGSR.

Yager, R. R. (1979h). On the measure of fuzziness and negation. Part 1 — Membership in the unit interval. *Int. J. General Systems*, Vol. 5, pp. 221–229.

Yager, R. R. (1980a). A logical on-line bibliographic searcher — an application of fuzzy sets. *IEEE Trans. Syst., Man, Cybern.*, Vol. 10, No. 1, Jan., pp. 51–53.

Yager, R. R. (1980b). A measurement — Informational discussion of fuzzy union and intersection. *Int. J. of Man-Machine Studies*, Vol. 11, pp. 189–200.

Yager, R. R. (1980c). On a general class of fuzzy connectives. *Fuzzy Sets and Systems*, Vol. 4, No. 3, pp. 235–242.

Yager, R. R. (1980d). A foundation for a theory of possibility. *J. of Cybernetics*, Vol. 10.

Yager, R. R. (1980e). A linguistic variable for importance of fuzzy sets. *J. of Cybernetics*, Vol. 10.

Yager, R. R. (1980f). Aspects of possibilistic uncertainty. *Int. J. Man-Machine Studies*, Vol. 12.

Yager, R. R. (1980g). Finite linearly ordered fuzzy sets with applications to decisions. *Int. J. Man-Machine Studies*, Vol. 12.

Yager, R. R. (1981a). Prototypical values for fuzzy subsets. *Kybernetes*, Vol. 10, No. 2.

Yager, R. R. (1981b). Approximate reasoning and possibilistic models in classification. *Int. J. of Computer and Info. Sci.*, Vol. 10, No. 2, pp. 141–175.

Yager, R. R. (1982a). Generalized probabilities of fuzzy events from fuzzy belief. *Information Science*, Vol. 28, No. 1, pp. 45–62.

Yager, R. R. (1982b). A new approach to the summarization of data. *Information Science*, Vol. 28, No. 1, p. 69.

Yager, R. R. (1982c). Measures of fuzziness based on *t*-norms. *Stochastica*, Vol. 6, No. 3, pp. 207–229.

Yager, R. R. (1982d). Fuzzy prediction based on regression models. *Information Science*, Vol. 26, No. 1, pp. 45–63.

Yager, R. R. (Ed.) (1982e). *Recent Advances in Fuzzy-Set and Possibility Theory*, Pergamon Press, Elmsford, NY.

Yager, R. R. (1983a). *Fuzzy Sets — A Bibliography*, Intersystems Publications, Seaside, CA.

Yager, R. R. (1983b). Some relationships between possibility, truth, and certainty. *Fuzzy Sets and Systems*, Vol. 11, No. 2, Oct., pp. 151–156.

Yager, R. R. (1983c). Presupposition in binary and fuzzy logics. *Kybernetes*, Vol. 12, pp. 135–140.

Yager, R. R. (1983d). Robot planning with fuzzy sets. *Robotica*, Vol. 1, Part 1, Jan., pp. 41–50.

Yager, R. R. (1984a). Approximate reasoning as a basis for rule-based expert systems. *IEEE Trans. Syst., Man, Cybern.*, Vol. SMC-14, No. 4, pp. 636–673.

Yager, R. R. (1984b). Fuzzy subsets with uncertain membership grades. *IEEE Trans. Syst., Man, Cybern.*, Vol. SMC-14, No. 2, pp. 271–275.

Yager, R. R. (1985a). φ-Projections on possibility distributions. *IEEE Trans. SMC*, Vol. SMC-15, No. 6, Nov.–Dec., pp. 775–777.

Yager, R. R. (1985b). Knowledge trees in complex knowledge bases. *Fuzzy Sets and Systems*, Vol. 15, No. 1, pp. 45–64.

Yager, R. R., and D. Basson (1975). Decision-making with fuzzy sets. *Decision Sciences*, Vol. 6, pp. 590–600.

Yao, J., M. Bowman, and M. Asce (1983). Fatigue damage assessment of Welder steel structures. *Proceedings of the W. H. Munse Symposium Behavior of Metal Structures*, Philadelphia, May 17.

Yingming, L. Pointwise characterization of complete regularity and imbedding theorem in fuzzy topological spaces. *Scientia Sincia*, Vol. 26, No. 2, pp. 138–147.

Zadeh, L. A. (1962). From circuit theory to system theory. *Proc. Institute of Radio Engineers*, Vol. 50, pp. 856–865.

Zadeh, L. A. (1965a). Fuzzy sets. *Inform. and Control*, Vol. 8, pp. 338–353.

Zadeh, L. A. (1965b). Fuzzy sets and systems. in *System Theory*, Microwave Research Institute Symposia Series XV (Fox, J. Ed.) Polytechnic Press, Brooklyn, New York, pp. 29–37.

Zadeh, L. A. (1966). Shadows of fuzzy sets. *Problems in Transmission of Information*, Vol. 2, pp. 37–44 (In Russian).

Zadeh, L. A. (1968a). Fuzzy algorithms, *Inform. and Control*, Vol. 12, pp. 94–102.

Zadeh, L. A. (1968b). Probability measures of fuzzy events. *J. Math. Anal. and Appl.*, Vol. 23, pp. 421–427.

Zadeh, L. A. (1969a). Biological applications of the theory of fuzzy sets and systems. In *Biocybernetics of the Central Nervous System* (Proctor, L. D., Ed.). Little, Brown, Boston, pp. 199–212.

Zadeh, L. A. (1969b). The concepts of system, aggregate and state in system theory. In *System Theory*. (Zadeh and Polak, Eds.) New York: McGraw-Hill, pp. 3–42.

Zadeh, L. A. (1971a). Quantitative fuzzy semantics. *Inform. Sci.*, Vol. 3, pp. 159–176.

Zadeh, L. A. (1971b). Similarity relations and fuzzy orderings. *Inform. Sci.*, pp. 177–200.

Zadeh, L. A. (1971c). Towards a theory of fuzzy systems. In *Aspects of Networks and Systems Theory* (Kalman, R. E., and R. N. DeClairis, Eds.). Holt, Rinehart & Winston, New York.

Zadeh, L. A. (1971d). Towards fuzziness in computer systems — Fuzzy algorithms and languages. In *Architecture and Design of Digital Computers*. (Boulaye, G. Ed.). Dunod, Paris, pp. 9–18.

Zadeh, L. A. (1972a). A fuzzy-set-theoretic interpretation of linguistic hedges. *Journ. of Cybernetics*, Vol. 2, pp. 4–34.

Zadeh, L. A. (1972b). A rationale for fuzzy control. *Journal of Dynamic Systems, Measurement and Control*, Vol. C94, pp. 3–4.

Zadeh, L. A. (1973a). A system-theoretic view of behaviour modification. In *Beyond the Punitive Society* (Wheeler, H., Ed.). W. H. Freeman, San Francisco, pp. 160–169.

Zadeh, L. A. (1973b). Outline of a new approach to the analysis of complex systems and decision processes. *IEEE Trans. Syst., Man, Cybern.*, Vol. SMC-2, pp. 28–44.

Zadeh, L. A. (1974a). A new approach to system analysis. In *Man and Computer,* (Marois, M. Ed.). North-Holland, Amsterdam, pp. 55–94.

Zadeh, L. A. (1974b). Fuzzy logic and its application to approximate reasoning. *Information Processing, Proc. IFIP Congress*, Vol. 74, No. 3, pp. 591–594.

Zadeh, L. A. (1974c). Numerical versus linguistic variables. *Newspaper of the Circuits and Systems Society*, Vol. 7, Feb., pp. 3–4.

Zadeh, L. A. (1974d). The concept of a linguistic variable and its application to approximate reasoning. In *Learning Systems and Intelligent Robots* (FU, K. S., and Tou, J. T., Eds.). Plenum Press, New York, pp. 1–10.

Zadeh, L. A. (1975a). Fuzzy logic and approximate reasoning. *Synthèse*, Vol. 30, pp. 407–428.

Zadeh, L. A. (1975b). Linguistic cybernetics. in *Advances in Cybernetics and Systems*, (Rose, J., Ed.). Gordon and Breach, London, Vol. 3, pp. 1607–1615.

Zadeh, L. A. (1976a). A fuzzy-algorithmic approach to the definition of complex or imprecise concepts. *Int. J. Man-Machine Studies*, Vol. 8, pp. 249–291. Also in *System Theory in the Social Sciences*. (Bossel, H., S. Klaczko, N. Moller, Eds.). Birkhauser-Verlag, Basel-Stuttgart, pp. 202–282.

Zadeh, L. A. (1976b). Semantic inference from fuzzy data by mathematical programming. *IEEE Man-System Cybern. Conference*.

Zadeh, L. A. (1976c). The linguistic approach and its application to decision analysis. In *Directions in Large-Scale Systems* (Ho, Y. C., and S. K. Mitter, Eds.). Plenum Press, New York.

Zadeh, L. A. (1977a). Fuzzy sets and their application to classification and clustering. In *Classification and Clustering* (J. Van Ryzin, Ed.) Acad. Press, New York, pp. 251–299.

Zadeh, L. A. (1977b). Linguistic characterization of preference relations as a basis for choice in social systems. *Erkenntnis*, Vol. 11, pp. 383–410.

Zadeh, L. A. (1978). Fuzzy sets as a basis for a theory of possibility. *Fuzzy Sets and Systems*, Vol. 1, No. 1, pp. 3–28.

Zadeh, L. A. (1980). Fuzzy sets versus probability. *IEEE Proceedings*, Vol. 68, No. 3.

Zadeh, L. A. (1981a). Possibility theory and soft data analysis. In *Mathematical Frontiers of the Social and Policy Sciences* (L. Cobb, and R. Mithrall, Eds.). Westview Press, Boulder, Colorado.

Zadeh, L. A. (1981b). Test-score semantics for natural languages and meaning representation via PRUF. In *Empirical Semantics*, Vol. 1 (B. B. Rieger, Ed.). Bochum: Brockmeyer.

Zadeh, L. A. (1982). A computational approach to fuzzy quantifiers in natural languages. *Memorandum No*. UCB-ERL M82-36, University of California, Berkeley.

Zadeh, L. A. (1983a). The role of fuzzy logic in the management of uncertainty in expert systems. *Fuzzy Sets and Systems*, Vol. 11, No. 3, Nov., pp. 199–228.

Zadeh, L. A. (1983b). Linguistic variables, approximate reasoning and disposition. *Med. Inform.*, Vol. 8, pp. 173–186.

Zadeh, L. A. (1983c). A fuzzy-set-theoretic approach to the compositionality of meaning: propositions, dispositions and canonical forms. *Memorandum* UCB/ERL M83-24, University of California, Berkeley. To appear in the *J. of Semantics*.

Zadeh, L. A. (1983d). Linguistic variables, approximate reasoning and dispositions. *Med. Inform.*, Vol. 8., No. 3, pp. 173–186. [Revised and expanded version of a paper published in the *Proc. 6th Annual Symp. on Computer Applications in Medical Care*, 30 Oct.-2 Nov. 1982, Washington D.C., pp. 787–791]

Zadeh, L. A. (1984). A theory of commonsense knowledge. In *Issues of Vagueness*, H. J. Skala, S. Termini, and E. Trillas (Eds.), Reidel Dordrecht, pp. 257–296.

Zadeh, L. A. (1984). A computational theory of disposition. In *Proc. 1984 Int. Conf. Computational Linguistics*, pp. 312–318.

Zadeh, L. A. (1985). Sollogistic reasoning in fuzzy logic and its application to usuality and reasoning with dispositions. *IEEE Trans. SMC*, Vol. SMC-15, No. 6, Nov.-Dec. pp. 754–763.

Zadeh, L. A., K. S. Fu, K. Tanaka, and M. Shimura, (Eds.) (1975). *Fuzzy Sets and Their Applications to Cognitive and Decision Processes*. Academic Press, New York.

Zeising, G., M. Wagenknecht, and K. Hartmann (1984). Synthesis of distillation trains with heat integration by a combined fuzzy and graphical approach. *Fuzzy Sets and Systems*, Vol. 12, No. 2, Feb., pp. 103–116.

Zeleny, M. (1974). A concept of compromise solutions and the method of displaced ideal computers. *O. R.*, Vol. 1, No. 4, pp. 479–496.

Zemankova-Leech, M., and A. Kandel, (1984). *Fuzzy Relational Data Bases — A Key to Expert Systems*. Verlag TÜV Rheinland, Köln, Germany.

Zhang, J. (1980). Some basic properties of the normal fuzzy-set structures. *Journal of Huazhong Institute of Technology*, English Edition, Vol. 2, No. 1.

Zheng, D., (1983). The extension theorem of *F*-measure spaces. *Fuzzy Sets and Systems*, Vol. 11, No. 1, Aug., pp. 19–30.

Zimmer, A. C. (1982). What really is turquoise? A note on the evaluation of color terms. *Psychology Research*, Vol. 44, pp. 213–230.

Zimmer, A. C. (1984). A model for the interpretation of verbal predictions. *Internat. Journal of Man-Machine Studies*, Vol. 20.

Zimmermann, H. J. (1975). Description and optimization of fuzzy systems. *Int. J. General Syst.*, Vol. 2, No. 4, pp. 209–215.

Zimmermann, H. J. (1977a). Fuzzy programming and linear programming with several objective functions. *Fuzzy Sets and Systems*, Vol. 1, No. 1, pp. 45–55.

Zimmermann, H. J. (1977b). Results of empirical studies in fuzzy-set theory. In *Applied General Systems Research — Recent Developments* (G. Klu, Ed.). Plenum Press, New York.

Zimmermann, H. J. (1985). *Fuzzy Set Theory — and Its Applications*. Kluwer, Nijhoff Publishing, Dordrecht.

Zimmermann, H. J., and H. Gehring (1975). Fuzzy information profiles for information selection. *Cong. Book*, Vol. II, 4th Internat. Congr., AFCET, Paris.

Zimmermann, H. J., L. A. Zadeh, B. R. Gaines, (Eds.) (1984). *Fuzzy Sets and Decision Analysis*. North Holland. Published by Elsevier Science Publishers B. V. as Vol. 20 in TIMS/Studies in the Management Sciences.

Zimmermann, H. J., and P. Zysno (1980). Latent connectives in human decision-making. *Fuzzy Sets and Systems*, Vol. 4, No. 1, July, pp. 37–52.

Zimmermann, H. J., and P. Zysno (1983). Decisions and evaluations by hierarchical aggregation of information. *Fuzzy Sets and Systems*, Vol. 10, No. 2, July, pp. 243–260.

JOURNALS, ETC., PROCEEDINGS, CONTAINING MAJOR PAPERS ON THE THEORY OF FUZZY SETS AND ITS APPLICATIONS

1) Multiple-Valued Logic, IEEE (Symposium Proceedings).
2) Cybernetics and Society (Conference Proceedings).
3) IEEE Conference on Decision and Control (Conference Proceedings).
4) Joint American Control, IEEE (Conference Proceedings).
5) BUSEFAL (Bulletins for Studies and Exchanges on Fuzziness and Its Applications), University of Paul Sabatier, Toulouse, France.
6) Beijing Working Group on Fuzzy Sets, People's Republic of China.
7) Collected papers on Fuzzy Mathematics (Abstracts), Huazhong Institute of Technology, People's Republic of China.
8) Fuzzy Logic Working-Group Reports, Queen Mary College, University of London, UK.
9) Meetings of the EWG on Fuzzy Sets.
10) Applied General Systems Research.
11) Summaries of Papers on General Fuzzy Problems, The Working Group on Fuzzy Systems, Tokyo, Japan.
12) Journal of Fuzzy Mathematics, Huazhong University of Science and Technology, People's Republic of China.
13) International Seminars on Fuzzy-Set Theory, Johannes Kepler University, Linz, Austria.
14) NAFIPS (North American Fuzzy Information Processing Society), Conference Reports and Newsletters.
15) International Conference on Systems, Man and Cybernetics (IEEE Proceedings 83 CH 1962-0).
16) Conference on Fuzzy Sets and Fuzzy Topology (Edited by S. E. Rodabaugh), Youngstown State University, Ohio, Dec. 1983.
17) 6th European Meeting on Cybernetics and Systems Research, Vienna, Austria, 1982.
18) Stochastica, Barcelona, Spain.
19) IFAC Symposium, New Delhi, India, 1982. (*Theory and Applications of Digital Control*; Ed. by A. K. Mahalonabis, Pergamon Press, 1983.)
20) IRIMS (International Research Institute of Management Sciences) and FORS (Finnish Operations Research Society) International Seminar, Moscow, 1981. (*Theory and Practice of Multiple-Criteria Decision-Making*; Ed. by C. Carlson and Y. Kochetkov, North Holland, 1983.)
21) IFSA (International Fuzzy Systems Association) Publications, including *Fuzzy Sets and Systems*, an international journal, North Holland, Amsterdam.

Index